SpringerWienNewYork

CISM COURSES AND LECTURES

The series presents lecture notes, monographs, edited works and proceedings in the field of Mechanics, Engineering, Computer Science and Applied Mathematics.
Purpose of the series is to make known in the international scientific and technical community results obtained in some of the activities organized by CISM, the International Centre for Mechanical Sciences.

INTERNATIONAL CENTRE FOR MECHANICAL SCIENCES

COURSES AND LECTURES - No. 501

VORTICES AND TURBULENCE AT VERY LOW TEMPERATURES

EDITED BY

CARLO F. BARENGHI
AND
YURI A. SERGEEV
NEWCASTLE UNIVERSITY, UK

SpringerWienNewYork

This volume contains 149 illustrations

SPIN 12325634

All contributions have been typeset by the authors.

ISBN 978-3-211-09446-4 SpringerWienNewYork

PREFACE

This book contains the lecture notes of the Advanced School "Vortices and Turbulence at Very Low Temperatures" organised at CISM, Udine, in July 2007. What motivated us in organising the school was the recent experimental and theoretical progress in the study of turbulence at very low temperatures near absolute zero using liquid and gaseous helium.

The aim of the school was to make more known to young turbulence researchers that the physical properties of helium allow experimenters to achieve extreme turbulence intensities, as measured by Rayleigh or Reynolds numbers. The second piece of information which we wanted to advertise is the observed striking similarity between ordinary turbulence and quantum turbulence (that is, turbulence in superfluid helium, in which the vortices are quantised). Quantum turbulence may thus represent a simplified version of the formidable turbulence problem.,

Why the need to advertise these results ? It is apparent that the problem of low temperature turbulence is at the intersection of physics communities which, normally, would not normally communicate much with each other: fluid dynamicists (who better understand the aspects of turbulence), low temperature physicists (who know the techniques to handle liquid helium) and atomic physicists (who can create quantised vortices in ultra-cold, Bose-Einstein condensed atomic gases). The school which we organised was thus an attempt to lower barriers between disciplines and stimulate young researchers with different backgrounds.

We are grateful to our colleagues Natalia Berloff, Shaun Fisher, Joe Niemela and Ladik Skrbek who gave the lectures on vortices and low temperature turbulence collected in this book. We are indebted to Professor Katepalli Sreenivasan who gave an exciting presentation of new experimental techniques to visualise liquid helium. We would like to thank CISM for hosting this event, in particular Cara Toros and the staff who helped with the day-to-day organization, Professor Paolo Serafini for the editorial activity, and CISM's Rector Professor Giulio Maier. Finally, we are grateful to Professor Alfredo Soldati for encouraging us in pursuing the idea of this advanced school.

Carlo Barenghi and Yuri Sergeev
Newcastle University
Newcastle upon Tyne, UK
January 2008

CONTENTS

Introduction to quantised vortices and turbulence

Carlo F. Barenghi*

School of Mathematics and Statistics, Newcastle University, Newcastle upon Tyne, U.K.

Abstract This chapter contains an introduction to liquid helium, the two–fluid model, quantised vortices and quantum turbulence. The last section gives a flavour of current research on quantum turbulence. At the end of the chapter a number of exercises test the reader's own understanding.

1 Liquid helium

1.1 The race toward absolute zero

During the XIX century physicists developed the science of thermodynamics and understood that there is a limit to the degree of cold which is possible. The temperature scale which starts from this limit is called the Kelvin scale of temperatures. One Kelvin degree is equal to one degree on the usual Centigrade scale, and absolute zero, $T = 0$ K, corresponds to $T = -273.15$ C. Physicists found that if the temperature is reduced there is less thermal disorder, thus the fundamental properties of matter become more apparent. Low temperature physics laboratories competed against each other, racing toward absolute zero, and attempted to liquefy all known gases, thus cooling matter to lower and lower temperatures. Oxygen become liquid at $T = 90$ K. Nitrogen required 77 K. In 1898 Dewar succeeded in liquefying hydrogen at $T = 20$ K. The only gas which resisted being liquefied was helium. Although helium is the second most common element in the Universe, it was identified only in the XIX century, first in the spectrum of solar radiation and then, by Ramsay, in rocks containing

*I am grateful to CISM for organizing the Advanced School on Vortices and Turbulence at Low Temperatures and for supporting the publication of these lecture notes. I also wish to thank Professor Alfredo Soldati for encouraging this Advanced School, and to Professor Yuri Sergeev for reading my manuscript. I also acknowedge the support of EPSRC, grants GR/T08876/01 and EP/D040892/1.

uranium. It was only in 1908 that Onnes succeeded in creating the first sample of liquid helium at $T = 4$ K. Few years later, in 1911, Onnes discovered superconductivity, the ability of some metals (e.g. mercury, tin, lead) and alloys to sustain electrical currents without any electrical resistance.

1.2 Engineering applications

Today the main practical application of liquid helium is cryogenic cooling. If we want to measure the low temperature properties of a substance (e.g. the specific heat or the thermal conductivity) it is necessary to cool a sample of that substance; the best way to extract heat it from the sample is to immerse it into a liquid, so that the area of thermal contact is maximised. At temperatures of few Kelvin degrees, the only existing liquid is helium: any other substance is solid.

A common application of helium is cooling superconducting magnets. The coils of these magnets are made of alloys which become superconducting if the temperature is less than a critical value. Superconducting magnets are routinely used in hospitals to make scans. They are also used in high energy physics laboratories to accelerate beams of elementary particles. An example is CERN's Large Hadron Collider. Along the 27 km long ring of the LHC there are more than one thousand superconducting magnets; to provide a magnetic field strength of $80,000$ Gauss, each magnet is held at the operating temperature of $T = 1.8$ K. Liquid helium is also used by astrophysicists to cool infrared detectors; for example, the IRAS satellite carried 720 litres of liquid helium held at $T = 1.6$ K.

1.3 Helium I and helium II

Figure 1.3.1 shows the typical phase diagram (pressure versus temperature) of an ordinary substance. In the diagram, the triple point marks the co-existence of gas, liquid and solid phases. The first experiments which investigated the properties of liquid helium were performed by Onnes and Dana in Leiden. Onnes and Dana found that liquid helium is transparent and has density equal to approximately $1/6$ of water's. They also noticed that, upon cooling the liquid helium by pumping on its vapour, the bubbling ceased when helium's temperature dropped below a critical value of approximately 2 K. Motivated by this strange effect, which suggests that some physical transformation takes place, Onnes and Dana measured helium's specific heat, C. They found that the temperature dependence of C has a remarkable peak at the same critical temperature 2 K. From the shape of the specific heat curve, they called T_λ this critical temperature. The value of T_λ on the current temperature scale is $T_\lambda = 2.1768$ K at saturated vapour pressure (SVP).

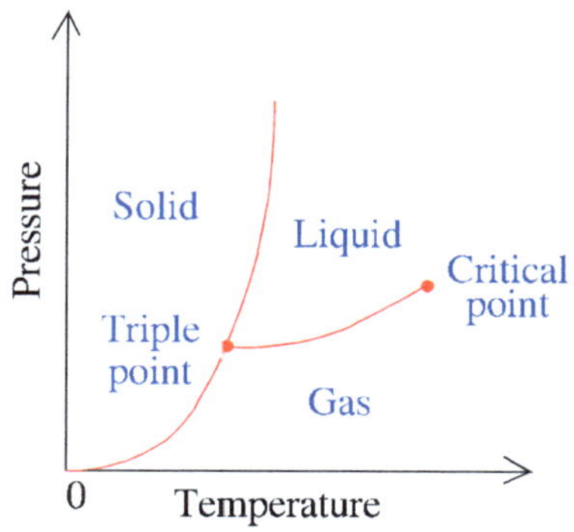

Figure 1.3.1. Phase diagram of an ordinary substance.

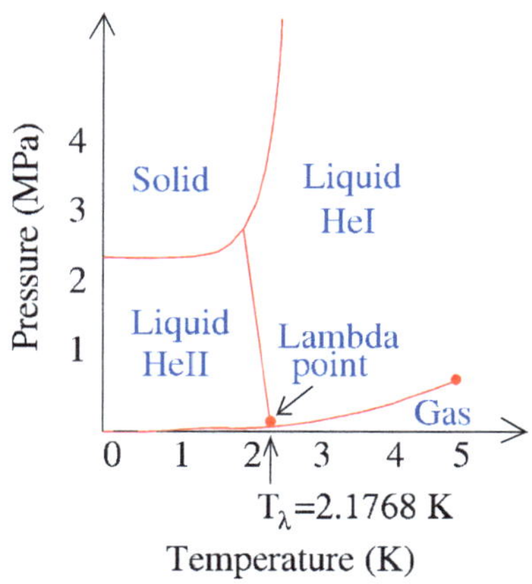

Figure 1.3.2. Phase diagram of liquid helium.

Soon it became clear that the properties of liquid helium above and below T_λ are very different. The low temperature liquid phase of helium was called helium II to distinguish it from the high temperature liquid phase called helium I. Figure 1.3.2 shows the phase diagram of liquid helium. Note the absence of a triple point and the fact that helium remains liquid down to absolute zero. To obtain solid helium, a pressure of about 25 bars must be applied. The boundary between helium I and helium II is called the lambda line; the intersection of the lambda line with the saturated pressure curve (along which most experiments are performed) is the lambda point.

The most striking property of helium II is superfluidity, which is the ability to flow without any viscous dissipation. Superfluidity was discovered independently by Kapitza and Allen in 1938, for which Kapitza was awarded the Nobel prize. Following the discovery of superfluidity, further experiments revealed that the flow of helium II has more strange properties if compared to the flow of ordinary liquids (including helium I). It was only in the 1940's that Landau and Tisza developed a theory, called the two–fluid model, which accounts for the observed flow of helium II, at least at small velocities. Landau was awarded the Nobel prize for his work on superfluidity. A prediction of the two–fluid model was an unusual mode of oscillation called second sound, which was observed by Peshkov in 1941.

In the 1940's, experiments on the rotational motion of helium II revealed more surprises. The quantisation of the circulation, predicted by Onsager (1948) and Feynman (1955), both Nobel prize winners, explained these experiments. The quantum of circulation was first observed by Vinen in 1961. Vinen also performed the first experimental investigations of quantum turbulence. Quantum turbulence limits the otherwise ideal properties of helium II to transfer heat, so it is important in the engineering applications of liquid helium. Current research in helium II is concerned with the similarities and difference between classical turbulence and quantum turbulence.

1.4 ^{4}He and ^{3}He

The nucleus of ordinary helium (^{4}He) consists of two protons and two neutrons. Naturally occurring helium gas contains a small fraction (approximately 1 part in 10^7) of the rare isotope ^{3}He, whose nucleus contains only one neutron. In 1972 Richardson, Lee and Osheroff were awarded the Nobel prize for the discovery that pure liquid ^{3}He becomes superfluid too, but at much colder temperatures (of the order of few milliKelvins) than ^{4}He.

1.5 Bose–Einstein condensation

The fundamental physical mechanism which is responsible for superfluidity is Bose – Einstein condensation (BEC), see Pethick and Smith (2001). Here it suffices to say that, according to quantum mechanics, a particle has also a characteristic wavelength, λ, associated with its momentum, p. In an ordinary gas, λ is much smaller than the average separation between the atoms, d. If the temperature is reduced, λ increases. At some critical temperature T_c, λ becomes of the order of d. In 1924 Bose and Einstein considered a sistem of bosons (particles with integer spin) and realized that, if the system consists of bosons, at this critical temperature it undergoes a phase transition, which corresponds to condensation in momentum phase. It was only in 1938 that London noticed that BEC is relevant to the superfluidity of liquid helium.

In the case of ^{3}He, the nucleus is not a boson (it has semi–integer spin), but paired atoms form Cooper pairs which are bosons and undergo BEC.

The study of BEC was greatly boosted by Wiemann, Cornell and Ketterlee's. They discovered BEC in trapped, ultra–cold alkali atoms in 1995, for which they were awarded the Nobel prize.

2 Two–fluid model

2.1 Thermal and mechanical effects

Early experiments showed that the motion of helium II has unusual properties. For example, consider a vessel A which contains helium and is linked to the helium in the bath B via a superleak S, as in Figure 2.1.1 left. A superleak is a very small hole (or holes); it can be realized, for example, by filling a channel with very fine powder, so fine that any ordinary fluid could not go through it. It was found that heating the helium in A with a resistor induces not only a temperature difference $\Delta T = T_A - T_B$, but also a flow from B to A through the superleak S, hence a pressure difference Δp, which is proportional to the height difference between the liquid in A and the liquid in B. This pressure difference can be large enough to create a small fountain, if A is open at the top (fountain effect). Note that the velocity (into A) opposes the flow of entropy (out of A), unlike what happens in an ordinary fluid.

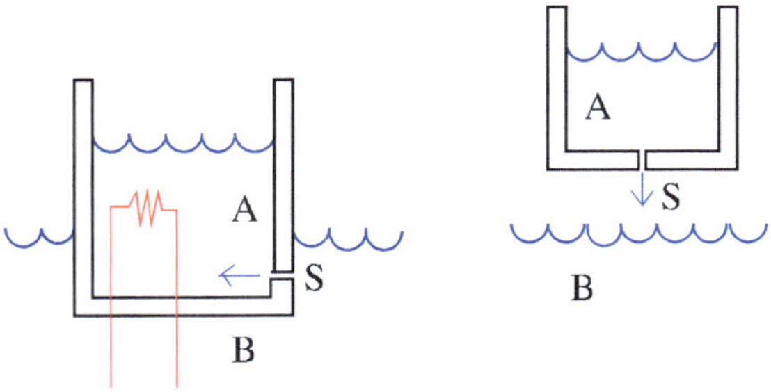

Figure 2.1.1. Left: thermo–mechanical effect. Right: mechano–thermal effect.

A second unusual effect, discovered by Daunt and Mendelsson and shown in Figure 2.1.1 right, is the following. If the vessel A is lifted above the bath B and helium flows out of the superleak S, the temperature in A increases, whereas the temperature is B decreases. This phenomenon is called the mechano–caloric effect.

Careful measurements by Kapitza of the chemical potential μ revealed that in these experiments μ remains the same in A and B: $\mu(p_A, T_A) = \mu(p_B, T_B)$. Since $d\mu = -sdT + dp/\rho$, where s is the specific entropy, we conclude that

$$\Delta p = \rho s \Delta T, \tag{2.1.1}$$

In another set of experiments it was found that helium's viscosity, η, seems to change at $T < T_\lambda$, depending on how it is measured.

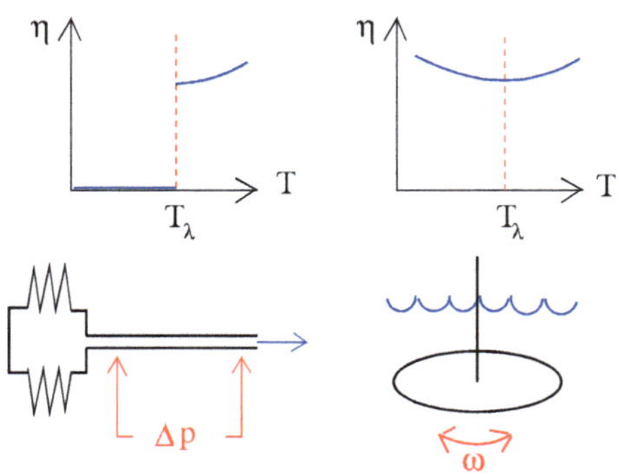

Figure 2.1.2. Left: the viscosity η, determined from the measurement of the pressure drop in a thin pipe, is discontinuous when plotted versus the temperature T. Right: if η is determined from the damping of an oscillating disk, it is continuous with T.

If the viscosity is measured by pushing helium along a capillary using bellows and detecting the pressure gradient along the channel, then $\eta = 0$ within experimental accuracy (see Figure 2.1.2 left). If the viscosity is measured by observing the damping of an oscillating disk, then $\eta \neq 0$ (see Figure 2.1.2 right).

2.2 Landau's equations

The apparently paradoxical results described in the previous subsection are explained by the two-fluid model of Landau and Tisza, see Landau and Lifshitz (1987) and Donnelly (1991). In this model, helium II is described as the intimate mixture of two fluids: the superfluid and the normal fluid. The first is related to the quantum ground state, and has zero viscosity and entropy. The second consists of thermal excitations and carries the total viscosity and entropy of the liquid. Each fluid has its own velocity and density fields, $\mathbf{v}_s$ and ρ_s for the superfluid and $\mathbf{v}_n$ and ρ_n for the normal fluid; the total density of helium II, $\rho = \rho_n + \rho_s$, is approximately temperature independent. The table below summarises the two–fluid model:

component	velocity	density	viscosity	entropy
normal fluid	$\mathbf{v}_n$	ρ_n	η	s
superfluid	$\mathbf{v}_s$	ρ_s	0	0

The two–fluid model accounts for the experimental observations. The superleak S is so small that the viscous normal fluid cannot move through it: only the superfluid flows through S. The observation that the chemical potential is constant across S both in the steady state (when $\mathbf{v}_s = 0$ in the superleak) and during transients (when $\mathbf{v}_s \neq 0$) led Landau to postulate that gradients of the chemical potential are responsible for the acceleration of the superfluid.

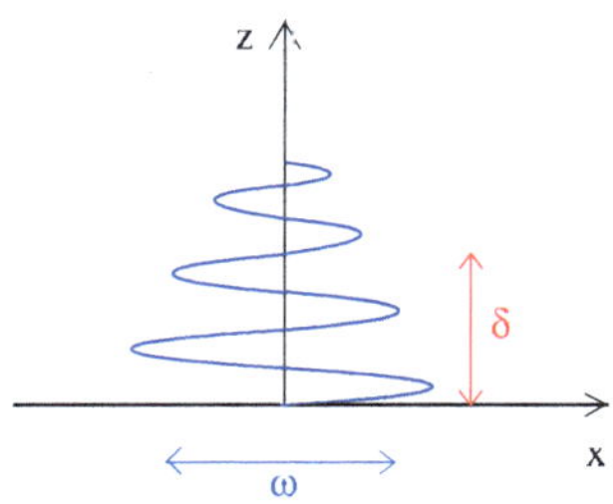

Figure 2.2.1. Penetration depth.

The relative proportion of superfluid and normal fluid at a given temperature was determined by Adronikashvili. He used the fact that the motion of an oscillating boundary penetrates into a viscous fluid only a distance of the order of $\sqrt{2\nu/\omega}$, where $\nu = \eta/\rho$, ν is the kinematic viscosity, η the viscosity, and ω the angular frequency of the oscillation - see Figure 2.2.1.

Adronikashvili's apparatus, shown schematically in Figure 2.2.2, was a special pendulum which consisted of a suspended stack of disks. Let Δz be the distance between the disks. If $\Delta z \ll \delta = \sqrt{2\eta/(\rho_n\omega)}$ the normal fluid is trapped between the disks and contributes to the moment of inertia of the pendulum, whereas the superfluid does not contribute (being inviscid,

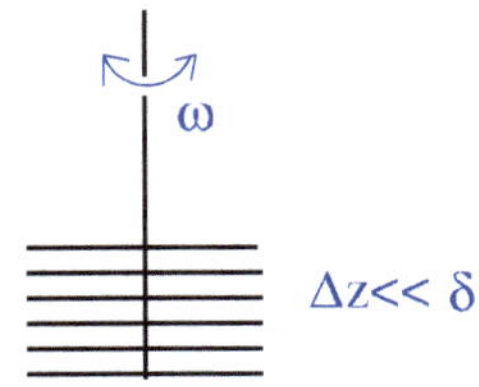

Figure 2.2.2. Adronikashvili's pendulum.

it moves freely between the disks). By measuring the damping rate of the torsional oscillations, Adronikashvili determined the ratios ρ_s/ρ and ρ_n/ρ as functions of the temperature T, which are shown schematically in Figure 2.2.3. Note that if the temperature is reduced the normal fluid fraction ρ_n/ρ decreases rapidly; below $T \approx 0.7$ K the normal fluid can be neglected.

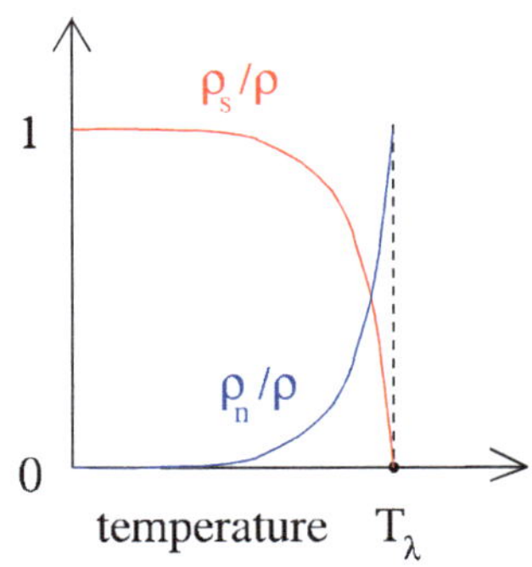

Figure 2.2.3.

The mathematical formulation of the two–fluid model consists of the equations of mass and entrophy conservation, and the equations of momentum conservation of the normal fluid and the superfluid, respectively. These equations are

$$\frac{\partial \rho}{\partial t} + \nabla \cdot (\rho_n \mathbf{v}_n + \rho_s \mathbf{v}_s) = 0, \tag{2.2.1}$$

$$\frac{\partial (\rho s)}{\partial t} + \nabla \cdot (\rho s \mathbf{v}_n) = 0, \tag{2.2.2}$$

$$\frac{\partial \mathbf{v}_n}{\partial t} + (\mathbf{v}_n \cdot \nabla)\mathbf{v}_n = -\frac{1}{\rho}\nabla p - \frac{\rho_s}{\rho_n} s \nabla T + \frac{\eta}{\rho_n}\nabla^2 \mathbf{v}_n, \tag{2.2.3}$$

$$\frac{\partial \mathbf{v}_s}{\partial t} + (\mathbf{v}_s \cdot \nabla)\mathbf{v}_s = -\frac{1}{\rho}\nabla p + s\nabla T. \tag{2.2.4}$$

Equation 2.2.2 states that entropy flows with the normal fluid. Note that, in isothermal conditions, the superfluid obeys the classical Euler equation, and the normal fluid obeys the classical Navier–Stokes equation.
Finally Landau recognised that, since $\mathbf{v}_s$ is proportional to the gradient of the phase of a quantum mechanical wavefunction, we must also have

$$\nabla \times \mathbf{v}_s = 0. \tag{2.2.5}$$

It must be stressed that Equation (2.2.3) and (2.2.4) are valid only at small velocities. In the presence of quantised vortices Landau's equation require modifications.

2.3 The spectrum of elementary excitations

The normal fluid consists of thermal excitations of energy ϵ and momentum $\mathbf{p}$. Landau showed that the shape of the dispersion curve $\epsilon = \epsilon(p)$, where $p = |\mathbf{p}|$, is responsible for the superfluid nature of helium II. Landau's spectrum, confirmed by neutron scattering experiments, is shown in Figure 2.3.1. Note the minimum at momentum p_0 and energy Δ. The excitations at low p (linear part of the spectrum) are called phonons; the excitations in the quadratic region near the minimum of the dispersion curve are called rotons.

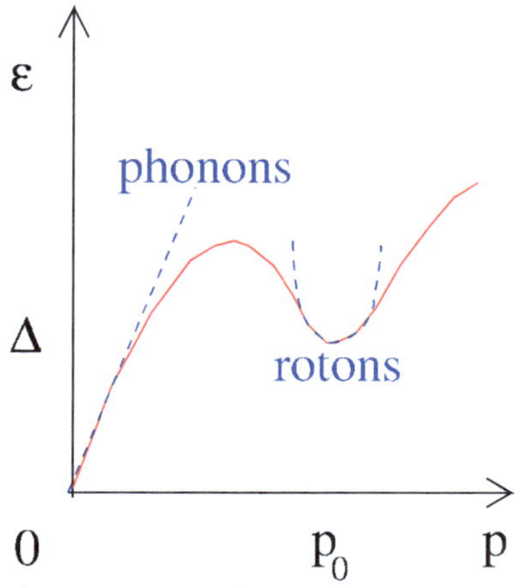

Figure 2.3.1. Landau's spectrum of the excitations. Note the roton minimum at (p_0, Δ_0).

Landau's argument for superfluidity is the following. Consider an object (e.g. an ion) of mass m moving with momentum $\mathbf{p}_1 = mV_1$ and energy E_1 which creates an excitation of energy ϵ and momentum $\mathbf{p}$ changing its own energy and momentum to E_2 and $\mathbf{p}_2$. Conservation of energy and momentum requires $E_1 = E_2 + \epsilon$ and $\mathbf{p}_1 = \mathbf{p}_2 + \mathbf{p}$, hence

$$P_1 p \cos\theta = m\epsilon + \frac{1}{2}p^2, \tag{2.3.1}$$

where θ is the angle between $\mathbf{p}_1$ and $\mathbf{p}$. Thus the object can lose energy and create an excitation if the initial velocity satisfies

$$V_1 > \frac{p}{2m} + \frac{\epsilon}{p} \approx \frac{\epsilon}{p}. \tag{2.3.2}$$

Let us minimise this velocity ϵ/p:

$$\frac{d}{dp}\left(\frac{\epsilon}{p}\right) = -\frac{1}{p^2}\epsilon + \frac{1}{p}\frac{d\epsilon}{dp} = 0. \tag{2.3.3}$$

We find:

$$\frac{d\epsilon}{dp} = \frac{\epsilon}{p}. \tag{2.3.4}$$

The minimum of ϵ/p thus corresponds to the line from the origin to a point slightly to the right of (p_0, Δ) on the dispersion curve; the critical velocity is $V_1 = V_c = 58$ m/s (at SVP). In conclusion, at sufficiently low temperature such that the normal fluid is negligible, we expect the ion to experience no drag for $0 < V_1 < V_c$.

At SVP, an ion moving in liquid helium creates a vortex ring at velocity smaller than V_c. Fortunately, at higher pressures the velocity of roton creation is smaller than the velocity required to create a vortex ring, and Landau's argument can be tested directly, as done by Allum *et al.* (1977).

3 Consequences of the two–fluid model

3.1 Second sound

The existence of two separate fluid components has a striking consequence on the oscillatory motion of helium II. Let us consider helium at rest ($\mathbf{v}_{n0} = 0$, $\mathbf{v}_{s0} = 0$) with density $\rho_0 = \rho_{s0} + \rho_{n0}$, pressure p_0, temperature T_0 and entropy s_0. We introduce small perturbations (indicated by primed quantities) $\rho = \rho_0 + \rho'$, $\rho_n = \rho_{n0} + \rho_n'$, $\rho_s = \rho_{s0} + \rho_s$, $\mathbf{v}_n = \mathbf{v}_n'$, $\mathbf{v}_s = \mathbf{v}_s'$, $p = p_0 + p'$, $T = T_0 + T'$ and $s = s_0 + s'$; neglecting quadratic terms in the perturbations, Landau's equations become

$$\frac{\partial \rho'}{\partial t} + \rho_{n0}\nabla \cdot \mathbf{v}'_n + \rho_{s0}\nabla \cdot \mathbf{v}'_s = 0, \tag{3.1.1}$$

$$\rho_0 \frac{\partial s'}{\partial t} + s_0 \frac{\partial \rho'}{\partial t} + \rho_0 s_0 \nabla \cdot \mathbf{v}'_n = 0, \tag{3.1.2}$$

$$\frac{\partial \mathbf{v}'_n}{\partial t} = -\frac{1}{\rho_0}\nabla p' - \frac{\rho_{s0}}{\rho_{n0}} s_0 \nabla T', \tag{3.1.3}$$

$$\frac{\partial \mathbf{v}'_s}{\partial t} = -\frac{1}{\rho_0}\nabla p' + \rho_{s0} s_0 \nabla T'. \tag{3.1.4}$$

In writing these equations we have neglected the viscous term $\eta\nabla^2 v_{\mathrm{n}}$, because we know already that its effect is to damp any motion. Assuming the solution in the form $e^{i\omega(t-x/c)}$, we find two values for the phase speed c:

$$c_1 = \sqrt{\left(\frac{\partial p}{\partial \rho}\right)_0}, \tag{3.1.5}$$

$$c_2 = \sqrt{\frac{\rho_{s0} s_0^2 T_0}{\rho_{n0} C_V}}. \tag{3.1.6}$$

We conclude that there are two modes of oscillation. The first mode is a pressure and density wave at (almost) constant temperature and entropy, in which $\mathbf{v}_n$ and $\mathbf{v}_s$ move in phase. In analogy with ordinary sound, we call this mode first sound. The second mode is a temperature and entropy wave at (almost) constant pressure and density, in which $\mathbf{v}_n$ and $\mathbf{v}_s$ move in anti–phase. We call this mode second sound. The speed of first sound is $c_1 \approx 200$ m/s at all temperatures; the speed of second sound, c_2, is approximately ten times less, and drops to zero as $T \to T_\lambda$.

It is interesting to notice that in second sound temperature perturbations obey the wave equation

$$\frac{\partial^2 T'}{\partial t^2} \approx c_2^2 \nabla^2 T',$$

whereas in ordinary fluids (e.g. helium I) they obey the heat equation

$$\frac{\partial T'}{\partial t} = \kappa \nabla^2 T'.$$

3.2 Thermal counterflow

Another consequence of the existence of two fluids is the unusual form of heat transfer. Consider Figure 3.2.1. A closed channel is open to the helium bath at one end; a resistor, placed at the closed end, dissipates a known heat flux $\dot{Q}$. This heat flux is carried away by the normal fluid, $v_n = \dot{Q}/(\rho ST)$. With the channel being closed, the mass flux is zero, $\rho_n v_n + \rho_s v_s = 0$, hence the superfluid moves towards the resistor, $v_s = (\rho_n/\rho_s)v_n$, setting up a counterflow velocity $v_{ns} = v_n - v_s$ which is proportional to the applied heat flux:

$$v_{ns} = v_n - v_s = \frac{\dot{Q}}{\rho_s ST}. \tag{3.2.1}$$

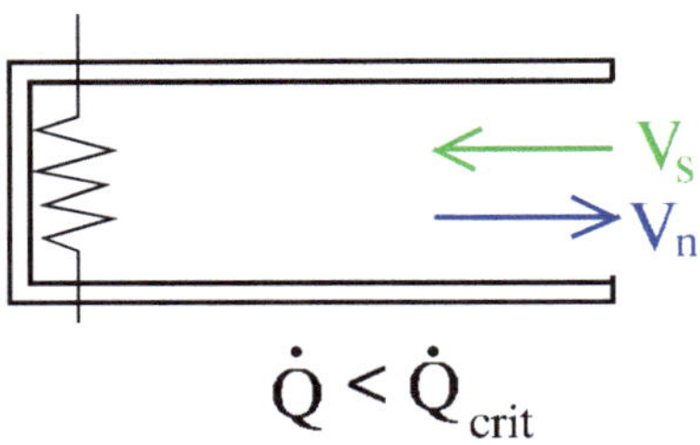

Figure 3.2.1. Laminar counterflow for $\dot{Q} < \dot{Q}_{crit}$.

Provided that $\dot{Q}$ is less than a critical heat flux $\dot{Q}_c$, this form of heat transfer is laminar.

4 Quantised vortex lines

4.1 Helium in rotation

Quantum mechanics introduces remarkable constraints on the rotational motion of helium II. It is instructive to consider the rotation of an ordinary, classical fluid first. A bucket of water which rotates at constant angular velocity Ω around the z axis has a height profile given by

$$z = \frac{\Omega^2 r^2}{2g}, \tag{4.1.1}$$

as shown in Figure 4.1.1 left; the water's velocity field is $\mathbf{v} = \Omega\hat{\boldsymbol{\phi}}$ (solid body rotation), and the vorticity is $\boldsymbol{\omega} = \nabla \times \mathbf{v} = 2\Omega\hat{\mathbf{z}}$, where $\hat{\mathbf{z}}$ and $\hat{\boldsymbol{\phi}}$ are the unit vectors along the axial and azimuthal direction respectively.

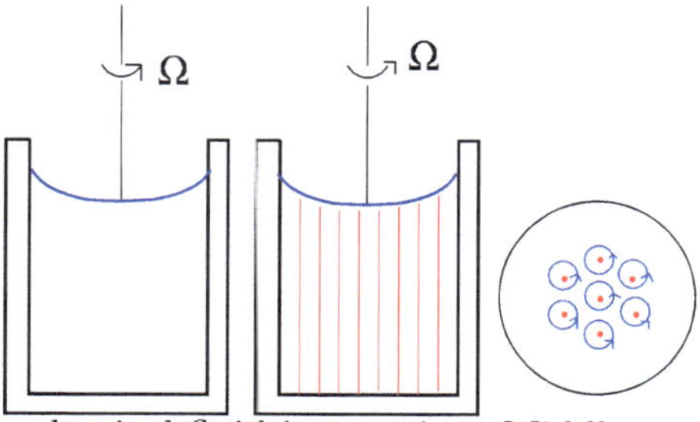

Figure 4.1.1. Left: classical fluid in rotation. Middle: rotating helium II. Right: top view of the vortex lattice.

The rotation of helium II is very different, because quantum mechanics introduces important constraints on the rotational motion. According to the two–fluid model, $\nabla \times \mathbf{v}_s = 0$, which means that the superfluid component cannot rotate; we expect that the profile of rotating helium II is

$$z = \left(\frac{\rho_n}{\rho}\right) \frac{\Omega^2 r^2}{2g}. \tag{4.1.2}$$

which is temperature dependent.

The observed profile did not agree with this prediction. The puzzle was solved by Onsager (1949) and Feynman (1955), who argued that the superfluid forms vortex lines, as in Figure 4.1.1 middle and right, around which the circulation κ is quantised, see Donnelly (1991):

$$\oint_C \mathbf{v}_s \cdot \mathbf{d}\boldsymbol{\ell} = \kappa, \tag{4.1.3}$$

where h is Planck's constant. The quantum of circulation (measured by Vinen in 1961) is

$$\kappa = \frac{h}{m} = 9.97 \times 10^{-4}\ cm^2\ s^{-1}, \tag{4.1.4}$$

where m is the mass of the helium atom.

Equation 4.1.3 can be used to determine the velocity field. Let C be a circle of radius r around the axis of the vortex; then the superfluid velocity is

$$v_\phi = \frac{\kappa}{2\pi r}, \tag{4.1.5}$$

as shown in Figure 4.1.3.

In Section 5 we shall see that the vortex core is hollow, so Equation 4.1.5 is valid only for $r \geq a_0$ where a_0 is the vortex core radius.

Figure 4.1.2. Vortex line

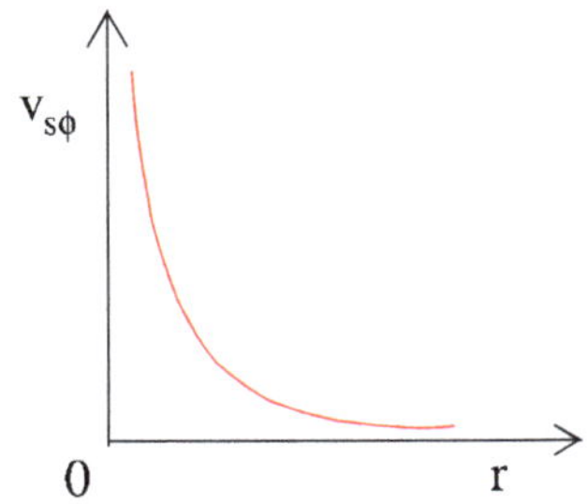

Figure 4.1.3. Velocity field around a vortex line.

4.2 The first vortex

The critical angular velocity Ω_c for the appearance of the first vortex line can be determined in the following way, see Donnelly (1991). Thermodynamical equilibrium requires minimisation of the free energy, $F = E - TS$, in the rotating frame of reference, which is

$$F' = F - \mathbf{\Omega} \cdot \mathbf{L} = (E - TS - \mathbf{\Omega} \cdot \mathbf{L}, \tag{4.2.1}$$

where E is internal energy, $\mathbf{\Omega}$ the angular velocity, and $\mathbf{L}$ the angular momentum. Let $T = 0$ and consider helium II contained in a rotating cylinder of radius R. The first vortex appears if

$$\Delta F' = F'_{vortex} - F'_{no\ vortex} = E - \Omega L < 0, \tag{4.2.2}$$

where the energy and the angular momentum (per unit length) are

$$E = \int_0^{2\pi} d\phi \int_{a_0}^{R} \frac{\rho_s v_s^2}{2} r dr, \tag{4.2.3}$$

$$L = \int_0^{2\pi} d\phi \int_{a_0}^{R} \rho_s r v_s r dr. \tag{4.2.4}$$

Substituting $v_s = \kappa/(2\pi r)$ we find that the critical velocity of vortex appearance is

$$\Omega_c = \frac{\kappa}{2\pi R^2} \ln(R/a_0), \tag{4.2.5}$$

where $a_0 \approx 10^{-8}$ cm is the vortex core radius.

4.3 Vortex lattice

If Ω is increased past Ω_c, more and more vortex lines appear in the flow. A bucket of helium rotating at constant angular velocity $\Omega > \Omega_c$ contains a lattice of quantised vortex lines aligned along the axis of rotation as shown in Figure 4.1.1. The lattice is steady in the rotating frame (see Figure 4.1.1 right; the number of vortex lines per unit area is given by Feynman's rule

$$n = \frac{2\Omega}{\kappa}. \tag{4.3.1}$$

Note that although the microscopic superflow is potential ($v_s \sim 1/r$), the macroscopically–averaged flow which results from the vortex lattice corresponds to solid body rotation ($v_s \sim \Omega r$). In other words, by creating n quantised vortices per unit area, helium II has the same (large–scale) vorticity of a classical rotating fluid ($2\Omega = n\kappa$).

Equation (4.3.1) has been tested by direct visualisation of quantised vortices at low temperatures by citeWP74; their technique consisted in trapping electrons along the vortex lines and then collecting them on electrodes at the top of the container. More recently, direct visualisation of quantised vorticity was achieved by Bewley, Lathrop, and Sreenivasan (2006), who trapped micron–size particles in the vortices and imaged the particles with a laser. The method, called Particle Image Velocimetry (PIV), is described in the Chapter by Sergeev in this book. Quantised vorticity in atomic Bose–Einstein condensates has also been achieved directly using lasers, see Madison, Chevy, Wohlleben, and Dalibard (2000).

5 The Bose–Einstein condensate model

5.1 The NLSE

The natural question raised by Equation (4.1.5) is: what happens if $r \to 0$? To answer this question we need a model for the vortex core.

Such model is provided by the Nonlinear Schrödinger Equation (NLSE), also called the Gross–Pitaevskii (GP) equation (see the Chapter by Berloff in this book). The NLSE describes accurately weakly interacting Bose particles in atomic Bose–Einstein condensates, In the case of helium II, the interaction between the bosons is strong, and our core model will be only qualitative, but sufficient for our purpose.

In the Hartree approximation, a condensate of weakly interacting Bose particles is described by a single particle wavefunction ψ for N bosons of mass m which obeys the NLSE

$$i\hbar\frac{\partial\psi}{\partial t} = -\frac{\hbar^2}{2m}\nabla^2\psi + V_0|\psi|^2\psi - E_0\psi. \tag{5.1.1}$$

where V_0 is the strength of the repulsive interaction between the bosons and E_0 is the energy increase upon adding one boson (chemical potential).

The simplest solution of Equation (5.1.1) is the uniform condensate at rest:

$$\psi = \psi_\infty = \sqrt{\frac{E_0}{V_0}}, \tag{5.1.2}$$

Another exact solution is the 1–dimensional solution in $0 \leq x < \infty$ near a wall at $x = 0$:

$$\psi = \psi_\infty = \tanh(x/a_0), \tag{5.1.3}$$

where the quantity

$$a_0 = \sqrt{\frac{\hbar^2}{mE_0}} \tag{5.1.4}$$

is called the healing length.

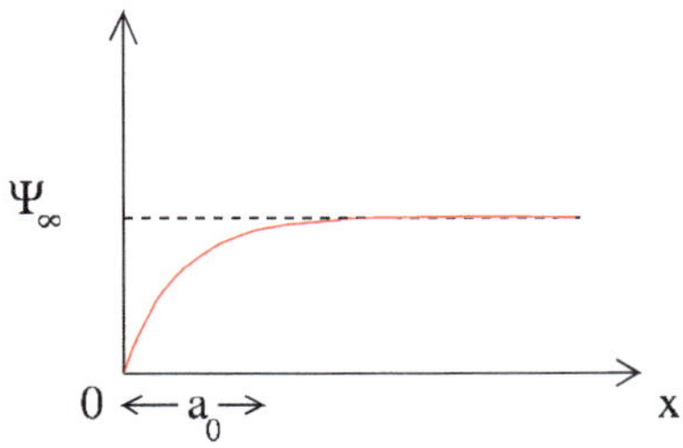

Figure 5.1.1. Condensate near the wall. The healing length is a_0.

It is easy to verify that small perturbations of the uniform solution obey the dispersion relation

$$\omega = k\sqrt{\frac{E_0}{m}\left(1 + \frac{\hbar^2 k^2}{4mE_0}\right)} \tag{5.1.5}$$

shown in Figure 5.1.2, where ω is the angular frequency and k is the wavenumber. Note that, if $k \ll 1$, then $\omega \approx ck$ where $c = \sqrt{E_0/m}$ is the speed of sound; if $k \gg 1$, then $\omega \approx \hbar^2 k^2/(2m)$, which is the dispersion relation of free particles. Note also that the NLSE dispersion relation – see Figure 5.1.2 – does not have a roton minimum like Landau's spectrum – see Figure 2.3.1.

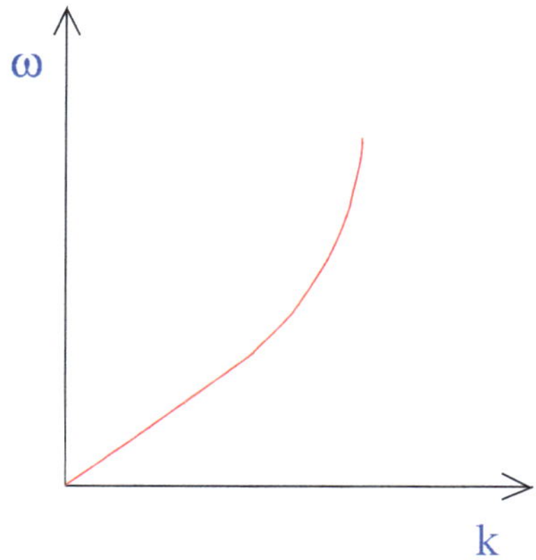

Figure 5.1.2. NLSE: dispersion relation $\omega = \omega(k)$.

5.2 Fluid dynamics interpretation of the NLSE

The Madelung transformation

$$\psi = Re^{iS}, \tag{5.2.1}$$

where R is the amplitude and S is the phase of ψ, provides us with a simple fluid dynamics interpretation of the NLSE. If we substitute Equation (5.2.1) into Equation (5.1.1), we obtain the continuity equation

$$\frac{\partial \rho_s}{\partial t} + \nabla \cdot (\rho_s \mathbf{v}_s) = 0 \tag{5.2.2}$$

and the (quasi) Euler equation

$$\rho_s \left(\frac{\partial v_{sj}}{\partial t} + v_{sk} \frac{\partial v_{sj}}{\partial x_k} \right) = -\frac{\partial p}{\partial x_j} + \frac{\partial \Sigma_{jk}}{\partial x_k} \tag{5.2.3}$$

where the usual convention applies of summation over repeated indices,

$$\rho_s = mR^2 \tag{5.2.4}$$

is the density,

$$\mathbf{v}_s = \frac{\hbar}{m} \nabla S \tag{5.2.5}$$

is the velocity (with Cartesian components v_{sk}, $k = 1, 2, 3$),

$$p = \frac{V_0}{2m^2} \rho_s^2 \tag{5.2.6}$$

is the pressure, and

$$\Sigma_{jk} = \left(\frac{\hbar}{2m} \right)^2 \rho_s \frac{\partial^2 \ln \rho_s}{\partial x_j \partial x_k} \tag{5.2.7}$$

is the quantum stress. Note that, without the quantum stress, the NLSE would describe a classical inviscid Euler fluid.

5.3 Vortex line solution of the NLSE

A vortex line solution of the NLSE is easily found. We use cylindrical coordinates (r, ϕ, z), substitute $S = \phi$ into $\psi = Re^{iS}$, and solve for the resulting velocity and density fields. We obtain

$$\mathbf{v}_s = \frac{\hbar}{m} \nabla S = \frac{\kappa}{2\pi r} \hat{\phi}, \tag{5.3.1}$$

which is the velocity field of a vortex line aligned along z shown in Figure 4.1.3.

To find the density field corresponding to this velocity field a numerical integration is required; the result is very similar to what shown in Figure 5.1.1: the density drops from its bulk value (away from the vortex axis) to zero (on the vortex axis) over a distance of the order of the healing length. This distance is thus called the vortex core radius. In helium II, it is approximately 10^{-8} cm. We conclude that that although the velocity $\kappa/(2\pi r) \to \infty$ for $r \to 0$, there is no problem because in the same limit $\rho_s \to 0$.

Finally we compute the circulation around the vortex axis along a closed path C. Taking for C a circle of radius r we recover

$$\oint_C \mathbf{v}_s \cdot \mathbf{d}\boldsymbol{\ell} = \int_0^{2\pi} v_s r d\phi = \kappa, \qquad (5.3.2)$$

as expected.

5.4 NLSE versus Euler

We have seen that the difference between the NLSE and the Euler equation is the quantum stress term. The natural question is thus in which sense a superfluid is similar to a classical inviscid Euler fluid.

Let D be the typical length scale of a problem. The ratio of the pressure term and the quantum stress term in the NLSE scales as $\hbar^2/(mE_0D^2)$, which is unity for $D \sim a_0$. We conclude that the quantum stress is important only at scales not larger than the healing length, $D \ll a_0$.

The quantum stress is indeed responsible for phenomena such as vortex nucleation, see Frish, Pomeau, and Rica (1992) and Winiecki and Adams (2000), and vortex reconnections, see Koplik and Levine (1993); these phenomena are outside the range of predictions of the classical Euler equations. Away from vortices, however, where the density is approximately constant, the quantum stress is zero, and the NLSE reduces to the Euler equation. Since the healing length is $a_0 \approx 10^{-8}$ cm and the typical vortex separation in quantum turbulence experiments is $\ell \approx 10^{-3}$ to 10^{-4} cm, there is a wide separation between the two scales. We conclude that, apart from relatively rare events such as vortex reconnections and nucleation, in most of the flow at most of the time the superfluid described by the NLSE is essentially a classical inviscid Euler fluid.

6 Two–fluid model with friction

6.1 Mutual friction

Quantised vortex lines interact with the phonons and rotons which make up the normal fluid, thus coupling the superfluid component with the normal fluid component, see Barenghi, Donnelly, and Vinen (1983). The coupling force $\mathbf{F}_{ns}$ is called mutual friction. It is proportional to the relative velocity between the two fluids, acting as a friction on each fluid. Thus, in the presence of quantised vorticity, the momentum conservation equations of Landau's two–fluid model become

$$\rho_n \left(\frac{\partial \mathbf{v}_n}{\partial t} + (\mathbf{v}_n \cdot \nabla)\mathbf{v}_n \right) = -\frac{\rho_n}{\rho} \nabla p - \rho_s s \nabla T, + \eta \nabla^2 \mathbf{v}_n + \mathbf{F}_{ns}, \qquad (6.1.1)$$

$$\rho_s\left(\frac{\partial \mathbf{v}_s}{\partial t}+(\mathbf{v}_s\cdot\nabla)\mathbf{v}_s\right)=-\frac{\rho_s}{\rho}\nabla p+\rho_s s\nabla T-\mathbf{F}_{ns}. \tag{6.1.2}$$

6.2 Attenuation of second sound

Consider a vessel which contains helium II and rotates at constant angular velocity Ω. A second sound pulse or resonance which moves across helium suffers a bulk attenuation. What concerns us here is the extra attenuation which arises due to the presence of vortices, shown in Figure 6.2.1. This extra attenuation can be used to measure the density of vortex lines.

Following Hall and Vinen (1956), the mutual friction force is

$$\mathbf{F}_{ns}=\frac{B\rho_n\rho_s}{\rho}\hat{\mathbf{\Omega}}\times(\mathbf{\Omega}\times\mathbf{q})+\frac{B'\rho_n\rho_s}{\rho}\mathbf{\Omega}\times\mathbf{q}, \tag{6.2.1}$$

where

$$\mathbf{q}=\mathbf{v}_n-\mathbf{v}_s, \tag{6.2.2}$$

and B and B' are temperature–dependent mutual friction coefficients which depend on the interaction of phonons and rotons with the quantised vortices. Substituting $\mathbf{q}$ and $\mathbf{F}$ into the Equation (6.1.1) and Equation (6.1.2), we obtain the following second sound wave equation:

$$\frac{d^2\mathbf{q}}{dt^2}+(2-B')\mathbf{\Omega}\times\frac{d\mathbf{q}}{dt}-B\hat{\mathbf{\Omega}}\times(\mathbf{\Omega}\times\frac{d}{dt}\mathbf{q})=c_2^2\nabla(\nabla\cdot\mathbf{q}) \tag{6.2.3}$$

Let us assume that the second sound propagates in the x direction:

$$\mathbf{q}=(q_x,q_y,0)e^{ikx-i\omega t}, \tag{6.2.4}$$

where k is the wavenumber and ω the angular frequency. In typical experimental conditions we have $\Omega/\omega \ll 1$, hence we obtain

$$k\approx\frac{1}{c_2}(\omega+i\frac{\Omega B}{2}). \tag{6.2.5}$$

The attenuation coefficient $\tilde{\alpha}$ is the imaginary part of k:

$$\tilde{\alpha}=\frac{B\Omega}{2c_2}. \tag{6.2.6}$$

In fact

$$\mathbf{q}=(q_x,q_y,0)e^{-\tilde{\alpha}x}e^{i\omega(x/c_2-t)}, \tag{6.2.7}$$

Note that angular velocity of rotation Ω is related to the vortex line density L via Feynman's rule, Equation (4.3.1), so, by measuring $\tilde{\alpha}$, we can recover the vortex line density.

It is therefore possible to perform an absolute measurement of the amount of vortex lines which are present in a turbulent flow. First the vessel is rotated, and the second sound signal is calibrated against the known vortex line density (number of vortices per unit area) $L = 2\Omega/\kappa$. Secondly, the vessel is stopped, the turbulence experiment is performed, and the second sound attenuation allows us to recover the vortex line density L (now to be interpreted as the vortex length per unit volume). Finally, it must be noticed that second sound is not attenuated by vortex lines which are parallel to the direction of propagation. If we assume that the turbulence is isotropic, only 2/3 of the vortices will attenuate the second sound wave.

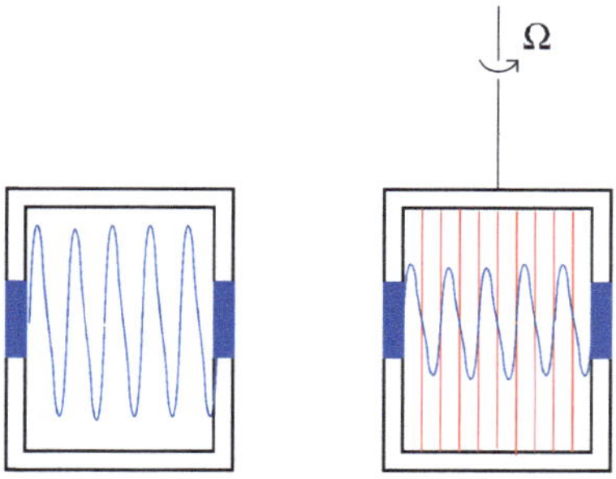

Figure 6.2.1. Left: second sound wave in a non–rotating vessel (no vortices). Right: second sound wave in the presence of vortices in a rotating vessel (vortices are present): note the reduced amplitude of the wave.

7 Vortex dynamics

7.1 The Biot–Savart law

The radius of the superfluid vortex core, approximately $a \approx 10^{-8}$ cm, is much smaller than any length scale of interest, so it is a good idea to approximate vortex lines as space curves of infinitesimal thickness. This approach was introduced by Schwarz (1985, 1998). The curves must be either closed loops or end at a boundary because a vortex cannot terminate in the middle of the flow.

Let $\mathbf{s} = \mathbf{s}(\xi)$ be the position of a point on such a curve, where ξ is the arc length. Following the classical theory of space–curves, we define the tangent $\hat{\mathbf{T}}$, normal $\hat{\mathbf{N}}$ and binormal $\hat{\mathbf{B}}$ unit vectors:

$$\mathbf{s}' = \frac{d\mathbf{s}}{d\xi} = \hat{\mathbf{T}}, \tag{7.1.1}$$

$$\frac{d\hat{\mathbf{T}}}{d\xi} = c\hat{\mathbf{N}}, \tag{7.1.2}$$

$$\hat{\mathbf{B}} = \hat{\mathbf{T}} \times \hat{\mathbf{N}}, \tag{7.1.3}$$

where $c = |\mathbf{s}''|$ is the curvature and $R = 1/c$ the local radius of curvature. The three vectors $\hat{\mathbf{T}}$, $\hat{\mathbf{N}}$ and $\hat{\mathbf{B}}$ form a right–handed system, as shown in Figure 7.1.1.

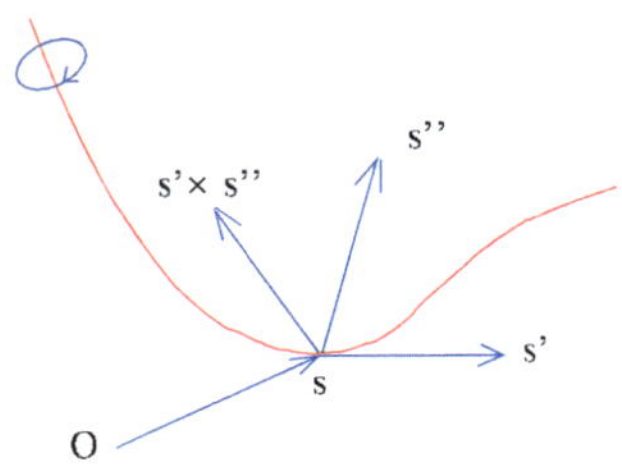

Figure 7.1.1.

The next step is to find the equation of motion of the vortex line. We start from classical definition of vorticity field $\boldsymbol{\omega}$ associated with a velocity field $\mathbf{v}$:

$$\boldsymbol{\omega} = \nabla \times \mathbf{v}. \tag{7.1.4}$$

Let us introduce the vector potential $\mathbf{v} = \nabla \times \mathbf{A}$; then $\mathbf{A}$ obeys the Poisson equation

$$\nabla^2 \mathbf{A} = -\boldsymbol{\omega}, \tag{7.1.5}$$

whose solution is

$$\mathbf{A}(\mathbf{x}) = \frac{1}{4\pi} \int \frac{\boldsymbol{\omega}(\mathbf{x}')d^3x'}{r}, \tag{7.1.6}$$

where $r = |\mathbf{x} - \mathbf{x}'|$. In our case the vorticity is formally concentrated on the vortex filament, $\boldsymbol{\omega}(\mathbf{x}')d^3x' = \kappa \mathbf{d}\boldsymbol{\ell}(\mathbf{x}')$, thus

$$\mathbf{A} = \frac{\kappa}{4\pi} \oint \frac{1}{r} \mathbf{d}\boldsymbol{\ell}', \tag{7.1.7}$$

from which we obtain the Biot–Savart law

$$\mathbf{v}(\mathbf{x}) = -\frac{\kappa}{4\pi} \oint \frac{(\mathbf{x} - \mathbf{x}')}{r^3} \times \mathbf{d}\boldsymbol{\ell}'. \tag{7.1.8}$$

The Biot–Savart law is often too difficult for analytical purpose, and is also computationally expensive. In many cases it is convenient to replace it with the following Local Induction Approximation (LIA):

$$\mathbf{v}(\mathbf{x}) \approx \beta \mathbf{s}' \times \mathbf{s}'', \tag{7.1.9}$$

where

$$\beta = \frac{\kappa}{4\pi R} \ln{(R/a_{eff})}, \tag{7.1.10}$$

where a_{eff} is an effective core radius and $R = 1/|\mathbf{s}''|$. It is apparent from the LIA that a vortex filament at a given position moves in the binormal direction with speed which is inversally proportional to radius of curvature at that position.

7.2 The Schwarz equation

Now we take into account friction. Let $\mathbf{v}_L$ be the vortex line velocity. The forces acting on unit length of vortex line are the Magnus force $\mathbf{f}_M$ and the drag force $\mathbf{f}_D$:

$$\mathbf{f}_M = \rho_s \boldsymbol{\kappa} \times (\mathbf{v}_L - \mathbf{v}_s), \tag{7.2.1}$$

$$\mathbf{f}_D = \rho_s \kappa \alpha \mathbf{s}' \times [\mathbf{s}' \times (\mathbf{v}_s - \mathbf{v}_n)] + \alpha' \mathbf{s}' \times (\mathbf{v}_s - \mathbf{v}_n), \tag{7.2.2}$$

where $\boldsymbol{\kappa} = \kappa \mathbf{s}' = \kappa \hat{\boldsymbol{\omega}}$. Both $\mathbf{f}_M$ and $\mathbf{f}_D$ are forces per unit length of vortex line. The Magnus force arises in general when there is a flow past a body with circulation (in this case the vortex core); on one side of the body the flow is faster, hence the pressure is lower, which causes a transverse force, see Figure 7.2.1.

The drag force consists of a parallel and transverse part, and is parametrised by friction coefficients α and α' which are related to the mutual friction coefficients B and B' already introduced by

$$\alpha = \frac{B\rho_n}{2\rho}, \qquad \alpha' = \frac{B'\rho_n}{2\rho}. \tag{7.2.3}$$

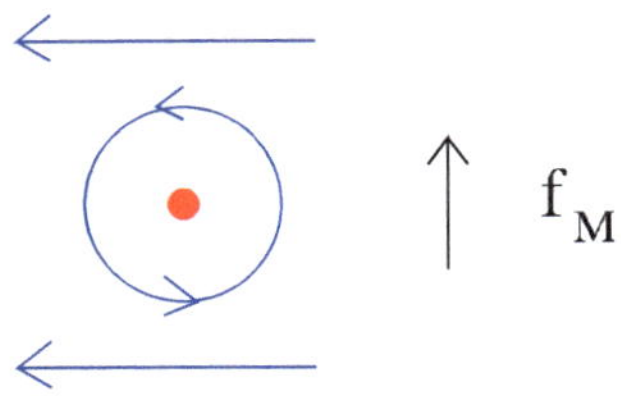

Figure 7.2.1. Magnus force.

Since the vortex core is very small, it has approximately no inertia, thus $\mathbf{f}_M + \mathbf{f}_D = \mathbf{0}$. We obtain Schwarz's equation

$$\mathbf{v}_L = \frac{d\mathbf{s}}{dt} = \mathbf{v}_s - \alpha \mathbf{s}' \times (\mathbf{v}_s - \mathbf{v}_n) + \alpha' \mathbf{s}' \times [\mathbf{s}' \times (\mathbf{v}_s - \mathbf{v}_n)], \qquad (7.2.4)$$

where we decompose

$$\mathbf{v}_s = \mathbf{v}_s^{self} + \mathbf{v}_s^{ext}. \qquad (7.2.5)$$

Here $\mathbf{v}_s^{self}$ is the self–induced velocity (Biot–Savart or LIA) and $\mathbf{v}_s^{ext}$ is an externally applied superflow.

Under LIA, we have $\mathbf{v}_s^{self} = \beta \mathbf{s}' \times \mathbf{s}''$ and Schwarz's equation reduces to

$$\frac{d\mathbf{s}}{dt} = \mathbf{v}_s^{ext} + \beta \mathbf{s}' \times \mathbf{s}'' + \alpha \mathbf{s}' \times (\mathbf{v}_{ns}^{ext} - \beta \mathbf{s}' \times \mathbf{s}'') \\ - \alpha' \mathbf{s}' \times [\mathbf{s}' \times (\mathbf{v}_{ns}^{ext} - \beta \mathbf{s}' \times \mathbf{s}'')],$$

where $\mathbf{v}_{ns}^{ext} = \mathbf{v}_n^{ext} - \mathbf{v}_s^{ext}$

The numerical method to move vortex filaments consists in dividing each filament into a large number of vortex points; each vortex points moves according to Schwarz's equation.

Schwarz was the first to recognise that an additional physical ingredient is required: vortex reconnections, shown schematically in Figure 7.2.2. His numerical algorithm performed reconnections when the distance between two vortex filaments is less then the discretization distance along each filament. Schwarz's insight was confirmed when Koplik and Levine (1993) demonstrated the existence of vortex reconnections using the NLSE.

The first numerical calculation of a vortex tangle was performed by Schwarz. The initial condition consisted of few vortex rings. In the presence of a counterflow velocity V_{ns}, the rings distorted each other, reconnected,

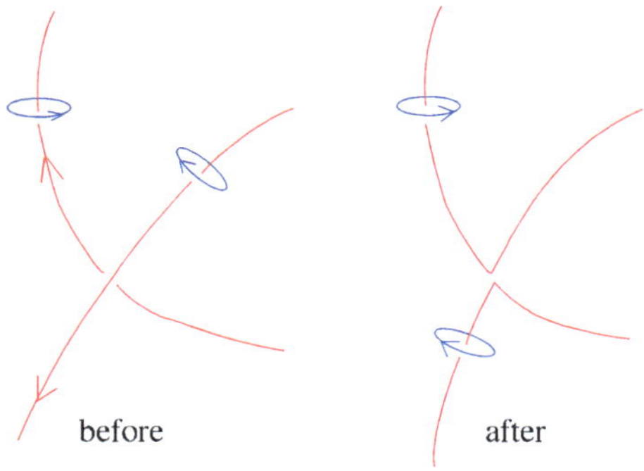

Figure 7.2.2. A vortex reconnection.

more loops were created, until a random–looking vortex tangle was generated in which the vortex line density oscillated around a non–zero statistical steady state.

7.3 Kelvin waves

Consider a vortex line which is straight and aligned in the z direction. Let $\mathbf{s} = \mathbf{s}(\xi)$ be the position of a point along the line. We use the LIA, Equation (7.1.9), to determine the self–induced motion of the vortex. Clearly if the vortex is straight, then $\mathbf{s}'' = 0$ and $\mathbf{v}_s^{self} = \beta \mathbf{s}' \times \mathbf{s}'' = 0$, that is to say the vortex does not move.

Now suppose that the vortex line is slightly perturbed away from the straight position in the form of a helix as in Figure 7.3.1:

$$\mathbf{s} = (\epsilon \cos\phi, \epsilon \sin\phi, z), \tag{7.3.1}$$

where $\phi = kz - \omega t$, ϵ is the amplitude of the helical wave, and ω_0 is the angular frequency. If $\epsilon \ll 1$ then $z \approx \xi$. We have $\mathbf{s}' = (-k\epsilon \sin\phi, k\epsilon \cos\phi, 1)$, $\mathbf{s}'' = (-k^2\epsilon \cos\phi, -k^2\epsilon \sin\phi, 0)$ and, neglecting terms proportional to ϵ^2, we conclude that the amplitude ϵ is constant and that the angular frequency is

$$\omega = \omega_0 = \beta k^2, \tag{7.3.2}$$

which is the dispersion relation for Kelvin waves.

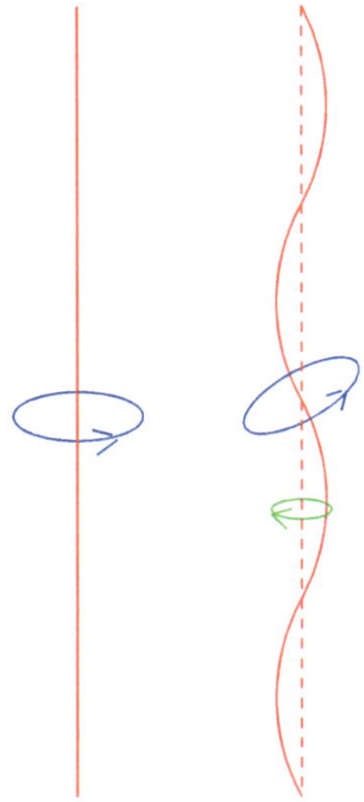

Figure 7.3.1. Kelvin wave on a vortex line.

7.4 The Donnelly–Glaberson instability

In the presence of friction, α and α' are not zero and the vortex line obeys Schwarz's equation

$$\frac{d\mathbf{s}}{dt} = \mathbf{v}_s^{self} + \alpha\mathbf{s}' \times (\mathbf{v}_{ns}^{ext} - \mathbf{v}_s^{self}) - \alpha'\mathbf{s}' \times [\mathbf{s}' \times (\mathbf{v}_{ns} - \mathbf{v}_s^{self})]. \qquad (7.4.1)$$

Let us assume that $\mathbf{v}_{ns} = \mathbf{v}_n - \mathbf{v}_s$ is in the direction parallel to the vortex line: $\mathbf{v}_{ns}^{ext} = (0; 0; V_{ns})$. We obtain

$$\omega = \omega_0 + \alpha'(kV_{ns} - \beta k^2), \qquad (7.4.2)$$

and

$$\frac{d\epsilon}{dt} = \sigma\epsilon. \qquad (7.4.3)$$

The first equation says that the friction changes the frequency of the Kelvin wave, ω_0. In the second equation $\sigma = \alpha(kV_{ns} - \beta k^2)$ is the growth rate of the Kelvin wave. The solution of Equation 7.4.3 is

$$\epsilon(t) = \epsilon(0)e^{\sigma t}, \qquad (7.4.4)$$

that is to say, if $\sigma > 0$ the amplitude of the Kelvin wave will grow exponentially, as shown in Figure 7.4.1.

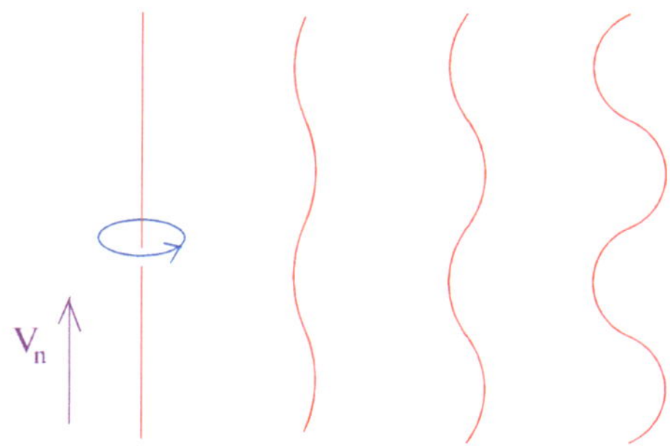

Figure 7.4.1. Donnelly–Glaberson instability.

In general, an arbitrary disturbances on a vortex line can be decomposed over modes with wavenumbers k. Initially, at small amplitude, each mode will grow or decay independently. We are interested in the mode which grows faster, because it will dominate all other modes. The maximum growth rate, σ_{max}, is obtained by setting $d\sigma/dk = 0$; we find

$$\sigma_{max} = \alpha \frac{V_{ns}^2}{4\beta}, \tag{7.4.5}$$

corresponding to the wavenumber $k_{max} = V_{ns}/(2\beta)$.

This instability, first observed by Cheng, Cromar, and Donnelly (1973) and then interpreted by Ostermeier and Glaberson (1975), plays an important role in the dynamics of a vortex tangle because it feeds energy into quantised vorticity.

8 Quantum Turbulence

8.1 The vortex tangle

Configurations of vortex lines can be either laminar or turbulent. An example of laminar vortex configuration is the rotating vortex lattice, shown in Figure 4.1.1 (middle). Another example is Taylor–Couette flow. Turbulent vortex configurations are easily created in the laboratory by applying a sufficiently large heat current, see Barenghi and Skrbek (2007), or by stirring the liquid helium using grids, see Smith, Donnelly, Goldenfeld, and Vinen (1993) or propellers, see Roche *et al.* (2007). The resulting quantum turbulence consists of a tangle of quantised vortices, as shown in Figure 8.1.1.

Figure 8.1.1. Vortex tangle.

The tangle can be characterised by the vortex line density, L, which is defined as the length of quantised vortex lines per unit volume. From the vortex line density we infer the typical distance between the vortices, $\ell \approx L^{-1/2}$. Thus $v \approx \kappa/(2\pi\ell)$ is the typical velocity inside the tangle.

8.2 Turbulent counterflow

Let us consider the quantum turbulence generated by a heat current (counterflow turbulence). We have seen that heat transfer in helium II takes the form of a counter current of $\mathbf{v}_n$ and $\mathbf{v}_s$. Experiments show that if $\dot{Q}$ (hence $V_{ns} = V_n - V_s$) exceeds a critical value $\dot{Q}_c$, then this laminar counterflow breaks down, and a tangle of vortex lines appears, see Figure 8.2.1.

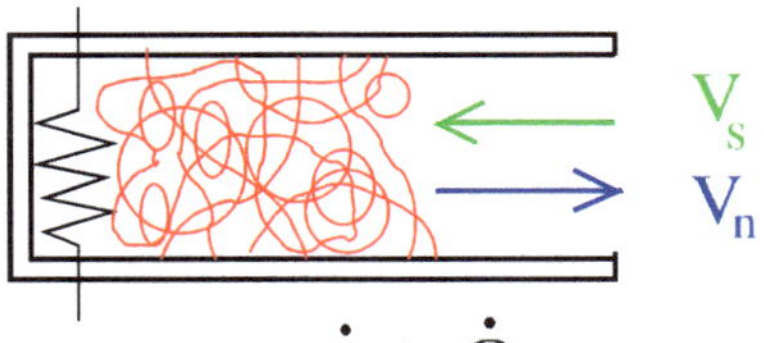

Figure 8.2.1. Turbulent counterflow for $\dot{Q} > \dot{Q}_{crit}$.

The vortex lines can be detected by monitoring the amplitude of a second sound signal across the channel; the measured vortex line density is

$$L = \gamma^2 V_{ns}^2, \tag{8.2.1}$$

where γ depends on temperature.

In the first approximation counterflow turbulence is homogeneous and the only important length scale is the intervortex distance $\delta \approx L_0^{-1/2}$, where L_0 is the vortex line density. Vinen argued that L_0 is due to the balance of production and destruction processes, which he modelled as the growth of vortex rings and the annihilation of opposite oriented vortex lines, obtaining

$$\frac{dL}{dt} = \frac{\chi_1 B \rho_n}{2\rho} V_{ns} L^{3/2} - \frac{\chi_2 \kappa}{2\pi} L^2, \tag{8.2.2}$$

where χ_1 and χ_2 are dimensionless constants of order one. The steady state solution to Equation (8.2.2) is indeed Equation (8.2.1) where

$$\gamma = \frac{\pi B \rho_n \chi_1}{\kappa \rho \chi_2}. \tag{8.2.3}$$

Tough discovered the existence of a second critical velocity, past which the vortex line density is larger. He called this second state of turbulence the T-2 state, to distinguish it from the less intense turbulent regime at smaller values of V_{ns}, called the T-1 state, see Figure 8.2.2.

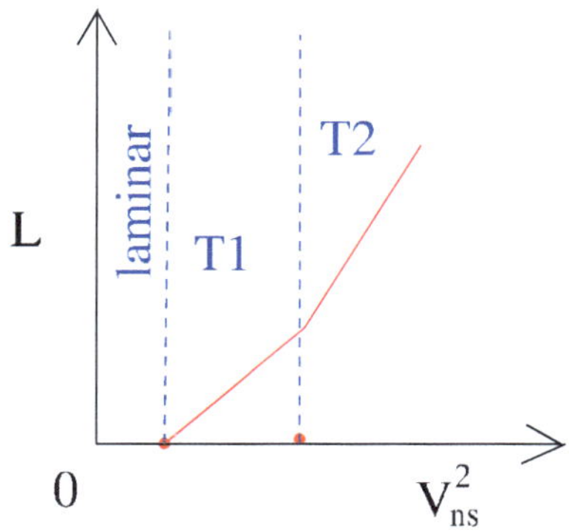

Figure 8.2.2. Laminar and turbulent regimesin thermal counterflow.

The natural question that arises is what is the nature of the two turbulent regimes. Melotte and Barenghi (1998) showed that the laminar normal fluid profile in a channel can become unstable at relatively small velocity if there are enough vortex lines. They considered a channel of circular cross section and radius R and a parabolic normal fluid profile of amplitude V_{ax} in the presence of the vortex line density L_0. Using the following model equation for the normal fluid,

$$\rho_n\left(\frac{\partial \mathbf{v}_n}{\partial t}+(\mathbf{v}_n\cdot\nabla)\mathbf{v}_n\right)=-\frac{\rho_n}{\rho}\nabla p-\rho_s S\nabla T+\eta\nabla^2\mathbf{v}_n \qquad (8.2.4)$$
$$-\left(\frac{B\rho_n\rho_s}{2\rho}\right)\left(\frac{2}{3}\right)\kappa L_0(\mathbf{v}_n-\mathbf{v}_s),$$

they performed a stability analysis. The resulting stability boundaries for the $m = 1$ azimuthal mode at various axial wavelengths k are shown in Figure 8.2.3, where

$$\beta=\frac{B\rho_s\kappa L_0R^2}{3\rho\nu_n} \qquad (8.2.5)$$

is the forcing parameter.

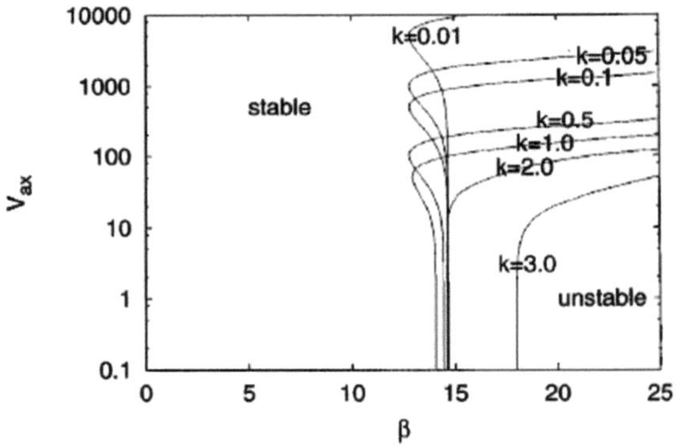

Figure 8.2.3. Normal fluid axial flow V_{ax} in the presence of friction forcing β: stability boundaries of perturbations with azimuthal wavenumber $m1 =$ and axial wavenumber k.

Figure 8.2.3 shows that many modes becomes unstable at approximately the same critical value of L_0, which suggests the onset of turbulence. The computed critical vortex line density corresponds to the observed values of the T-1 to T-2 transition. This suggests that in the T-1 state the superfluid is turbulent but the normal fluid is not, and in the T-2 state both superfluid and normal fluid are turbulent.

8.3 Classical aspects of quantum turbulence

Our understanding of ordinary homogeneous, isotropic turbulence (such as the turbulence away from walls in a wind tunnel) is based on the idea that energy is fed into the turbulence at large scale ℓ_0, and transferred to smaller and smaller scales by inertial instabilities (with no role played by viscous forces). At sufficiently small scale ℓ_1, called the Kolmogorov length scale, viscous forces become of the same order of magnitude as inertial forces, and, at scales smaller than ℓ_1, viscosity dissipates kinetic energy into heat. It is convenient to consider this process in the wavenumber space. Let k be the magnitude of the wave vector $\mathbf{k}$. It is found that the energy spectrum, $E(k)$, in the inertial range $1/\ell_0 \ll k \ll 1/\ell_1$ obeys the Kolmogorov scaling $E(k) \sim k^{-5/3}$.

A number of recent experiments have revealed many classical aspects of quantum turbulence. For example, Walstrom, Weisend, Maddocks, and Van Sciver (1998) forced helium II at high velocities along pipes and channels and observed the same pressure drops which are detected in ordinary turbulence. Smith, Hilton, and Van Sciver (1999) observed the same drag crisis in helium II which is seen in an ordinary fluid when a sphere moves at high velocity. Maurer and Tabeling (1998), who used counter–rotating propellers to continually excite turbulence in helium II, measured the energy spectrum, and found the classical Kolmogorov $-5/3$ scaling over the entire temperature range explored, from T_λ down to $T = 1.4$ K. Numerical simulations of vortex tangles in the absence of friction produced similar energy spectra, see Nore, Abid, and Brachet (1997), Araki, Tsubota, and Nemirowskii (2002), and Kobayashi and Tsubota (2005).

Smith, Donnelly, Goldenfeld, and Vinen (1993) found that the temporal decay of turbulence in helium II behind a towed grid has the same $t^{-3/2}$ power law which is expected in an ordinary fluid from the Kolmogorov spectrum. The same time dependence of turbulence decay was found in counterflow turbulence, see Barenghi and Skrbek (2007).

The current interpretation of these experiments is that at length scales larger than the intervortex spacing, the normal fluid and the superfluid are coupled by the friction and behave like a classical, ordinary fluid, as explained by Vinen and Niemela (2002).

Hulton, Barenghi, and Samuels (20020(@) introduced a simple model which illustrates the coupling of the normal fluid and superfluid vorticity. Using spherical coordinates (r, θ, ϕ), consider for simplicity a straight vortex segment which is initially in the plane $\theta = \pi/2$. For the sake of simplicity, we also neglect the component of the friction corresponding to the parameter α'. A normal fluid eddy, modelled by $\mathbf{v}_n = (0, 0, \Omega r \sin\theta)$ is also present, see Figure 8.3.1. The governing equations

$$\frac{d\theta}{dt} = -\alpha\Omega \sin\theta, \tag{8.3.1}$$

$$\frac{dr}{dt} = 0, \tag{8.3.2}$$

$$\frac{d\phi}{dt} = 0 \tag{8.3.3}$$

have the solution

$$\theta(t) = 2tan^{-1}(e^{-\alpha\Omega t}). \tag{8.3.4}$$

Thus the vortex segment will align with the normal fluid vorticity ($\theta \to 0$) for $t \to \infty$. Equation (8.3.4) shows that the vorticity of the superfluid tends to align with the vorticity of the normal fluid.

We must also take into account the fact that the lifetime τ of the normal fluid eddy is not infinite: τ is only of the order of the turnover time, $\tau \approx 1/\Omega$, so the segment will turn only by the angle $\theta(\tau) \approx \pi/2 - \alpha$ where α is the friction coefficient. Although this angle is small, the effect is sufficient to create a net polarisation of the vortex tangle in the direction of the normal fluid so that the superfluid and normal fluid vorticities match.

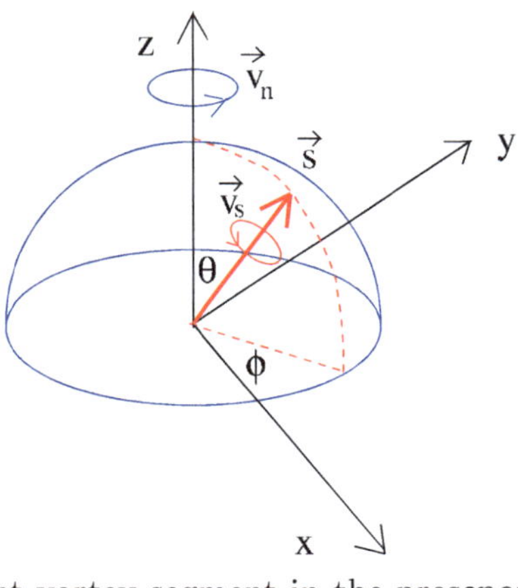

Figure 8.3.1. Straight vortex segment in the presence of normal fluid rotation.

8.4 Quantum turbulence at absolute zero

At temperatures below 0.7 K, the normal fluid is negligible and helium II can be considered a pure superfluid. The question is what should be the properties of this special kind of turbulence. For example, it is apparent from numerical calculations of Tsubota that, without friction, the vortex tangle looks more "crinkled". Quantum turbulence experiments at temperatures as low as few mK were performed by Davis, Hendry, and McClintock (2000) using a vibrating grid; they found that the turbulence decays rapidly. This result was at first surprising. In ordinary fluids, turbulence decays without a continuous supply of kinetic energy because energy is dissipated by viscous forces at very small scales. In helium II, if the temperature is small enough that friction and viscous effects are negligible, it was not clear what this energy sink should be.

A numerical calculation by Nore, Abid, and Brachet (1997) shed light onto the problem. They computed the evolution of a tangle of vortices using the NLSE model, starting from an arbitrary, large–scale Taylor–Green flow. They observed that the kinetic energy decreases with time, while the sound energy increases (the total energy being constant). The sink of kinetic energy was thus found: it the generation of sound. Samuels and Barenghi (1998) showed that the kinetic energy of vortices in a typical quantum turbulence experiments, if turned into phonons, would create a temperature increase which, because of the small heat capacity of the system, for $T < 0.3$ K would be large enough to be detected in a suitably built cell.

Further work revealed more details of the generation of sound energy by vortices and the classical aspect of the problem, i.e. vortex sound is well–known in classical fluid dynamics. In the context of superfluids, numerical simulations performed using the NLSE revealed that quantised vortices radiate sound energy when they accelerate. For example, Barenghi *et al.* (2005) found that a 2–dimensional vortex–antivortex pair which interacts with an isolated vortex changes suddenly the direction of motion and emits a ripple of sound waves, as shown in Figure 8.4.1. After the interaction, the vortex–antivortex is smaller because it has less kinetic energy.

Parker *et al.* (2004) showed that the energy–sound interaction can be studied in great details in an atomic Bose–Einstein condensate confined in a special trap; the trap consists of the usual harmonic potential with a Gaussian dimple at the bottom. By tuning the depth of the dimple, the sound emitted by the vortex which moves in the trap escapes and cannot be re absorbed by the vortex, which is confined in the dimple.

By studying numerically the collision of vortex rings in the NLSE model as in Figure 8.4.2, Leadbeater *et al.* (2001) discovered a more fundamental aspect of the transformation of kinetic energy into sound: a rarefaction pulse is emitted at each vortex reconnection event. The pulse is short (few healing lengths), intense (initially, at the reconnection, the density drops to zero), and localized with respect to the angle of the reconnection. The

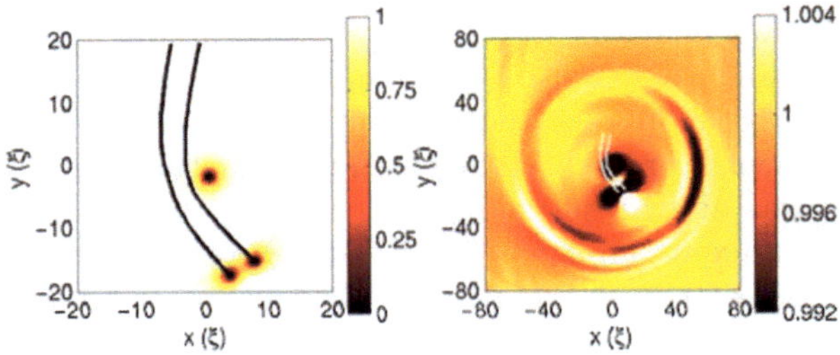

Figure 8.4.1. Left: vortex–antivortex pair which approaches a third isolated vortex and becomes deflected. Right: the sound ripple (small density oscillations) which are generated. The trajectory of the vortex pair is superimposed.

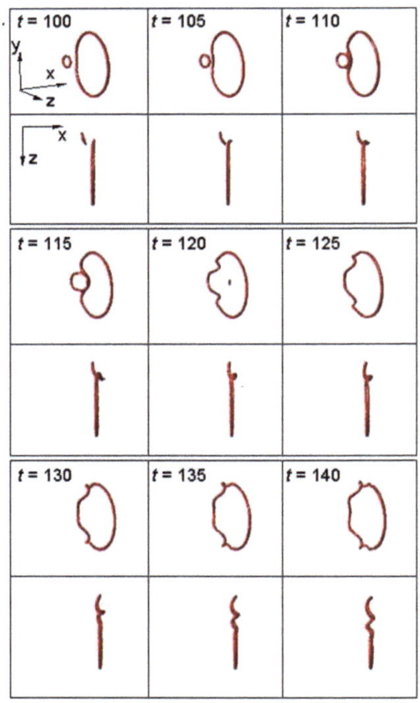

Figure 8.4.2. Collision of vortex rings. The time sequence shows two views of the colliding rings. The dot visible at $t = 120$ is the rarefaction pulse.

pulse removes kinetic energy from the vortex configuration. Unlike vortex sound, this effect has no classical counterpart: vortices in a classical inviscid Euler fluid cannot reconnect (changes in topology are forbidden by the conservation of helicity); vortices in a classical viscous Navier–Stokes fluid can reconnect, but, being viscous, are not relevant to our problem. Figure 8.4.2 shows the collision of vortex rings, and Figure 8.4.3 the sound pulse which is generated. Within a dense vortex tangle, quantised vortices lose energy both ways, through radiation and reconnection pulses, as shown in Figure 8.4.4.

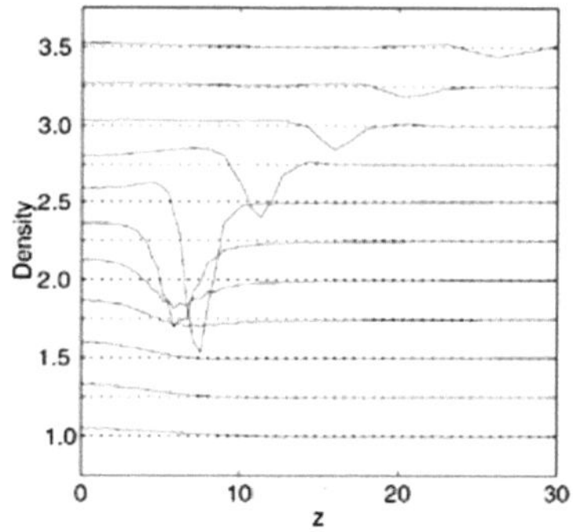

Figure 8.4.3. Density along the z–axis for a collision of two vortex rings initially offset with respect to each other. The curves correspond to different times and are offset with respect to each other for clarity. Just before the reconnection (bottom curve) the density is uniform except for a slight increase near the origin indicating the approaching rings. At the reconnection a rarefaction pulse is created at the centre of which the depth drops to zero. As the pulse moves away, the depth decreases and the pulse becomes more shallow

By studying numerically the collision of vortex rings in helium II, in typical experimental conditions, the vortex line density is not large enough for the decay of kinetic energy to be explained by vortex reconnections alone. Vinen considered the sound classically radiated by simple vortex configurations such as vortex pairs and Kelvin waves. He found that the power which is radiated per unit length by a co–rotating vortex–vortex pair separated by the distance ℓ is proportional to ℓ^{-6}. Taking for ℓ the average intervortex spacing deduced from the observed vortex line density, Vinen concluded that sound radiation by moving vortices cannot account for the

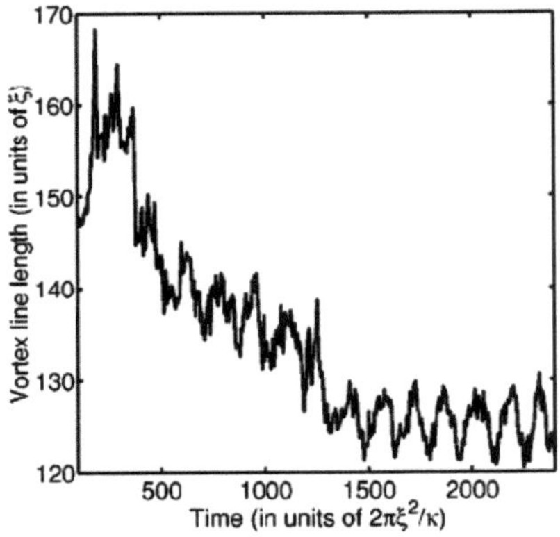

Figure 8.4.4. Vortex length vs time. The sudden drop corresponds to the emission of a rarefaction pulse, the decaying oscillations to sound radiation.

observed decay of superfluid turbulence: a much shorter length scale is necessary to radiate enough sound and explain the measurements.

The mechanism to shift kinetic energy to shorter and shorter length scales, short enough that sound can be radiated away, is the Kelvin wave cascade, see Svistunov (1995), Kivotides *et al.* (2001), and Kozik and Svistnov (2005). The Kelvin wave cascade has some analogy with the classical Richardson cascade. The following numerical calculation by Kivotides *et al.* (2001) showed how vortex reconnections trigger the cascade. Figures 8.4.5, 8.4.6, 8.4.7 and 8.4.8 show four vortex rings which collide and undergo vortex reconnections. The cusps produced at the reconnections relax, and the nonlinear interaction of large amplitude Kelvin waves generate Kelvin waves at shorter and shorter scales, until the resulting energy spectrum saturates - see Figure 8.4.9. Current theoretical work by L'vov, Nazarenko, and Rudenko (2007) on the problem is concerned with the possibility of a bottleneck between the classical Kolmogorov spectrum (at wavenumbers $k \ll 1/\ell$) and the Kelvin wave spectrum (at wavenumbers $k \gg 1/\ell$).

8.5 Rotating quantum turbulence

Quantum turbulence in a rotating system is particularly interesting, because of the conflict between the rotation (which tends to order the vortex configuration) and the mechanism which generates the turbulence (e.g. heat flow, stirring) which tends to randomise the vortices. The most studied rotating turbulence experiment in helium II was performed by Swan-

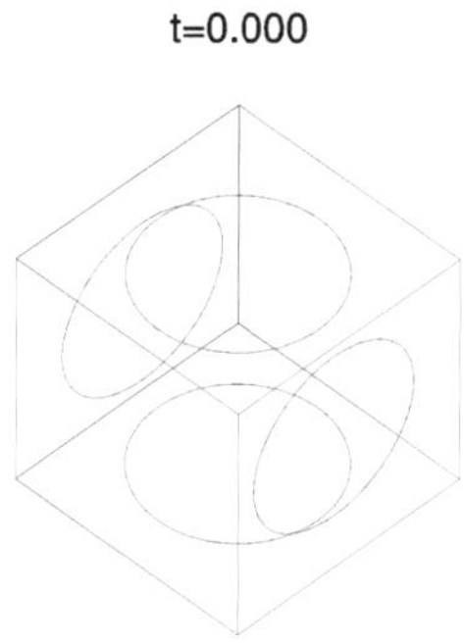

Figure 8.4.5. Kelvin wave cascade, $t = 0.0$ s: just before vortex reconnections.

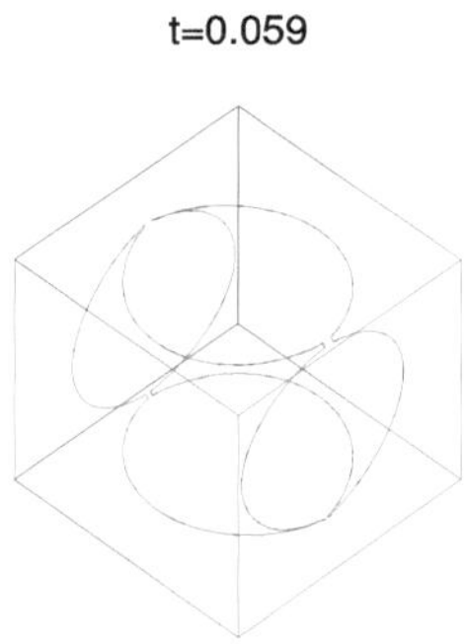

Figure 8.4.6. Kelvin wave cascade, $t = 0.059$ s: just after vortex reconnections.

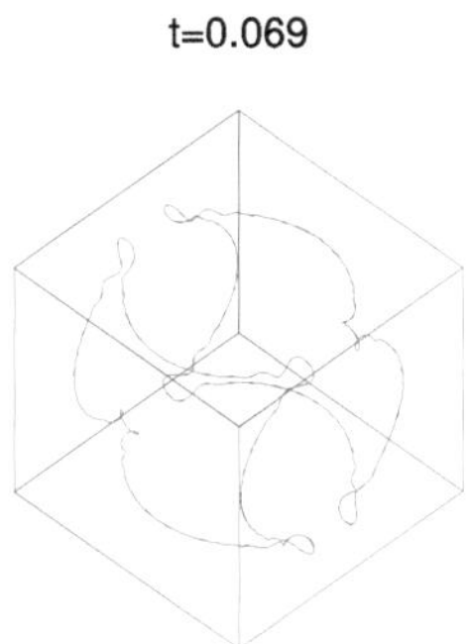

Figure 8.4.7. Kelvin wave cascade, $t = 0.129$ s: note the large amplitude Kelvin waves.

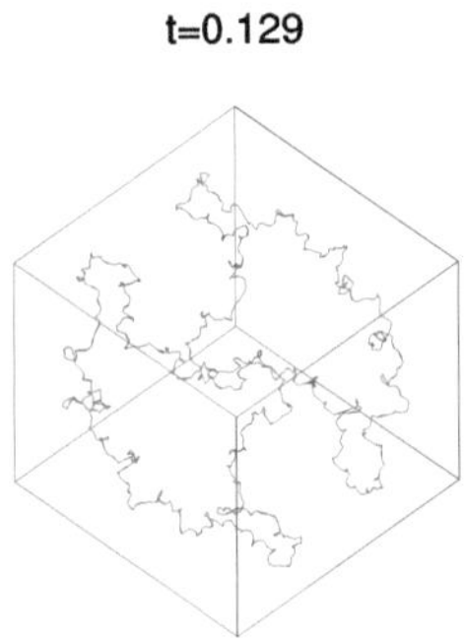

Figure 8.4.8. Kelvin wave cascade, $t = 0.129$ s: note the very short wavelengths.

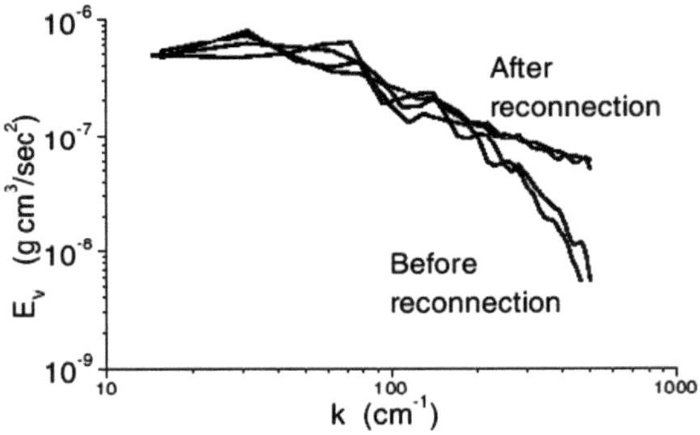

Figure 8.4.9. Kelvin wave cascade: the energy spectrum before the vortex reconnections and after saturation.

son, Barenghi, and Donnelly (1983). They applied higher and higher heat fluxes to a rotating channel, and found three regimes: a vortex lattice for $0 < \dot{Q} < \dot{Q}_{c1}$, a first state of turbulence for $\dot{Q}_{c1} < \dot{Q} < \dot{Q}_{c2}$, and a second state of stronger turbulence for $\dot{Q} > \dot{Q}_{c2}$. They identified $\dot{Q}_{c1}$ as the critical heat flux of the Glaberson–Donnelly instability, but the nature of $\dot{Q}_{c2}$ is not clear yet.

Numerical calculations performed by BarenghiTsubota (Araki) confirmed the findings at small values of heat flux. Figure 8.5.1 shows how a vortex lattice becomes unstable via the generation of Kelvin waves.

Rotating turbulence in ^{3}He-B is currently studied in Helsinki by Krusius and collaborators. In this superfluid, the viscosity is very large, so the normal component is expected to remain in solid body rotation and to

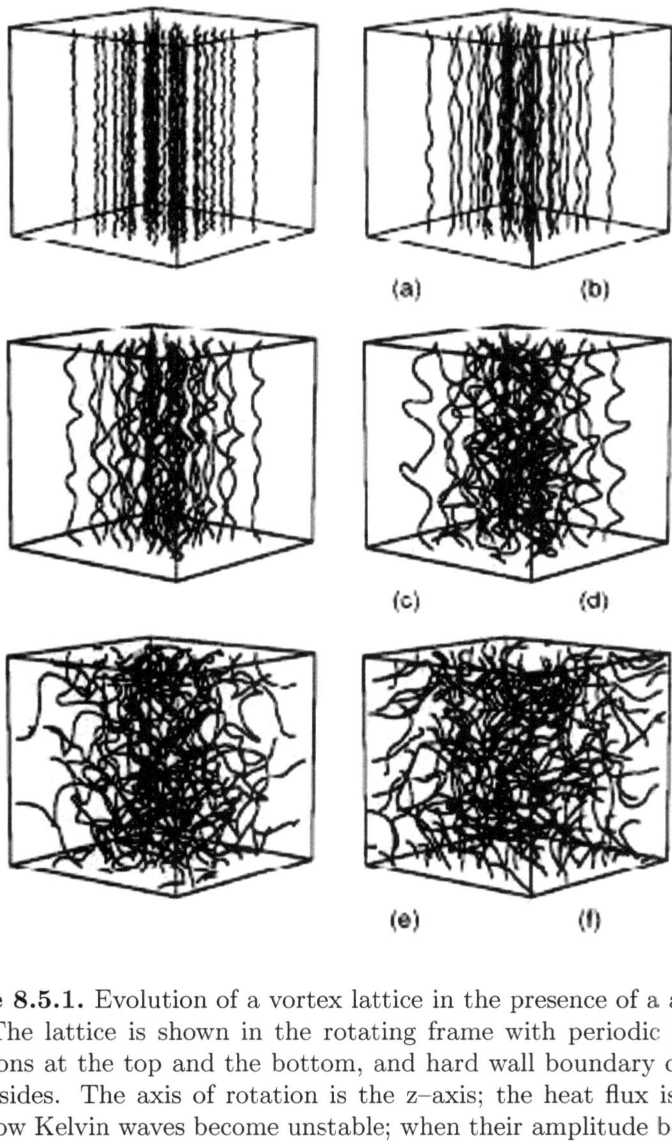

Figure 8.5.1. Evolution of a vortex lattice in the presence of a axial heat flux. The lattice is shown in the rotating frame with periodic boundary conditions at the top and the bottom, and hard wall boundary conditions at the sides. The axis of rotation is the z–axis; the heat flux is along z. Note how Kelvin waves become unstable; when their amplitude becomes of the order of the vortex separation, vortex reconnections take place and the vortex configuration becomes turbulent.

provide a sink of energy for the superfluid via the friction. By injecting a vortex loop in a rotating but still vortex–free system, Finne *et al.* (20030) discovered that the vortex configuration evolves to turbulence only if the parameter $q = \alpha/(1-\alpha')$ is small enough. At large enough values of q (high temperature), the initial vortex relaxes and aligns along the axis of rotation without creating turbulence. The parameter q was identified as the ratio of friction and inertial forces, which is the Reynolds number in ordinary fluids. Unlike the Reynolds number, q depends only on the temperature, not on the velocity.

%newpage

9 Exercises

9.1 Exercise 1

Consider a channel which is closed at one end and open to a bath of liquid helium II at the other end. At the closed end a resistor dissipates a known heat flux $\dot{Q}$. This heat is carried away from the resistor by the normal fluid according to $\dot{Q} = \rho S T v_n$ where S is the specific entropy. By imposing the condition that $\rho_s v_s + \rho_n v_n = 0$ (zero mass flux) and using the fact that $\rho = \rho_n + \rho_s$, show that

1. The relative velocity $v_{ns} = v_n - v_s$ between the normal fluid and the superfluid is proportional to the applied heat flux: $v_{ns} = \dot{Q}/(\rho_s S T)$.
2. $|v_s|/|v_n| \to 0$ for $T \to 0$ and $|v_s|/|v_n| \to \infty$ for $T \to T_\lambda$.
3. What happens at large values of the heat flux ?

9.2 Exercise 2

Use the quantisation of the circulation to show that the energy per unit length, $E' = E/h$, of a straight vortex line in a container of radius b is

$$E' = \frac{\rho_s \kappa^2}{4\pi} \ln{(b/a)},$$

where h is the height of the container and a is the vortex core radius.

9.3 Exercise 3

Use the LIA

$$\mathbf{v}_{si} = \beta \mathbf{s}' \times \mathbf{s}'',$$

where a prime denotes derivative with respect to arclength, and

$$\beta = \frac{\kappa}{4\pi} \ln{(1/(|\mathbf{s}'|a_{eff}))}$$

to find the self–induced velocity of a vortex ring of radius R.

9.4 Exercise 4

In an experiment it is found that the vortex line density is $L = 10^6\ cm^{-2}$. What is the typical separation δ between the quantised vortices ? Is δ of the order of the vortex core radius $a_0 \approx 10^{-8}\ cm$ or much larger ? Estimate the typical velocity within the turbulent tangle.

9.5 Exercise 5

Consider the two–fluid equations in the presence of superfluid turbulence:

$$\rho_s \frac{D\mathbf{v}_s}{Dt} = -\frac{\rho_s}{\rho}\nabla p + \rho_s S \Delta T - \mathbf{F}_{ns},$$

$$\rho_n \frac{D\mathbf{v}_n}{Dt} = -\frac{\rho_n}{\rho}\nabla p + \rho_s S \Delta T + \mu \nabla^2 \mathbf{v}_n + \mathbf{F}_{ns},$$

where $D/Dt = \partial/\partial t + \mathbf{v}\cdot\nabla$. Neglect the $(\mathbf{v}\cdot\nabla)\mathbf{v}$ terms and assume a steady state regime. Taking suitable averages show that the average pressure gradient is insensitive on whether the superfluid is turbulent or not.

9.6 Exercise 6

Consider a vortex ring of radius R, vortex core radius a ($a \ll R$), and circulation κ, in an inviscid fluid of density ρ. The energy, velocity and impulse of the ring are respectively

$$E = \frac{\rho\kappa R^2}{2}\left(\ln(8R/a_0) - \frac{3}{2}\right),$$

$$V = \frac{\kappa}{4\pi R}\left(\ln(8R/a_0) - \frac{1}{2}\right),$$

$$P = \rho\kappa\pi R^2.$$

Show that the vortex ring obeys Hamilton's equation

$$V = \frac{dE}{dP}.$$

Now consider an ordinary particle of mass m moving at velociy V. Its energy and momentum are respectively

$$E = \frac{1}{2} m V^2,$$

and

$$P = mV.$$

Show that the particle obeys the same Hamilton's equation $V = dE/dP$.

Finally, sketch the velocity V versus energy E for both the vortex ring and the particle. What is the striking difference between the vortex ring and the particle in terms of the V vs E relation ?

9.7 Exercise 7

Consider a superfluid vortex ring of radius $R \ll a_0$ (where a_0 is the vortex core) located on the (x,y)–plane and moving in the z–direction. The motion of the ring is determined by the balance of the Magnus force and the drag force: $2\pi R(\mathbf{f}_M + \mathbf{f}_D) = 0$.

The Magnus force on the unit length of vortex is

$$\mathbf{f}_M = \rho_s \kappa \times (\mathbf{v}_L - \mathbf{v}_{s,tot}),$$

where the magnitude of the vector κ is the quantum of circulation κ and the direction of κ is along the vortex. The quantity $\mathbf{v}_{s,tot} = \mathbf{v}_s + \mathbf{v}_{si}$ is the total superfluid velocity, sum of $\mathbf{s}_{si}$, which is the self–induced velocity of the vortex (which arises from its curvature) and $\mathbf{v}_s$, which is any imposed background superfluid flow.

The drag force on the unit length of vortex can be written as

$$\mathbf{f}_D = \gamma_0(\mathbf{v}_n - \mathbf{v}_L) + \gamma_0' \hat{\kappa} \times (\mathbf{v}_n - \mathbf{v}_L),$$

where γ_0 and γ_0' are know temperature–dependent mutual friction coefficients and $\hat{\kappa}$ is the unit vector in the direction of κ.

Use cylindrical coordinates (r, θ, z) so that $\kappa = (0, \kappa, 0)$, $\mathbf{v}_L = (\dot{R}, 0, V)$ with $\dot{R} = dR/dt$, the imposed superfluid and normal fluid flows are $\mathbf{v}_s = (0, 0, V_s)$ and $\mathbf{v}_n = (0, 0, V_n)$ respectively, and the self–induced velocity of the vortex ring is $\mathbf{v}_{si} = (0, 0, V_R)$ with

$$V_R = \frac{\kappa}{4\pi R}\left(\ln(8R/a_0) - \frac{1}{2}\right).$$

At sufficiently low temperature the normal fluid is negligible and there is no friction ($\gamma_0 = \gamma_0' = 0$), so, if $\mathbf{v}_s = 0$, the ring will move with velocity V_R in the z–direction keeping the radius constant ($\dot{R} = 0$).

At finite temperature γ_0 and γ_0' are different from zero, the velocity of the ring in the z–direction, V is not V_R and the ring's radius R can change. Show that indeed

$$\frac{dR}{dt} = \frac{\gamma}{\rho_s \kappa}(V_n - V_s - V_R),$$

where

$$\gamma = \frac{\gamma_0 {\rho_s}^2 \kappa^2}{\gamma_0^2 + (\rho_s \kappa - \gamma_0')^2},$$

Finally, integrating the equation for dR/dt from $R = R_0$ (initial radius) at $t = 0$ to $R = a$ (final radius) at $t = T$, show that, if $V_n = V_s = 0$, the life–time T of the vortex ring is

$$T = \frac{2\pi \rho_s R_0^2}{\gamma \mathcal{L}},$$

Hint: take into account that $a \ll R_0$ and, for the sake of simplicity, assume that the quantity $\mathcal{L} = \ln(8R/a) - 1/2$, which depends on R only weakly, is a constant.

Bibliography

D. R. Allum, P. V. E. McClintock, A. Phillips, and R. M. Bowley, *Phil. Trans. Roy. Soc. London A* **284**:179, 1977.

T. Araki, M. Tsubota, and S. K. Nemirovskii, *Phys. Rev. Lett.* **89**: 145301, 2002.

C. F. Barenghi, *Phys. Rev. B* **45**: 2290, 1992.

C. F. Barenghi, R. J. Donnelly, and W. F. Vinen, *J. Low Temp. Physics* **52**:189, 1983.

C. F. Barenghi, N. G. Parker, N. P. Proukakis, and C. S. Adams, *J. Low Temp. Physics* **138**:629, 2005.

C. F. Barenghi and L. Skrbek, *J. Low Temp. Physics* **146**:5, 2007.

G. P. Bewley, D. P. Lathrop, and K. R. Sreenivasan, *Nature* **44**:588, 2006.

D. K. Cheng, M. W. Cromar, and R. J. Donnelly, *Phys. Rev. Lett.* **31**:433, 1973.

S. I. Davis, P. C. Hendry, and P. V. E. McClintock, *Physica B* **280**:43, 2000.

R. J. Donnelly, *Quantized Vortices In Helium II*, Cambridge University Press, 1991.

R. P. Feynman. In C. J. Gorter, editor. *Progress in Low Temperature Physics, Vol. 1.* North-Holland, Amsterdam, p. 17, 1955.

A. P. Finne, T. Araki, R. Blaauwgeers, V. B. Eltsov, N. B. Kopnin, M. Krusius, L. Skrbek, M. Tsubota, and G. E. Volovik, *Nature* **424**:1022, 2003.

T. Frish, Y. Pomeau, and S. Rica, *Phys. Rev. Lett.* **69**:1644, 1992.

H. E. Hall and W. F. Vinen, *Proc. R. Soc. London, Ser. A* **238**:215, 1956.

S. Hulton, C. F. Barenghi, and D. C. Samuels, *Phys. Rev. Lett.* **89**:275301, 2002.

D. Kivotides, J. C. Vassilicos, D. C. Samuels, and C. F. Barenghi, *Phys. Rev. Lett.*, **86**:3080, 2001.

M. Kobayashi and M. Tsubota, *Phys. Rev. Lett.* **94**: 065302, 2005.

J. Koplik and H. Levine, *Phys. Rev. Lett.* **71**:1375, 1993.

E. Kozik and B. Svistunov, *Phys. Rev. B* **72**:172505, 2005.

L. D. Landau and E. M. Lifshitz, *Fluid Mechanics. Course of Theoretical Physics, Vol. 6.* Butterworth-Heinemann, 1987.

M. Leadbeater, T. Winiecki, D. C. Samuels, C. F. Barenghi, and C. S. Adams, *Phys. Rev. Lett.* **86**: 1410, 2001.

V. S. L'vov, S. Nazarenko, and O. Rudenko, *Phys. Rev. B* **76**:024520, 2007.

K. W. Madison, F. Chevy, W. Wohlleben, and J. Dalibard, *Phys. Rev. Lett.*, **84**: 806, 2000.

J. Maurer and P. Tabeling, *Europhys. Lett.* **43**,:29, 1998.

D. J. Melotte and C. F. Barenghi, *Phys. Rev. Lett.* **80**:4181, 1998.

C. Nore, M. Abid, and M.-E. Brachet, *Phys. Rev. Lett.* **78**:3896, 1997.

R. M. Ostermeier and W. I. Glaberson, *J. Low. Temp. Phys.* **21**:191, 1975.

L. Onsager, *Nuovo Cim. Suppl. 2* **6**:249, 1949.

N. G. Parker, N. P. Proukakis, C. F. Barenghi, and C. S. Adams, *Phys. Rev. Lett.* **92**:160403, 2004.

C. J. Pethick and H. Smith, *Bose-Einstein Condensation In Dilute Gases*, Cambridge University Press, Cambridge 2001.

P.-E. Roche, P. Diribarne, T. Didelot, O. Francais, L. Rousseau, and H. Willaime, *Europhys. Lett.* **77**:66002, 2007.

D.C. Samuels and C.F. Barenghi, *Phys. Rev. Lett.* **81**:4381, 1998).

K. W. Schwarz, *Phys. Rev. B* **31**, 5782, 1985.

K. W. Schwarz, *Phys. Rev. B* **38**:2398, 1988.

M. R. Smith, R. J. Donnelly, N. Goldenfeld, and W. F. Vinen, *Phys. Rev. Lett.* **71**:2583, 1993.

M. R. Smith, D. K. Hilton, and S. V. Van Sciver, *Phys. Fluids* **11**:751, 1999.

B. Svistunov, *Phys. Rev. B* **52**:3647, 1995.

C. E. Swanson, C. F. Barenghi, and R. J. Donnelly, *Phys. Rev. Lett.* **50**:190 (1983).

M. Tsubota, T. Araki, and C. F. Barenghi, *Phys. Rev. Lett.* **90**,:205301, 2003.

W. F. Vinen and J. J. Niemela, *J. Low Temp. Phys.* **128**:167, 2002, and Erratum, **129**:213, 2002.

P. L. Walstrom, J. G. Weisend, J. R. Maddocks, and S. V. Van Sciver, *Cryogenics* **28**:101, 1998.

G. A. Williams and R. E. Packard, *Phys. Rev. Lett.* **33**:280, 1974.

T. Winiecki and C. S. Adams, *Europhysics Lett.* **52**:257, 2000.

Motion of Solid Particles and Flow Visualization in Helium II

Yuri A. Sergeev*

School of Mechanical and Systems Engineering, Newcastle University, Newcastle upon Tyne, U.K.

Abstract Recent experiments have shown that it is possible to implement Particle Image Velocimetry (PIV) in liquid helium. However, to interpret the PIV data in the superfluid phase, it is necessary to understand how the particles are affected by the two components, the viscous normal fluid and the inviscid superfluid, as well as by the quantized vortex lines that may exist in the superfluid component. The chapter is supplemented by exercises aiming to assist the reader in better understanding this rapidly developing area of research.

1 Introduction

In this Chapter we will be concerned with the motion of small solid particles in turbulent helium II. This new and rapidly developing area of research has been initiated by the recent success of implementation of the Particle Image Velocimetry (PIV) technique in superfluid helium, see Donnelly *et al.* (2002), Zhang and Van Sciver (2002, 2005a, 2005b), Celik and Van Sciver (2002), Zhang, Celik, and Van Sciver (2004), and Bewley *et al.* (2006, 2007). This is perhaps the only technique which can identify the flow patterns in turbulent superfluid helium. PIV, based on tracking of small solid particles, has been a standard technique in classical fluid dynamics for several decades (there are many books written on this subject, see, for example Raffel, Willert, and Kompenhaus (1998)). In experimental studies of classical turbulence, one can be confident that the motion of sufficiently small particles will reveal the details of turbulent motion of the viscous fluid. In superfluids, the two-fluid

*I am grateful to CISM for organizing the Advanced School "Vortices and Turbulence at Very Low Temperature", and to Professor A. Soldati (University of Udine) for encouraging this School. On behalf of C. F, Barenghi (Newcastle University), D. Kivotides (University of California Santa Barbara), and myself, who contributed to the original research described in this Chapter, I wish to thank EPSRC for support provided by Grants GR/T08876/01 and EP/D061601. I am also grateful to Dr. D. C Swailes, who was the first reader of the manuscript, for useful comments and suggestions.

nature of helium II makes an interpretation of PIV measurements much more difficult: a solid particle interacts with both the normal fluid and the superfluid; moreover, the particle may interact strongly with quantized vortex lines in the superfluid component.

Basic properties of helium II were already described by C. F. Barenghi in the previous Chapter. Below we will revise briefly those properties which will be essential for the understanding of particulate motion in helium II.

Two-Fluid Model

According to the Landau's Two-Fluid Model, at the macroscopic level helium II can be modelled as the intimate mixture of two fluid components: the normal fluid and the superfluid. Their proportion depends on temperature, see Figure 1.0.1; here $\rho_{\rm n}$ and $\rho_{\rm s}$ are the normal and the superfluid density, respectively; the total density, $\rho = \rho_{\rm n} + \rho_{\rm s}$ is approximately temperature independent; $T_\lambda \approx 2.1768\,\mathrm{K}$ is the temperature of λ transition. The superfluid corresponds to the quantum ground state and has zero viscosity and zero entropy. The normal fluid consists of thermally excited states and, at a macroscopic level, can be modelled as a viscous Navier-Stokes fluid.

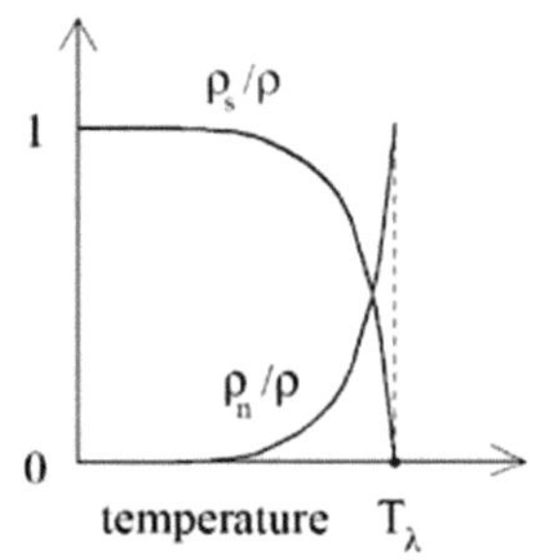

Figure 1.0.1. $\rho_{\rm s}$ and $\rho_{\rm n}$ as functions of temperature.

Heat Transfer Near Absolute Zero

Many experiments and technological applications of helium II are related to the ideal ability of superfluid helium to conduct heat. At temperatures below T_λ any other substance is solid. Liquid helium is important as cryogenic coolant of solid state physics samples and superconducting magnets (in hospital scanners, particle physics accelerators, infrared satellites, etc.).

Consider the channel filled with liquid helium at temperature below T_λ.

The applied heat is carried away by the normal fluid,

$$q = \rho S T v_{\mathrm{n}}\,, \tag{1.0.1}$$

where q is the heat flux, and S specific entropy. But to conserve mass we have $\rho_{\mathrm{n}} v_{\mathrm{n}} + \rho_{\mathrm{s}} v_{\mathrm{s}} = 0$, so that a superflow is generated in the opposite direction, $v_{\mathrm{s}} = -(\rho_{\mathrm{n}}/\rho_{\mathrm{s}}) v_{\mathrm{n}}$. Therefore, a counterflow arises (see Figure 1.0.2) proportional to the heat flux:

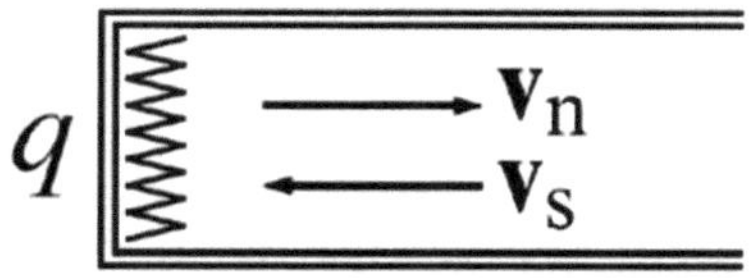

Figure 1.0.2. Thermal counterflow.

$$v_{\mathrm{ns}} = v_{\mathrm{n}} - v_{\mathrm{s}} = \frac{q}{\rho_{\mathrm{s}} S T}\,. \tag{1.0.2}$$

Quantized Vortices

The rotational motion in superfluid component is constrained: if v_{s} exceeds a small critical value, the superfluid forms vortex filaments, see Figure 1.0.3 for illustration. Their main features are: 1) inviscid; 2) the atomic

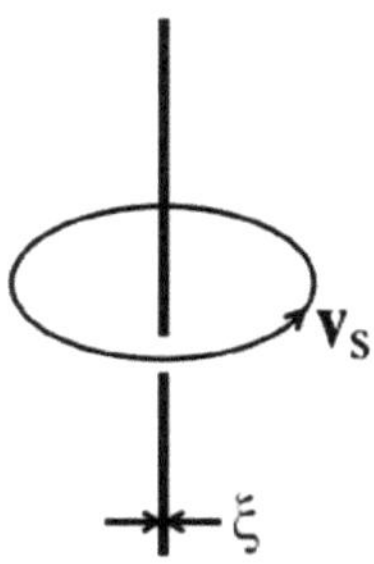

Figure 1.0.3. Superfluid vortex.

size core ($\xi \approx 10^{-8}$ cm); 3) fixed circulation:

$$\oint \mathbf{v}_{\mathrm{s}} \cdot d\mathbf{l} = \frac{h}{m} = \kappa\,, \tag{1.0.3}$$

where h is Planck's constant, m the mass of helium atom, $\kappa = 0.997 \times 10^{-8}\,\mathrm{cm}^2/\mathrm{s}$ the quantum of circulation.

The extra dissipation introduced by the superfluid vortices spoils the otherwise ideal ability of liquid helium to conduct heat, and, in the case where the heat flux is sufficiently large, the counterflow illustrated by Figure 1.0.2 looks different, see Figure 1.0.4: a tangle of superfluid vortices fills the channel. The tangle of quantized vortices constitutes the turbulence

Figure 1.0.4. Turbulent thermal counterflow.

in the superfluid component (the so-called "quantum turbulence"). The normal flow can be either laminar (T-1 turbulence), or turbulent (T-2 turbulence), see Melotte and Barenghi (1998); in the latter case the turbulence in the normal fluid closely resembles classical turbulence.

Numerical simulations, in the periodic box, of the development of the vortex tangle were performed first by Schwarz (1998) and many authors afterwards; time sequence of the tangle's development is illustrated by Figure 8.5.1 of the previous Chapter.

Flow Visualization Techniques and PIV

The aim of experimental physicists working in the area of superfluid turbulence is to understand the vortex tangle (quantum turbulence) and to measure velocity fluctuations in both the normal fluid and the superfluid. Until recently the major difficulty was lack of direct flow visualization technique near absolute zero. Existing methods, such as measurements of temperature differences to detect extra dissipation, ion trapping, measurements of second sound attenuation by quantized vortices, etc. probed only the vortex line density averaged over an experimental cell and did not reveal turbulent velocity fluctuations (although a remarkable resolution has been recently achieved in second sound measurements by Roche *et al.* (2007) of the local tangle density).

Experimental breakthrough was made in 2002 when the PIV (Particle Image Velocimetry) technique was successfully introduced in helium II (see references in the first paragraph of this Section). This technique, which has been standard in classical fluid dynamics for many years, is based on injecting many small particles into the liquid. Two images are produced using short laser pulses of different frequencies (corresponding, for example, to the green and red colours) focused into a narrow sheet and separated in time by a few milliseconds. Software then analyses the images and identifies green and red dots corresponding to the same particle at the two different times, see Figure 1.0.5. In this way the observed distance between the

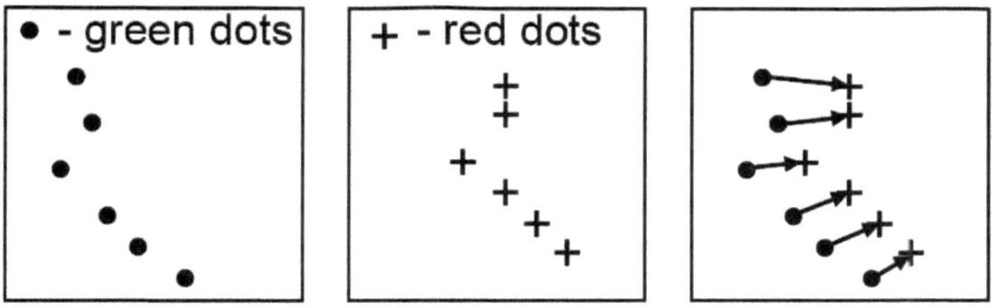

Figure 1.0.5. Particle Image Velocimetry. Images at times t (left) and $t + \Delta t$ (centre). Right – particle displacements.

corresponding dots yields the component of the local velocity in the plane of the light sheet.

In classical fluids, provided the particle size is sufficiently small (in turbulence studies, "small" means smaller than the Kolmogorov length), the dominating force acting on the particle is the viscous drag force, so that a researcher can be confident that small particles trace the fluid motion (in particular, turbulent velocity fluctuations) rather well.

Figure 1.0.6[1] illustrates the results of PIV measurements by Zhang and Van Sciver (2005b) of particle velocity field around the cylinder in helium II counterflow (here L, cm^{-2} is the vortex line density).

Because the viscosity of liquid helium is very low the Kolmogorov scale in the turbulent normal fluid can be very small, so that it is essential that the particles used in PIV measurements are as small as possible. In the cited experimental works the typical particle diameter was of the order of $1\,\mu$m.

What do tracer particles trace in superfluid helium? One may expect that, although the viscosity of helium II is low, the dominating force exerted on the particle by the fluid will still be the viscous drag, so that solid particles should trace the normal fluid. However, this is not always true: due to the two-fluid nature of helium II, the particles interact not only with the

[1]Reprinted by permission from Macmillan Publishers Ltd: T. Zhang and S. W. Van Sciver, *Nature Physics*, **1**:36-38, copyright 2005.

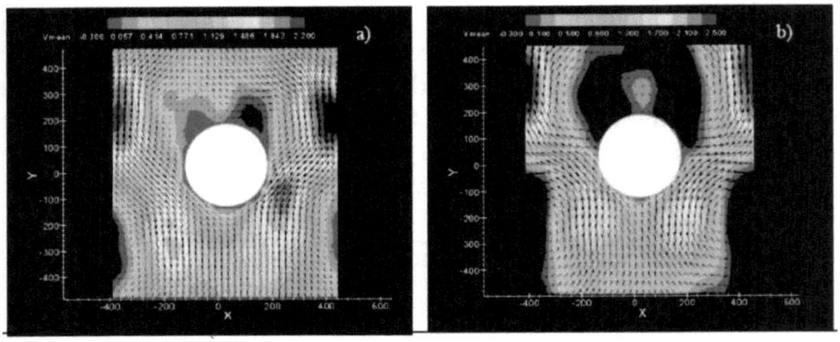

Figure 1.0.6. PIV measurements by Zhang and Van Sciver (2005b) of particle velocity field around the cylinder in helium II counterflow. Particles – polymer microspheres, $a_{\mathrm{p}} = 0.85\,\mu\mathrm{m}$, $\rho_{\mathrm{p}} = 1.1\,\mathrm{g/cm^3}$. Left: $T = 1.6\,\mathrm{K}$, $q = 400\,\mathrm{mW/cm^2}$, $L = 1 \times 10^6\,\mathrm{cm^{-2}}$. Right: $T = 2.03\,\mathrm{K}$, $q = 1120\,\mathrm{mW/cm^2}$, $L = 2.6 \times 10^6\,\mathrm{cm^{-2}}$.

normal fluid, but also with the superfluid component through inertial and added mass forces; moreover, the particles interact strongly with quantized vortices in superfluid and, as we will see later, may even be trapped on superfluid vortex lines. Therefore, if we want to interpret results of PIV measurements correctly, we must answer first the question asked in the beginning of this paragraph.

"One-Way Interaction" Model

We begin answering this question with a relatively simple, "One-Way Interaction" model which follows the approach standard in classical two-phase turbulence studies. (The detailed analysis of the "One-Way Interaction" model of particle motion in helium II will be given later in Section 3.) We will generalize to the two-fluid model of superfluid helium the equations of motion of a solid spherical particle of radius a_{p} in a nonuniform flow of classical fluid. We assume that 1) the presence of particles does not modify the turbulence, 2) the flow velocities vary little over a distance of the order of particle size, and 3) particles do not interact strongly with quantized vortex lines and are certainly not trapped on these lines. These assumptions require that the particle size be much smaller than both the Kolmogorov lengthscale, b_η in the normal fluid, and the mean intervortex distance, $\ell = L^{-1/2}$ in the superfluid, as illustrated in Figure 1.0.7.

To formulate, under these assumptions, the equations of particle motion,

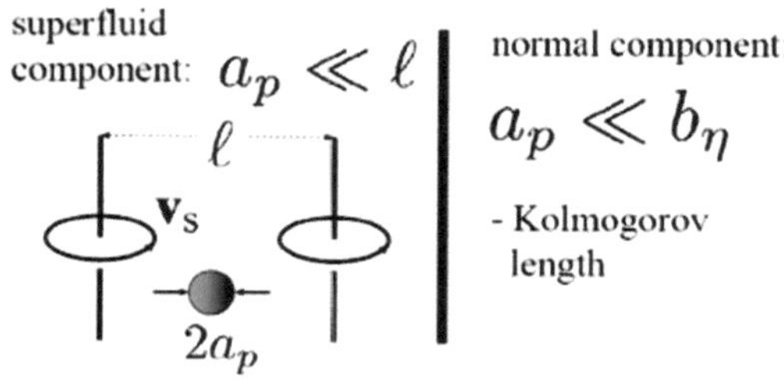

Figure 1.0.7. Particle size must be smaller than the smallest scales of turbulence in both the normal fluid and the superfluid.

we will take into account all forces exerted on the particle by both the inviscid superfluid and the viscous normal fluid. This is not a trivial task, and the correct expressions for the forces acting on the spherical particle in the inviscid and viscous nonuniform flows were formulated only in the last two decades. Until recently, many articles, monographs and textbooks contained fundamental errors where e.g. added mass and inertial forces were concerned. This is why in the next Section we will analyze the forces acting on the particle in classical nonuniform flows in detail.

2 Fluid-Particle Interaction in Classical Liquids

We consider a spherical solid particle of radius a_{p} in the nonuniform flow. Let the ambient (that is, in the absence of the particle) fluid velocity field be $\mathbf{v}_0(\mathbf{r}, t)$, see Figure 2.0.8. The relative velocity between the ambient

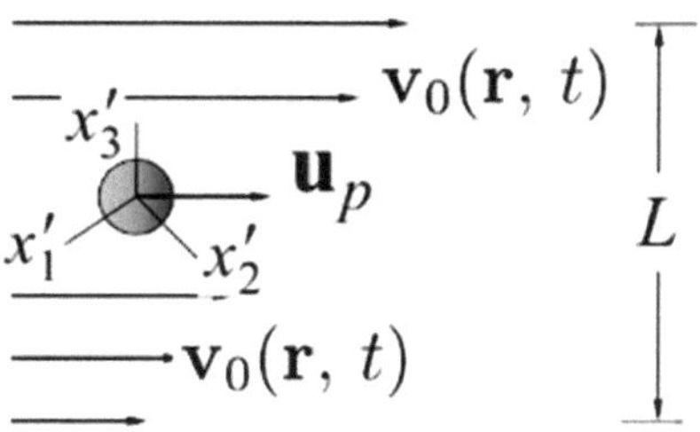

Figure 2.0.8. Spherical particle in the nonuniform flow.

flow and the particle is $\mathbf{W} = \mathbf{v}_0(\mathbf{R}(t), t) - \mathbf{u}_{\mathrm{p}}$, where $\mathbf{R}(t)$ is the current position of the sphere's center. According to the assumptions formulated in

Section 1, the size of the particle is much smaller than the flow lengthscale, L, i.e. $a_{\mathrm{p}} \ll L$, so that we can introduce a small parameter

$$\varepsilon = a_{\mathrm{p}} \|\nabla \mathbf{v}_0\| / W \ll 1 \,. \tag{2.0.1}$$

Spherical Particle in the Inviscid Nonuniform Flow

Irrotational flow. The following derivation is based on the work of Auton, Hunt, and Prud'homme (1988) (see also summary of this derivation in Poole *et al.* (2005)). We start with what seems as the simplest situation: a spherical particle in the slightly nonuniform (i.e. $a_{\mathrm{p}} \ll L$), irrotational flow of the inviscid fluid. It is convenient to consider the fluid flow in the reference frame (x_1', x_2', x_3') of the particle, see Figure 2.0.8. To analyze the flow field around the particle, we will use the perturbation theory with the small parameter ε defined by formula (2.0.1).

The flow field can be decomposed as

$$\mathbf{w} = \mathbf{W} + \Delta \mathbf{w}^{(\mathrm{U})} + \Delta \mathbf{w}^{(\mathrm{E})} \,, \tag{2.0.2}$$

where $\Delta \mathbf{w}^{(\mathrm{U})}$ is the perturbation caused by the presence of the particle (this contribution is clearly of the same order of magnitude as the uniform flow velocity $\mathbf{W}$), and $|\Delta \mathbf{w}^{(\mathrm{E})}|/W \sim O(\varepsilon)$ is the perturbation due to nonuniformity of the ambient flow field; to the leading order, $O(\varepsilon)$, this contribution can be represented as an extensional flow.

Since the flow field is irrotational then $\mathbf{w} = \nabla \phi$, and the potential ϕ itself can be decomposed as

$$\phi = \phi_0 + \phi^{(\mathrm{U})} + \phi^{(\mathrm{E})} \,, \tag{2.0.3}$$

where ϕ_0, $\phi^{(\mathrm{U})}$ and $\phi^{(\mathrm{E})}$ are the contributions provided by the uniform flow, the flow caused by the presence of the sphere, and the extensional flow:

$$\phi_0 = \mathbf{W} \cdot \mathbf{r}' \,, \qquad \phi^{(\mathrm{U})} = C^{(\mathrm{U})}(r') \mathbf{W} \cdot \mathbf{r}' , \qquad \phi^{(\mathrm{E})} = e_{ij} x_i' x_j' \,, \tag{2.0.4}$$

where $C^{(\mathrm{U})} = a_{\mathrm{p}}^3 / (2 r'^3)$ for the spherical particle, and

$$e_{ij} = \frac{1}{2} \left(\frac{\partial v_{0i}}{\partial x_j'} + \frac{\partial v_{0j}}{\partial x_i'} \right) \tag{2.0.5}$$

are the components of the rate-of-strain tensor; here and below the usual convention applies of summation over repeating indices.

In the non-inertial reference frame moving with the particle, the Euler equation for the inviscid flow around the particle can be formulated as

$$\frac{\partial \mathbf{w}}{\partial t} + (\mathbf{w} \cdot \nabla) \mathbf{w} = -\frac{1}{\rho} \nabla p' - \frac{d \mathbf{u}_{\mathrm{p}}}{dt} \,, \tag{2.0.6}$$

where $\mathbf{u}_{\mathrm{p}}(t)$ is the particle velocity, and p' the fluid pressure in the reference frame of the particle.

The force acting on the particle is

$$\mathbf{F} = -\int_S p'\mathbf{n}\,dS\,, \tag{2.0.7}$$

where S is the surface of the particle with $\mathbf{n}$ being the normal unit vector. In order to calculate p', we note that, for potential flow $\mathbf{w}$, Equation (2.0.6) possesses the Cauchy-Lagrange integral

$$\rho\left(\frac{\partial\phi}{\partial t} + \frac{1}{2}\mathbf{w}\cdot\mathbf{w} + \frac{d\mathbf{u}_{\mathrm{p}}}{dt}\cdot\mathbf{r}'\right) + p' = f(t)\,, \tag{2.0.8}$$

where $f(t)$ is the function which can be determined from the conditions at $|\mathbf{r}'| \to \infty$. Applying the perturbation theory up to and including the terms $O(\varepsilon)$, the force acting on the sphere can be calculated from Equations (2.0.7) and (2.0.8) as

$$\begin{aligned} \frac{\mathbf{F}}{\rho} &= \int_S \left(\frac{\partial\phi_0}{\partial t} + \frac{\partial\phi^{(\mathrm{U})}}{\partial t}\right)\mathbf{n}\,dS \\ &+ \int_S \left[\left(\mathbf{W} + \nabla\phi^{(\mathrm{U})}\right)\cdot\nabla\phi^{(\mathrm{E})}\right]\mathbf{n}\,dS + \vartheta\,\frac{d\mathbf{u}_{\mathrm{p}}}{dt}\,, \end{aligned} \tag{2.0.9}$$

where $\vartheta = \frac{4}{3}\pi a_{\mathrm{p}}^3$ is the particle volume. Using potentials (2.0.4), for the components of the force Equation (2.0.9) yields

$$F_i = \rho\vartheta\left\{(1 + C)\left(\dot{W}_i + e_{ij}W_j\right) + (\dot{u}_p)_i\right\}, \tag{2.0.10}$$

where $(\dot{\ }) \equiv d/dt$ and $C = C^{(\mathrm{U})}(r')\big|_{r'=a} = \frac{1}{2}$ is the added mass coefficient for a spherical particle. Now,

$$\begin{aligned} \dot{W}_i + e_{ij}W_j &= \frac{\partial v_{0i}}{\partial t} + e_{ij}v_{0j} - (\dot{u_{\mathrm{p}}})_i \\ = \frac{\partial v_{0i}}{\partial t} &+ v_{0j}\,\frac{\partial v_{0j}}{\partial x_j} - (\dot{u_{\mathrm{p}}})_i = \frac{Dv_{0i}}{Dt} - \frac{d(u_{\mathrm{p}})_i}{dt}\,, \end{aligned} \tag{2.0.11}$$

where

$$\frac{D}{Dt} = \frac{\partial}{\partial t} + (\mathbf{v}\cdot\nabla) \tag{2.0.12}$$

is the substantial derivative.

Combining Equations (2.0.10) and (2.0.11), we obtain the following representation of the total force acting on the particle in the nonuniform, irrotational, inviscid flow:

$$\mathbf{F} = \mathbf{F}^{(\mathrm{i})} + \mathbf{F}^{(\mathrm{a})}\,, \tag{2.0.13}$$

where

$$\mathbf{F}^{(\mathrm{i})} = \rho\vartheta\,\frac{D\mathbf{v}}{Dt} \qquad \text{and} \qquad \mathbf{F}^{(\mathrm{a})} = C\rho\vartheta\left(\frac{D\mathbf{v}}{Dt} - \frac{d\mathbf{u}_\mathrm{p}}{dt}\right) \tag{2.0.14}$$

are the inertial and the added mass force, respectively (here and below the subscript "0" referring to the ambient flow is omitted).

Rotational flow. Assume now that $\nabla \times \mathbf{v}_0 = \boldsymbol{\omega}_0 \neq 0$. Using the perturbation theory, with ε defined by (2.0.1) being the small parameter, we write

$$\mathbf{w} = \mathbf{W} + \Delta\mathbf{w}^{(\mathrm{U})} + \Delta\mathbf{w}^{(\mathrm{E})} + \Delta\mathbf{w}^{(\omega)}\,, \tag{2.0.15}$$

where $\Delta\mathbf{w}^{(\omega)} \sim O(\varepsilon)$ is the rotational contribution. Calculation by Auton, Hunt, and Prud'homme (1988), similar to that in the previous section, yields:

$$\mathbf{F} = \rho\vartheta\,\frac{D\mathbf{v}}{Dt} + C\rho\vartheta\left(\frac{D\mathbf{v}}{Dt} - \frac{d\mathbf{u}_\mathrm{p}}{dt}\right) + \mathbf{F}^{(\omega)}\,, \tag{2.0.16}$$

where

$$\mathbf{F}^{(\omega)} = \frac{1}{2}\,\rho\vartheta(\mathbf{v} - \mathbf{u}_\mathrm{p}) \times \boldsymbol{\omega} \tag{2.0.17}$$

is the lift force arising due to stretching of vortex lines in the vicinity of the sphere's surface.

Particle in the Nonuniform Viscous Flow

We will consider the motion of a solid particle in the viscous fluid assuming that the particle Reynolds number is small:

$$\mathrm{Re_p} = \frac{a_\mathrm{p}|\mathbf{v} - \mathbf{u}_\mathrm{p}|}{\nu} \ll 1\,, \tag{2.0.18}$$

where ν is the kinematic viscosity. Note that small particle Reynolds numbers are typical of PIV experiments, both in classical fluids and in superfluid helium. We will consider the particle motion under the assumptions formulated in Sections 1 and 2 (in particular that the parameter ε introduced by formula (2.0.1) is small).

Detailed analysis of the forces acting on the particle in the nonuniform viscous flow can be found in works by Maxey and Riley (1983), Mei (1994), and Kim, Elghobashi, and Sirignano (1998). The total force acting on the particle can be represented as a sum of several contributions:

$$\mathbf{F} = \mathbf{F}^{(\mathrm{g})} + \mathbf{F}^{(\mathrm{d})} + \mathbf{F}^{(\mathrm{F})} + \mathbf{F}^{(\mathrm{i})} + \mathbf{F}^{(\mathrm{a})} + \mathbf{F}^{(\mathrm{B})} + \mathbf{F}^{(\omega)} + \mathbf{F}^{(\mathrm{LM})} . \qquad (2.0.19)$$

The main contribution, dominating in most particulate flows, is the viscous drag force (in the case of small particle Reynolds numbers it is usually called the Stokes force):

$$\mathbf{F}^{(\mathrm{d})} = 6\pi a_{\mathrm{p}} \rho \nu (\mathbf{v} - \mathbf{u}_{\mathrm{p}}) . \qquad (2.0.20)$$

Derivation of this force can be found in any textbook on fluid dynamics, e.g. Landau and Lifshitz (1987) or Batchelor (1973). The force $\mathbf{F}^{(\mathrm{F})} \sim O(a_{\mathrm{p}}^2/L^2)\mathbf{F}^{(\mathrm{d})}$ is known as the Faxén correction, due to the local nonuniformity of the ambient flow, to the Stokes force and, provided $a_{\mathrm{p}} \ll L$, is very small.

Surprisingly, in the viscous flow the inertial and added mass forces, $\mathbf{F}^{(\mathrm{i})}$ and $\mathbf{F}^{(\mathrm{a})}$ are determined by the same formulae (2.0.13)-(2.0.14) as for the inviscid flow, with the same added mass coefficient, $C = \frac{1}{2}$ for the spherical particle.

The force acting on the particle in the viscous flow depends on the whole history of the relative fluid-particle motion. The contribution $\mathbf{F}^{(\mathrm{B})}$ is known as the Basset, or history force, and is determined by the integral

$$\mathbf{F}^{(\mathrm{B})} = 6\pi a_{\mathrm{p}} \rho \nu \int_{-\infty}^{t} \frac{a_{\mathrm{p}}}{\sqrt{\pi\nu(t-\tau)}} \frac{d(\mathbf{v} - \mathbf{u}_{\mathrm{p}})}{d\tau} \, d\tau . \qquad (2.0.21)$$

$\mathbf{F}^{(\omega)}$ is the so-called Saffman lift force acting on the particle due to the local shear of the ambient flow field (see Figure 2.0.9):

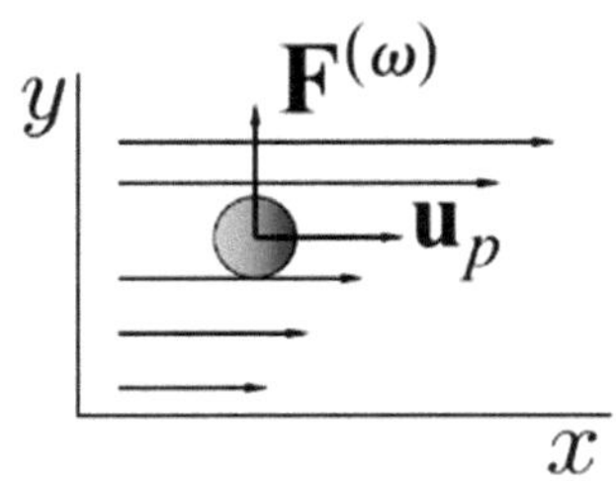

Figure 2.0.9. Saffman lift force in a simple shear flow.

$$\mathbf{F}_y^{(\omega)} = 6.46\, \rho \nu^{1/2} a_{\mathrm{p}}^2 |\boldsymbol{\omega}|^{-1/2} (\mathbf{v} - \mathbf{u}_{\mathrm{p}}) \times \boldsymbol{\omega} . \qquad (2.0.22)$$

The Magnus lift force, $\mathbf{F}^{(\mathrm{LM})}$ is caused by the particle rotation in the viscous fluid:

$$\mathbf{F}^{(\mathrm{LM})} = \pi a_{\mathrm{p}}^{3} \rho \boldsymbol{\Omega} \times (\mathbf{u}_{\mathrm{p}} - \mathbf{v}) , \qquad (2.0.23)$$

where $\boldsymbol{\Omega}$ is the angular velocity of particle rotation.

3 Particle Motion in Superfluid Helium: "One-Way Interaction" Model

Forces Acting on the Particle

To determine the force acting on the particle in helium II, we simply add together all the forces exerted by the normal fluid and the superfluid:

$$\begin{aligned} \mathbf{F} &= \mathbf{F}^{(\mathrm{g})} \\ &+ \mathbf{F}_{\mathrm{n}}^{(\mathrm{d})} + \mathbf{F}_{\mathrm{n}}^{(\mathrm{i})} + \mathbf{F}_{\mathrm{n}}^{(\mathrm{a})} + \mathbf{F}_{\mathrm{n}}^{(\mathrm{B})} + \mathbf{F}_{\mathrm{n}}^{(\mathrm{F})} + \mathbf{F}_{\mathrm{n}}^{(\omega)} + \mathbf{F}_{\mathrm{n}}^{(\mathrm{LM})} \\ &+ \mathbf{F}_{\mathrm{s}}^{(\mathrm{i})} + \mathbf{F}_{\mathrm{s}}^{(\mathrm{a})} + \mathbf{F}_{\mathrm{s}}^{(\omega)} , \end{aligned} \qquad (3.0.1)$$

where the subscripts "n" and "s" refer to the normal fluid and the superfluid, respectively. Here $\mathbf{F}^{(\mathrm{g})}$ is the combination of gravity and buoyancy, and $\mathbf{F}_{\mathrm{n}}^{(\mathrm{d})}$ is the viscous drag force exerted by the normal fluid:

$$\mathbf{F}^{(\mathrm{g})} = \vartheta(\rho_{\mathrm{p}} - \rho)\mathbf{g}, \qquad \mathbf{F}_{\mathrm{n}}^{(\mathrm{d})} = 6\pi a_{\mathrm{p}} \mu_{\mathrm{n}} (\mathbf{v}_{\mathrm{n}} - \mathbf{u}_{\mathrm{p}}) , \qquad (3.0.2)$$

where μ_{n} is the viscosity of helium II.

Substantial derivatives required for determining the inertial and added mass forces in the normal and superfluid components, respectively, are now defined as

$$\frac{D\mathbf{v}_{\mathrm{n}}}{Dt} = \frac{\partial \mathbf{v}_{\mathrm{n}}}{\partial t} + (\mathbf{v}_{\mathrm{n}} \cdot \nabla)\mathbf{v}_{\mathrm{n}} \quad \text{and} \quad \frac{D\mathbf{v}_{\mathrm{s}}}{Dt} = \frac{\partial \mathbf{v}_{\mathrm{s}}}{\partial t} + (\mathbf{v}_{\mathrm{s}} \cdot \nabla)\mathbf{v}_{\mathrm{s}} . \qquad (3.0.3)$$

Under the assumptions formulated in Sections 1 and 2 some of the contributing forces can be neglected, see Exercise 1-3.

Exercise 1. Explain why the lift force in the superfluid component, $\mathbf{F}_{\mathrm{s}}^{(\omega)}$ is identically zero, provided the particle does not become trapped on the superfluid vortex lines.

Exercise 2. Assuming that the particle size a_{p} is much smaller than the smallest length scale of the turbulence in the normal fluid, L (i.e. the Kolmogorov length b_η), and that the particle Reynolds number, $\mathrm{Re}_{\mathrm{p}} = a_{\mathrm{p}}|\mathbf{v} - \mathbf{u}_{\mathrm{p}}|/\nu$ is small, estimate the ratios of the Saffman and Magnus lift forces to the viscous drag force and hence show that these forces can be neglected. [You may find helpful making use of the small parameter $\varepsilon = a\|\nabla \mathbf{v}_0\|/W$.]

Exercise 3. Show that the Basset (history) force can be neglected compared with the viscous drag force provided the particle size is much smaller than $b_\eta(\rho_n/\rho)^{1/2}$, where b_η is the Kolmogorov length.

Lagrangian Equations of Particle Motion

Having solved Exercises 1-3, we arrive at the following equation of particle motion:

$$\begin{aligned} \rho_p \vartheta \frac{d\mathbf{u}_p}{dt} &= 6\pi a_p \mu_n (\mathbf{v}_n - \mathbf{u}_p) + \vartheta(\rho_p - \rho)\mathbf{g} \\ &+ \rho_n \vartheta \frac{D\mathbf{v}_n}{Dt} + C\rho_n \vartheta \left(\frac{D\mathbf{v}_n}{Dt} - \frac{d\mathbf{u}_p}{dt} \right) \\ &+ \rho_s \vartheta \frac{D\mathbf{v}_s}{Dt} + C\rho_s \vartheta \left(\frac{D\mathbf{v}_s}{Dt} - \frac{d\mathbf{u}_p}{dt} \right). \end{aligned} \tag{3.0.4}$$

This equation must be considered together with the kinematic equation

$$\frac{d\mathbf{r}}{dt} = \mathbf{u}_p , \tag{3.0.5}$$

where $\mathbf{r} = \mathbf{r}(t) = (x_1(t), x_2(t), x_3(t))$ $(\equiv \mathbf{r}_p(t))$ should be regarded as a Lagrangian trajectory of the solid particle. Equations (3.0.4) and (3.0.5) constitute the closed system for the unknown particle position and velocity, $\mathbf{r}(t)$ and $\mathbf{u}_p(t)$, respectively.

Very often (although not always) researchers performing PIV experiments use neutrally buoyant particles (with $\rho_p = \rho$) in order to eliminate unwanted effects of gravity on the particle motion. For neutrally buoyant particles, Equation (3.0.4) can be written in a more concise form

$$\frac{d\mathbf{u}_p}{dt} = \frac{1}{\tau}(\mathbf{v}_n - \mathbf{u}_p) + \frac{3}{2\rho_o}\left(\rho_n \frac{D\mathbf{v}_n}{Dt} + \rho_s \frac{D\mathbf{v}_s}{Dt} \right), \tag{3.0.6}$$

where

$$\rho_o = \rho_p + \frac{\rho}{2} = \frac{3}{2}\rho \quad \text{and} \quad \tau = \frac{2a_p^2 \rho_o}{9\mu_n}. \tag{3.0.7}$$

The parameter τ plays an important rôle in the study of particle motion in fluids, and is commonly known as either the particle response time, or viscous relaxation time (it shows how quickly the particle adjusts its motion to the fluid flow).

Exercise 4. Assume that the normal fluid and the superfluid are both stationary ($\mathbf{v}_\mathrm{n} = \mathbf{v}_\mathrm{s} = \mathbf{0}$). Consider, in the presence of gravity, the particle sedimentation from $\mathbf{u}_\mathrm{p} = \mathbf{0}$. Calculate the terminal speed of sedimentation, and find the time scale, τ_∞ of the transient motion. Calculate these quantities at temperature $T = 1.5\,\mathrm{K}$ for polymer microspheres of size $a_\mathrm{p} = 10^{-5}\,\mathrm{cm}$ and density $\rho_\mathrm{p} = 1.1\,\mathrm{g/cm^3}$.

Exercise 5. Assume that the velocities of the normal fluid and the superfluid have no spatial dependence but have harmonic time dependence:

$$\mathbf{v}_\mathrm{n} = \mathbf{V}_\mathrm{n}\exp(i\omega t), \qquad \mathbf{v}_\mathrm{s} = \mathbf{V}_\mathrm{s}\exp(i\omega t)\,.$$

Find the solution of Lagrangian equations of motion for the neutrally buoyant particle and hence show that, after a short transient,

a) if $\omega\tau \gg 1$, the particle moves with a velocity corresponding to the total current density, $\mathbf{j} = \rho_\mathrm{n}\mathbf{v}_\mathrm{n} + \rho_\mathrm{s}\mathbf{v}_\mathrm{s}$;

b) if $\omega\tau \ll 1$, the particle tracks the normal fluid.

In fact, the result *b)* of Exercise 5 remains valid for general, spatially dependent flows as well, provided that the effects associated with close interactions between particles and quantized vortices can be neglected. More generally, the condition in the result *b)* should be written as $\tau/\tau_\mathrm{f} \ll 1$, where τ_f is the timescale of the fluid motion (e.g. the Kolmogorov time).

Although Equations (3.0.4) or (3.0.6) cannot be used for the analysis of close interaction between the particle and the vortex (e.g. for the analysis of vortex reconnection to the particle surface and particle trapping by the vortex core), they can still be used for the analysis of an approach of the particle to the quantized vortex, see the following exercise.

Exercise 6. Particle-vortex approach at finite temperatures. Neglecting any influence of the particle on the straight quantized vortex, and assuming that the normal fluid velocity $\mathbf{v}_\mathrm{n} \equiv 0$, use the Lagrangian equations of motion for the neutrally buoyant particle,

$$\frac{d\mathbf{u}_\mathrm{p}}{dt} = -\frac{1}{\tau}\,\mathbf{u}_\mathrm{p} + \frac{3\rho_\mathrm{s}}{2\rho_\mathrm{o}}(\mathbf{v}_\mathrm{s}\cdot\nabla)\mathbf{v}_\mathrm{s}\,, \qquad \frac{d\mathbf{r}_\mathrm{p}}{dt} = \mathbf{u}_\mathrm{p}$$

to describe the approach of the particle to the vortex (see Figure 3.0.10). Assume that, due to viscosity of the normal fluid, any orbital motion of the particle will be strongly damped. In polar cylindrical coordinates, consider the motion of the particle in the radial direction, with $u_\mathrm{p}(t)$ and $r_\mathrm{p}(t)$ being

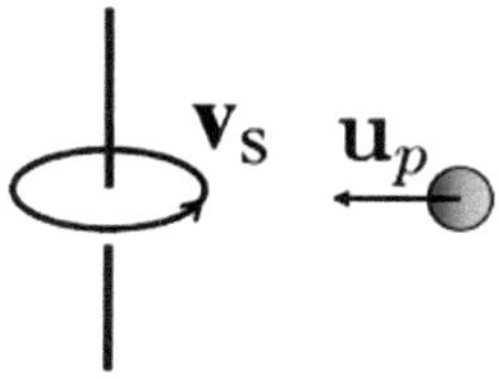

Figure 3.0.10. Particle-vortex approach.

the particle's velocity and radial coordinate. Taking into account that, in cylindrical polar coordinates, the velocity field of the superfluid vortex is $\mathbf{v}_s = [0,\, \kappa/(2\pi r),\, 0]$, you may find it useful to represent the last term in the equation of motion as a radial pressure gradient force.
Find $r_p(t)$ and $u_p(t)$. For neutrally buoyant particle of size $a_p = 1\,\mu$m, using numerical values for the properties of He II at $T = 1.85\,$K, recover the curve (a) of the graph shown in Figure 4.0.30 of Section 4. Calculate the time required for particle to travel from $r = r_0$ to $r = 2a_p$, and find numerical estimate for this time assuming $r_0 = 5a_p$.

Two important issues may invalidate the conclusions drawn from the "One-Way Interaction" model: 1) Instability of particle trajectories, and 2) Trapping of particles on superfluid vortex lines. The analysis of particle trapping will require more elaborate, self-consistent, Mutual Interaction model which would account for deformation of the vortex filament by the approaching particle, including possible reconnection of the vortex with the particle surface (see below Section 4).

Instability of Particle Trajectories at Finite Temperature.

Instability of Lagrangian trajectories of the neutrally buoyant particle in the classical viscous fluid was discovered and studied relatively recently by Babiano *et al.* (2000). To illustrate such an instability in turbulent superfluid at finite temperature such that both the normal fluid and the superfluid component are present, we consider the simplest case assuming that $\mathbf{v}_n \approx \mathbf{v}_s =: \mathbf{v}$ down to the length scales comparable to the vortex line spacing. The Lagrangian equations of particle motion, which become

$$\frac{d\mathbf{u}_p}{dt} = \frac{\mathbf{v} - \mathbf{u}_p}{\tau} + \frac{D\mathbf{v}}{Dt}, \qquad \frac{d\mathbf{r}_p}{dt} = \mathbf{u}_p, \tag{3.0.8}$$

have, provided $\mathbf{u}_p(0) = \mathbf{v}(0)$ and $\mathbf{r}_p(0) = \mathbf{r}_f(0)$, a formal solution $\mathbf{u}_p(t) = \mathbf{v}(\mathbf{r}_f,\, t)$, $\mathbf{r}_p(t) = \mathbf{r}_f(t)$, where $\mathbf{r}_f(t)$ is a trajectory of the fluid point. Therefore, it would seem natural to conclude that neutrally buoyant particles follow the fluid exactly.

However, let us consider now the particle motion in the so-called ABC (Arnold-Beltrami-Childress) flow, frequently used as the simplest model of turbulence, whose velocity field is

$$\begin{aligned} v_x &= A\sin(2\pi z) + C\cos(2\pi y)\,, \\ v_y &= B\sin(2\pi x) + A\cos(2\pi z)\,, \\ v_z &= C\sin(2\pi y) + B\cos(2\pi x)\,. \end{aligned} \tag{3.0.9}$$

The path lines of this flow in the $(x,\ z)$-plane are illustrated in Figure 3.0.11. The time sequence illustrating the position of large number of neutrally

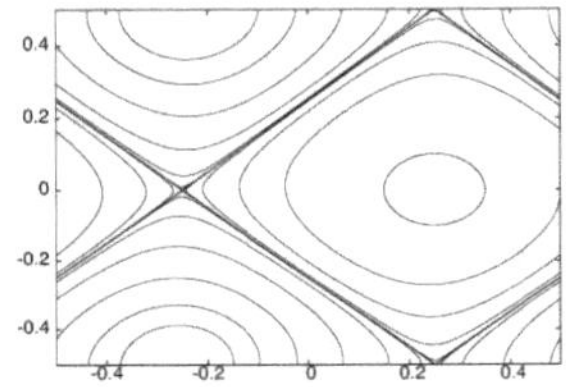

Figure 3.0.11. Path lines of the ABC flow.

buoyant tracer particles, starting from their uniform distribution, is shown in Figure 3.0.12[2]. It can be seen that particles do not follow the fluid, but instead, due to instabilities of their trajectories, travel from the regions of high vorticity to the regions of high rate of strain. (In classical multiphase fluid dynamics similar phenomenon for particles heavier than the fluid has been studied in detail. However, the mechanism of segregation of heavy particles is quite different from that of neutrally buoyant particles.) Note, though, that this instability develops rather slowly: the last frame of Figure 3.0.12 corresponds to 3 times of turnover of the ABC flow.

Instability of Particle Trajectories at Very Low Temperatures ($T \to 0$, $\rho_{\mathbf{n}} = 0$).

Instability of particle trajectories is more pronounced in the case $T \to 0$ when the normal fluid is absent. The equations of motion of neutrally

[2]Reprinted, by permission from the American Physical Society, from D. R. Poole, C. F. Barenghi, Y. A. Sergeev, and W. F. Vinen, *Phys. Rev. B*, **71**:064514, 2005. Copyright 2005 by the American Physical Society.

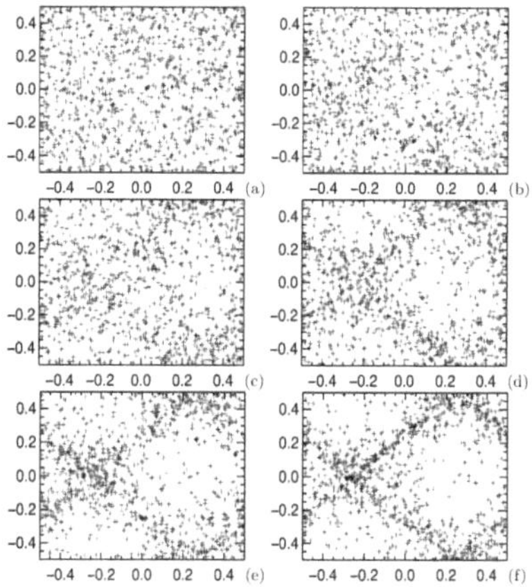

Figure 3.0.12. Position of tracer particles at different times (Poole *et al.* (2005)).

buoyant particle become

$$\frac{d\mathbf{u}_\mathrm{p}}{dt} = \frac{\partial \mathbf{v}_\mathrm{s}}{\partial t} + (\mathbf{v}_\mathrm{s} \cdot \nabla)\mathbf{v}_\mathrm{s} = -\nabla p\,, \qquad \frac{d\mathbf{r}_\mathrm{p}}{dt} = \mathbf{u}_\mathrm{p}\,, \tag{3.0.10}$$

and have a formal solution $\mathbf{u}_\mathrm{p}(t) = \mathbf{v}_\mathrm{s}(\mathbf{r}_\mathrm{s}(t),\, t)$, $\mathbf{r}_\mathrm{p}(t) = \mathbf{r}_\mathrm{s}(t)$, where $d\mathbf{r}_\mathrm{p}/dt = \mathbf{v}_\mathrm{s}(\mathbf{r}_\mathrm{s},\, t)$, so that, in the case of very low temperature when the normal fluid is absent, it can be expected that neutrally buoyant particles trace the superfluid. However, we will see that it is not the case either. Consider, as shown in Figure 3.0.13, the simplest case of the neutrally buoyant particle

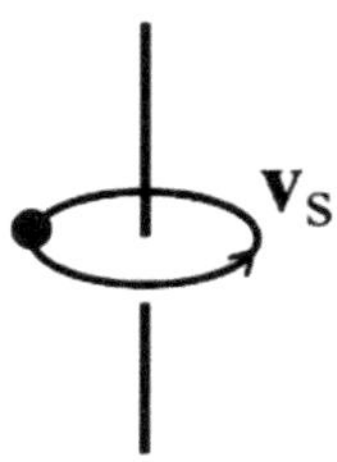

Figure 3.0.13. Particle motion around a stationary single vortex.

moving around a single, stationary, straight vortex line. In cylindrical polar coordinates $(r_\mathrm{p},\, \theta_\mathrm{p})$ the equations of particle motion are (see Exercises 7 and 8 below):

$$\ddot{r}_{\mathrm{p}} - r_{\mathrm{p}}\omega_{\mathrm{p}}^2 = -\frac{\kappa^2}{2\pi^2 r_{\mathrm{p}}^3}\,, \qquad 2\omega\dot{r}\mathrm{p} + r_{\mathrm{p}}\dot{\omega}_{\mathrm{p}} = 0\,, \tag{3.0.11}$$

where $\omega_{\mathrm{p}} = \dot{\theta}_{\mathrm{p}}$.

Exercise 7. Derive these equations from the Lagrangian equations of particle motion at very low temperatures (in the absence of normal fluid). [Use the flow field of a single straight vortex line to represent $\partial \mathbf{v}_{\mathrm{s}}/\partial t + (\mathbf{v}_{\mathrm{s}}\nabla)\mathbf{v}_{\mathrm{s}}$ as a two-dimensional central potential force.]

Exercise 8. Solve these equations and hence show that, if at the initial moment the particle velocity does not coincide exactly with the fluid velocity, the particle will spiral either outwards (towards $r_{\mathrm{p}} \to \infty$), or inwards (towards $r_{\mathrm{p}} \to 0$). [If you have difficulty solving these equations, try linearizing them around a circular trajectory, and then solve simpler, linearized equations. You will arrive at the same conclusion.]

Having solved Exercises 7 and 8, we arrive at the conclusion which remains valid in the general case of particle motion at temperature $T \to 0$ (Sergeev *et al.* (2006a)): unless the initial velocity of neutrally buoyant particle matches exactly the velocity of the fluid point, the trajectory of the solid particle deviates significantly from the trajectory of the fluid point. Such an instability is amplified by any macroscopic mismatch between the velocity of the superfluid and the velocity of the particle at the beginning of the experiment. This is further illustrated by Figure 3.0.14[3] showing the trajectories of solid and fluid particles around three vortices.

Trapping of Particles on Quantized Vortices: Mechanism and Experimental Evidence

Why can the superfluid vortex trap the particle? To answer this question, we have to take into account a possibility of reconnection of the quantized vortex to the surface of moving particle (a more detailed analysis of the mechanism of this process will be given below in Section 4).

Below three different versions, or, rather, three different ways of explaining the reason for particle trapping are suggested.

[3]Reprinted, by permission from the American Physical Society, from Y. A. Sergeev, C. F. Barenghi, D. Kivotides, and W. F. Vinen, *Phys. Rev. B*, **73**:052502, 2006. Copyright 2006 by the American Physical Society.

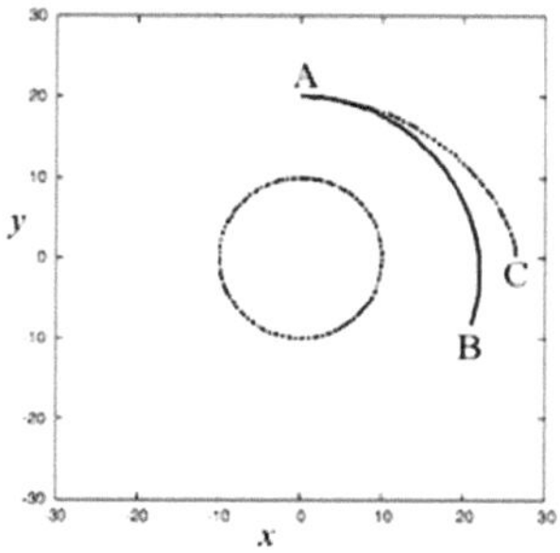

Figure 3.0.14. Trajectories of neutrally buoyant inertial particle (dashed line A to C) and superfluid particle (solid line A to B) around three vortices moving along the closed orbit shown (Sergeev *et al.* (2006a)).

1°. Imagine that the quantized vortex reconnects symmetrically to the surface of spherical particle, as shown in Figure 3.0.15. The kinetic energy of the flow field created by the straight quantized vortex can be easily

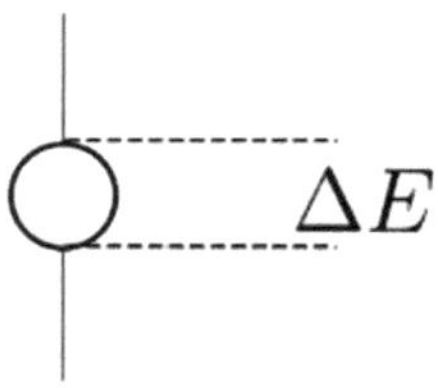

Figure 3.0.15. Reconnection of the vortex to the particle surface reduces kinetic energy of helium by ΔE.

calculated, and in the configuration shown in Figure 3.0.15 is reduced by

$$\Delta E = \frac{\rho_{\rm s}\kappa a_{\rm p}}{4\pi} \ln \frac{a_{\rm p}}{\xi} . \tag{3.0.12}$$

Formula (3.0.12) determines the kinetic energy which the particle would require to break free from the configuration shown in Figure 3.0.15.

2°. The flow field of the vortex (see Figure 3.0.16) creates a pressure gradient

$$\nabla p = -\rho_{\rm s}(\mathbf{v}_{\rm s} \cdot \nabla)\mathbf{v}_{\rm s} = \frac{\rho_{\rm s}\kappa^2}{8\pi^2} \nabla \left(\frac{1}{r^2}\right) \tag{3.0.13}$$

attracting the particle to the vortex.

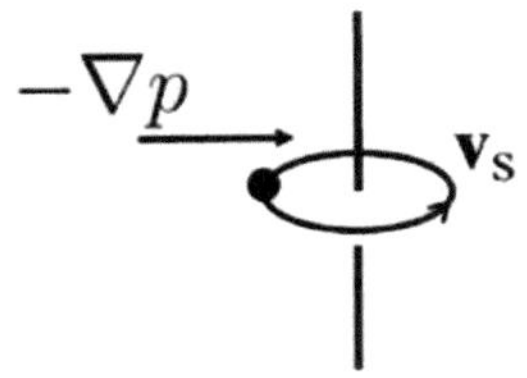

Figure 3.0.16. Flow field of the vortex creates a pressure gradient force acting on the particle.

3°. If the particle-vortex configuration is symmetric, as in the right part of Figure 3.0.17[4], then, obviously, the force acting on the particle is zero.

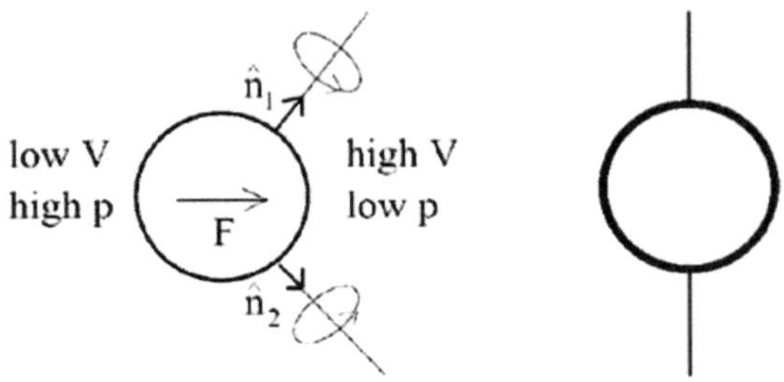

Figure 3.0.17. Asymmetric reconnection of the vortex to the particle surface creates a force restoring the symmetric particle-vortex configuration (Sergeev, Barenghi, and Kivotides (2006b)).

Now imagine that this symmetry is perturbed as shown in the left part of this figure. The superfluid component is inviscid, so that the following arguments apply based on Bernoulli's integral: the fluid velocity on the right side of the particle surface, where the two vortex strands are closer, is greater than that on the left side, and, therefore, the pressure is bigger on the left side of the sphere. This provides a net force restoring the symmetric particle-vortex configuration.

Is there any evidence of particle trapping by quantized vortices? Yes, and having provided three explanations for trapping phenomenon, below we will also provide three evidences.

[4]Reprinted, by permission from the American Physical Society, from Y. A. Sergeev, C. F. Barenghi, and D. Kivotides, *Phys. Rev. B*, **74**:184506, 2006. Copyright 2006 by the American Physical Society.

Evidence 1° comes from the recent publication in Nature by Bewley, Lathrop, and Sreenivasan (2006). Figure 3.0.18[5] shows that the researchers,

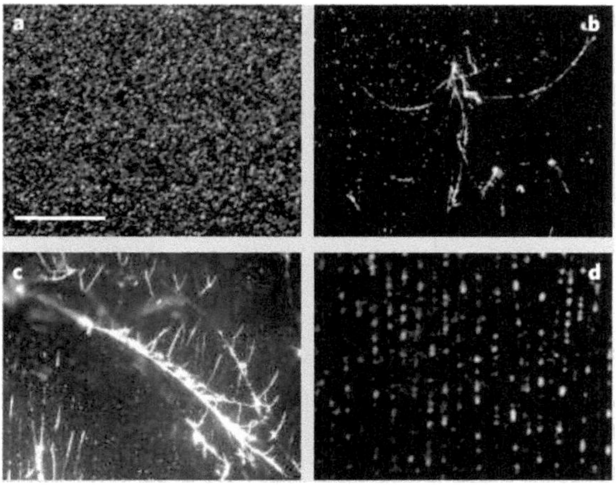

Figure 3.0.18. PIV visualization of quantized vortex cores (Bewley, Lathrop, and Sreenivasan (2006)): (a) – above λ transition; (b) and (c) – branching filaments tenth of mK below λ transition; (d) – regrouping along vertical lines in steady rotating apparatus.

using the PIV technique, actually "painted" quantized vortices by tracer particles.

Evidence 2° comes from the experimental results of Zhang and Van Sciver (2005a) who studied, by the PIV technique, the sedimentation of heavy (not neutrally buoyant) particles in the vertical thermal counterflow with the heat source at the bottom of the experimental apparatus. In such a counterflow the normal flow is directed upwards. Assuming that the dominating force of the fluid-particle interaction is the viscous drag, it could have been expected that the particle speed will be just the difference between the normal fluid velocity and the terminal velocity of particle sedimentation,

$$v_{\mathrm{slip}} = \frac{2a_{\mathrm{p}}^2}{9\mu_{\mathrm{n}}} (\rho_{\mathrm{p}} - \rho) \tag{3.0.14}$$

(see Exercise 4). Were this correct, on the graph showing the adjusted particle speed, $v_{\mathrm{pa}} = u_{\mathrm{p}} + v_{\mathrm{slip}}$ the experimental data would have collapsed

[5]Reprinted by permission from Macmillan Publishers Ltd: G. P. Bewley, D. P. Lathrop, and K. R. Sreenivasan, *Nature*, **44**:588, copyright 2006

on the solid straight line shown in Figure 3.0.19. However, as seen from this figure, the measured particle speed was considerably lower than pre-

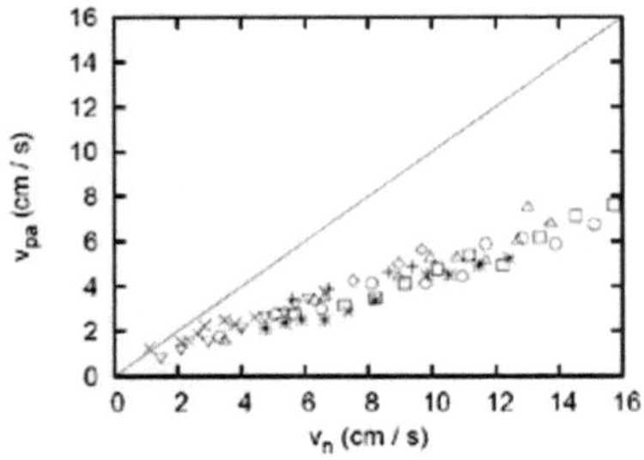

Figure 3.0.19. PIV measurement by Zhang and Van Sciver (2005a) of $v_{\text{pa}} = u_{\text{p}} + v_{\text{slip}}$ for particles sedimenting in turbulent thermal counterflow.

dicted by the simple model described above. Instead, it was found that $u_{\text{p}} = v_{\text{n}} - v_{\text{slip}} - v_{\text{add}}$, where the additional speed, v_{add} turned out to be proportional to the counterflow heat flux, $v_{\text{add}} \sim q = \rho S T v_{\text{n}}$. The presence of this additional velocity can be explained only by strong interactions (e.g. sequences of trapping-untrapping events) between particles and quantized vortices.

Evidence 3° comes not from experiments, but from the quantum calculation based on the Gross-Pitaevskii equation. Berloff and Roberts (2000) studied an interaction between the negative ion and the quantized vortex. In superfluid helium, the negative ion forms around itself a bubble of diameter approximately 16×10^{-8} cm (an order of magnitude higher than the size of the vortex core, $\xi \approx 10^{-8}$ cm) and, therefore, can be treated, for our purpose, as a particle (albeit very small). Results of Berloff and Roberts' calculation are illustrated in Figure 3.0.20[6]: the ion bubble approaches the vortex, which deforms ((b) and (c)) and then reconnects to the particle surface (d); the reconnection excites Kelvin waves ((f), (g), and (h)) which carry away the energy, and, eventually, the bubble-vortex configuration relaxes and the ion becomes trapped on the vortex core (i).

[6]Reprinted, by permission from the American Physical Society, from N. G. Berloff and P. H. Roberts, *Phys. Rev. B*, **63**:024510, 2000. Copyright 2000 by the American Physical Society.

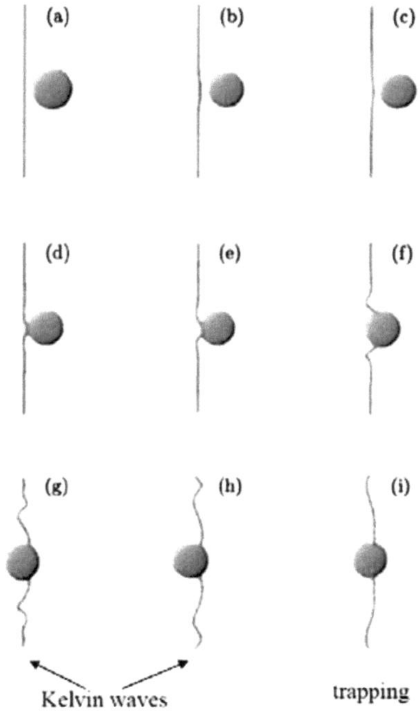

Figure 3.0.20. Trapping of the ion bubble by the quantized vortex: microscopic calculation by Berloff and Roberts (2000) based on the Gross-Pitaevskii equation.

Mutual Friction and Trapping of Particles on Vortex Cores

At this point, the question can be asked whether it is worth or not further exploiting the "One-Way Interaction" model which neglects any influence of the particle on the vortex evolution. The answer is "yes". We will see below that there are cases where trapping events are not very frequent so that we can learn a lot about the behaviour of tracer particles by just ignoring their trapping on vortex cores. We may also try to find out whether there is any phenomenon that can prevent trapping. In fact, such a phenomenon is mutual friction, described by C. F. Barenghi in the previous Chapter, between superfluid vortices and the normal fluid.

In the vicinity of the vortex core, the mutual friction induces, in the normal fluid, the vortex dipole whose typical lengthscale is expected to be about 100 μm. The results of numerical calculation by Idowu *et al.* (2000b) of the dipole-like normal fluid disturbance are shown in Figure 3.0.21[7]. This nor-

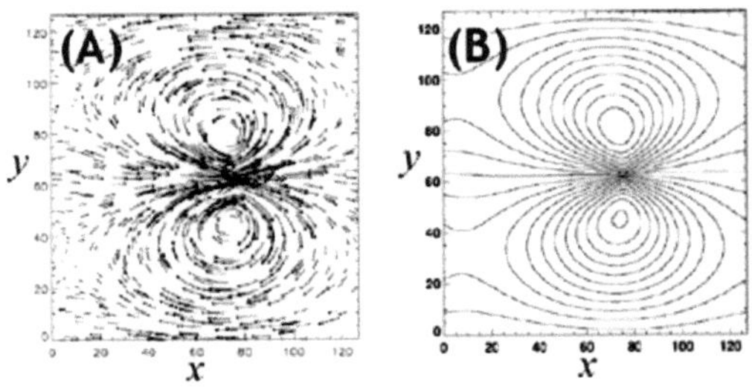

Figure 3.0.21. Velocity field of the normal fluid due to the mutual friction forcing of a single superfluid vortex (Idowu *et al.* (2000b)). (A): velocity arrow plot; (B): streamlines.

mal flow disturbance can deflect the tracer particle which otherwise would have collided with, and possibly trapped by the vortex. Typical trajectory (Sergeev *et al.* (2007c)) of the particle interacting with the superfluid vortex and normal fluid disturbance is shown in Figure 3.0.22[8] by the solid line. For comparison, the trajectory calculated without taking into account the

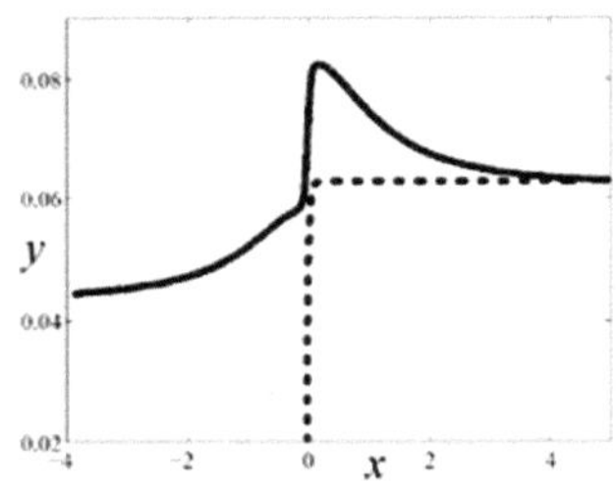

Figure 3.0.22. Particle trajectories in the presence (solid lines) and absence (dashed line) of the normal fluid disturbances (Sergeev *et al.* (2007c)).

[7]Reprinted, by permission from the American Physical Society, from O. C. Idowu, A. Willis, C. F. Barenghi, and D. C. Samuels, *Phys. Rev. B*, **62**:3409-3415, 2000. Copyright 2000 by the American Physical Society.

[8]Reproduced, with kind permission from Springer Science and Business Media, from Fig. 2 (right) of the original publication by Y. A. Sergeev, S. Wang, E. Meneguz, C. F. Barenghi, Influence of Normal Fluid Disturbances on Interactions of Solid Particles with Quantized Vortices, *J. Low Temp. Phys.*, **146**:417-434, 2007. Copyright 2007 by Springer.

normal fluid disturbance is shown by the dashed line. This trajectory leads to the collision with the vortex core located at the origin.

Mutual friction was discovered theoretically by Onsager (1949) and Feynman (1955). Normal fluid vortical structures induced by the mutual friction were predicted by Hall and Vinen (1956), but so far, because of low resolution of existing experimental techniques, there was no direct experimental proof of existence of the normal fluid disturbances. Owing to the much higher resolution, the PIV technique, recently implemented in helium II, can provide such an evidence. Perhaps the first experimental confirmation of existence of normal flow structures induced by the mutual friction has already been found. Figure 3.0.23 shows a typical particle trajectory observed by Bewley *et al.* (2007)[9], by means of the PIV technique, in the turbulent

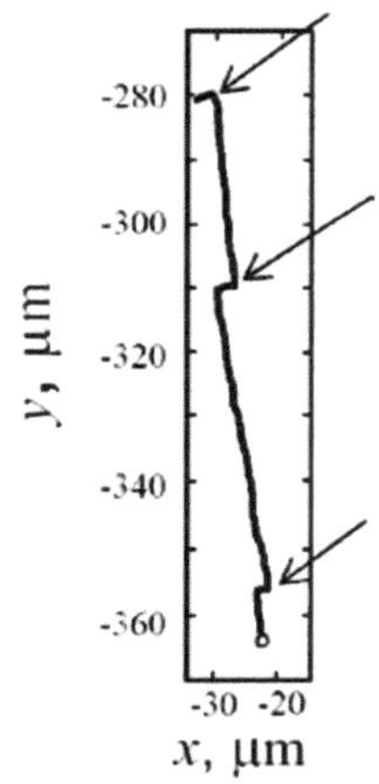

Figure 3.0.23. PIV observation by Bewley *et al.* (2007) of particle trajectories possibly indicating an influence of normal fluid disturbances induced by the mutual friction.

thermal counterflow. The particle trajectory has characteristic 'kinks' resembling the 'deflected' trajectory shown by the solid line in Figure 3.0.22.

In the conclusion of this Section we will analyze, neglecting trapping, two examples of particle motion in helium II at temperatures $1\,\mathrm{K} < T < T_\lambda$ when both the normal and the superfluid components are present.

[9]I am grateful to Prof. K. K. Sreenivasan for permission to use, prior to their publication, the experimental results by Bewley, Lathrop, Sreenivasan, and Paoletti.

Vortex Ring Propagating Against a Particular Sheet

This study (Kivotides, Barenghi, and Sergeev (2005)) led to a proposal of a simple experiment (not performed yet) which could, in principle, justify the use of PIV technique for measuring instantaneous normal fluid velocity patterns. We consider a single vortex ring propagating normally to a plane sheet of neutrally buoyant particles, see Figure 3.0.24[10] (left).

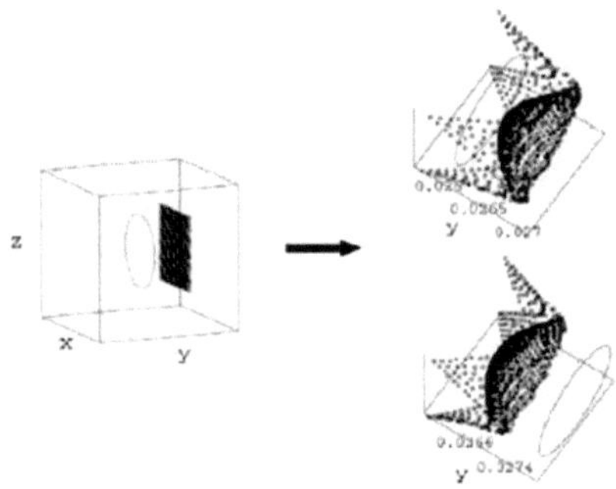

Figure 3.0.24. Vortex ring and solid particles' configurations before and after the ring has passed through the particulate sheet (Kivotides, Barenghi, and Sergeev (2005)).

In the considered model, the mutual friction between the normal fluid and the superfluid vortex is taken into account leading to emergence of the normal fluid disturbances described earlier. The motion of the superfluid vortex whose core is defined parametrically by $\mathbf{X}(s,\, t)$ is governed by the following equation derived by Idowu *et al.* (2000a):

$$\frac{\partial \mathbf{X}}{\partial t} = \mathbf{v}_\ell = h\mathbf{V}^{\mathrm{s}} + h_* \mathbf{X}' \times (\mathbf{v}_{\mathrm{n}} - \mathbf{V}^{\mathrm{s}}) - h_{**} \mathbf{X}' \times (\mathbf{X}' \times \mathbf{v}_{\mathrm{n}}) \,, \qquad (3.0.15)$$

where $\mathbf{X}' = \partial \mathbf{X}/\partial s$, $h(T)$, $h_*(T)$ and $h_{**}(T)$ are the known mutual friction coefficients, and the vortex-induced superfluid velocity $\mathbf{V}^{\mathrm{s}}$ is given by the Biot-Savart integral

$$\mathbf{V}^{\mathrm{s}}(\mathbf{x}) = -\frac{\kappa}{4\pi} \int ds \, \frac{\mathbf{X}' \times (\mathbf{X} - \mathbf{x})}{|\mathbf{X} - \mathbf{x}|^3} \,. \qquad (3.0.16)$$

[10] Figures 3.0.24, 3.0.25, and 3.0.26 are reprinted, by permission from the American Physical Society, from D. Kivotides, C. F. Barenghi, and Y. A. Sergeev, *Phys. Rev. Lett.*, **95**:215302, 2005. Copyright 2005 by the American Physical Society.

The motion of incompressible ($\nabla \cdot \mathbf{v}_n = 0$) normal fluid is governed by the equation

$$\frac{\partial \mathbf{v}_n}{\partial t} + (\mathbf{v}_n \cdot \nabla)\mathbf{v}_n = -\frac{1}{\rho}\nabla p + \nu \nabla^2 \mathbf{v}_n + \frac{1}{\rho}\mathbf{F}\,, \tag{3.0.17}$$

where $\mathbf{F}$ is the mutual friction force per unit volume. This force is determined as the sum of the drag force and the Iordanskii force, $\mathbf{f}$, per unit length:

$$\mathbf{f} = \rho_s \kappa [d_{**}\mathbf{X}' \times (\mathbf{X}' \times (\mathbf{v}_n - \mathbf{V}^s)) - d_* \mathbf{X}' \times (\mathbf{v}_n - \mathbf{V}^s)]\,. \tag{3.0.18}$$

Here the new mutual friction coefficients, $d_*(T)$ and $d_{**}(T)$ are known and can be expressed explicitly through $h(T)$, $h_*(T)$ and $h_{**}(T)$.

Trajectories of neutrally buoyant particles were found by numerical integration of Equations (3.0.5) and (3.0.6). Configurations of solid particles before and after the ring has passed the particulate sheet are shown in Figure 3.0.24 (right). A typical trajectory, illustrating the deflection of solid particle by the normal fluid structures induced by the mutual friction, is shown in Figure 3.0.25. Figure 3.0.26 shows the histograms of the angle

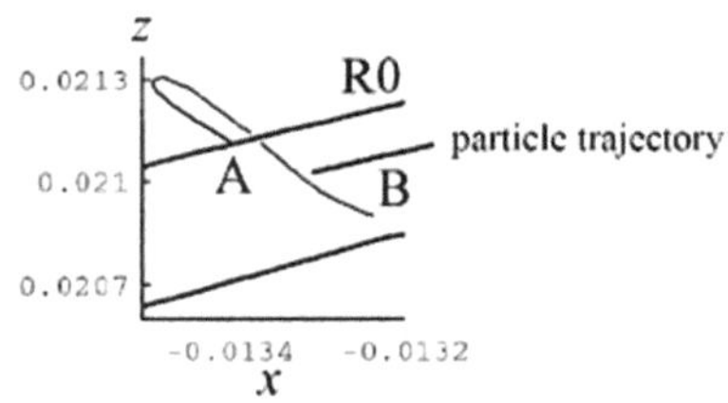

Figure 3.0.25. Example of particle trajectory close to the vortex ring (Kivotides, Barenghi, and Sergeev (2005)).

and of the relative difference between $\mathbf{v}_n$ and $\mathbf{u}_p$; it can be seen that $\mathbf{v}_n$ and $\mathbf{u}_p$, to a very good degree of accuracy, are identical both in magnitude and direction. Moreover, the calculation showed that trapping events are relatively rare – only 42 out of 900 particles approached the ring to a distance smaller than three particle diameters. This enables us to conclude that, in the proposed experiment, the measurement of particle velocities can provide direct information about instantaneous normal flow patterns.

Particle Motion in Thermal Counterflow

In this example (Kivotides, Barenghi, and Sergeev (2006a)) we will be concerned with the T-I state of helium II turbulence such that the vortex

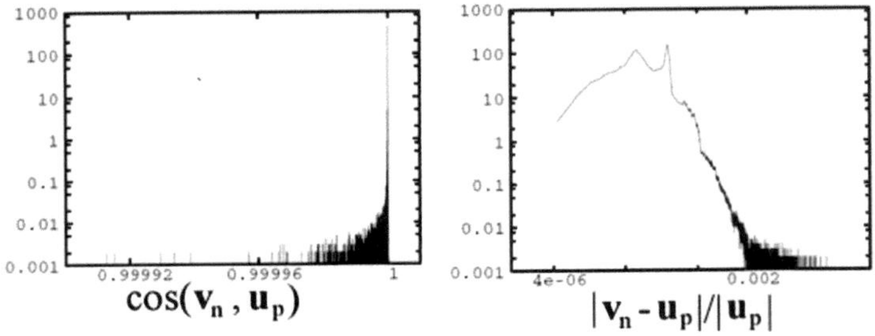

Figure 3.0.26. Histograms (Kivotides, Barenghi, and Sergeev (2005)) of the cosine of the angle between $\mathbf{v}_{\mathrm{n}}$ and $\mathbf{u}_{\mathrm{p}}$ (left) and of the magnitude of relative velocity, $|\mathbf{v}_{\mathrm{n}} - \mathbf{u}_{\mathrm{p}}|$ (right). The vertical axes are divided by 1000.

tangle in the superfluid component is present but the normal flow is laminar, see Figure 1.0.4. For simplicity, the normal flow is assumed uniform, $\mathbf{v}_{\mathrm{n}} = \mathbf{const}$. The superfluid velocity can be represented as $\mathbf{v}_{\mathrm{s}} = \mathbf{v}_{\mathrm{s}}^{C} + \mathbf{V}^{\mathrm{s}}$, where the mean (counterflow) superfluid velocity, $\mathbf{v}_{\mathrm{s}}^{C}$ is linked with the normal fluid velocity by the relation $\rho \mathbf{v}_{\mathrm{s}}^{C} + \rho_{\mathrm{n}} \mathbf{v}_{\mathrm{n}} = \mathbf{0}$, and $\mathbf{V}^{\mathrm{s}}$ is the fluctuating superfluid velocity induced by the vortex tangle. The dynamic vortex tangle is modelled, in the periodic box, taking into account the mutual friction between the normal fluid and quantized vortices, as well as the Biot-Savart interaction between vortex filaments. An influence of superfluid vortices on the motion of normal fluid is neglected. Based on Equations (3.0.5) and (3.0.6), the motion of neutrally buoyant tracer particles of diameter $d_{\mathrm{p}} = 6.25 \times 10^{-3}\,\mathrm{cm}$ was calculated after the vortex tangle has reached the statistically steady state. Calculations were performed at temperature $T = 1.3\,\mathrm{K}$ for $|\mathbf{v}_{\mathrm{n}}| = 1.1417$ and 06058 cm/s (corresponding values of the counterflow heat flux are $q = 1.07 \times 10^{-3}$ and $4.57 \times 10^{-4}\,\mathrm{J/(cm^2 \cdot s)}$), and at temperature $T = 2.171\,\mathrm{K}$ for $|\mathbf{v}_{\mathrm{n}}| = 0.01183\,\mathrm{cm/s}$ ($q = 0.125\,\mathrm{J/(cm^2 \cdot s)}$). The histogram of particle velocity magnitudes, illustrated for the latter case by Figure 3.0.27[11], shows that the particle velocity is very narrowly peaked around the constant normal velocity.

Three histograms of Figure 3.0.28, corresponding to the three cases considered above, show the absence of alignment between particle velocities and the superfluid velocity, $\mathbf{V}^{\mathrm{s}}$ induced by the vortex tangle, and clearly illustrate anisotropy in $\mathbf{V}^{\mathrm{s}}$.

[11] Figures 3.0.27 and 3.0.28 are reprinted, by permission from EDP Sciences, from D. Kivotides, C. F. Barenghi, and Y. A. Sergeev, *Europhys. Lett.*, **73**:733-739, 2006. Copyright 2006 by EDP Sciences

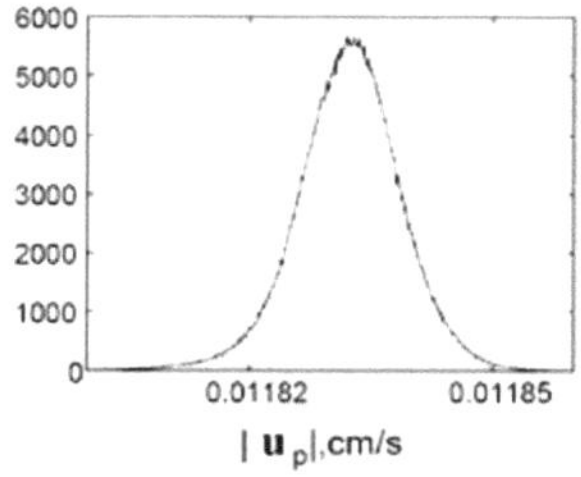

Figure 3.0.27. Histogram of particle velocity magnitudes at $T = 2.171\,\mathrm{K}$ (Kivotides, Barenghi, and Sergeev (2006a)).

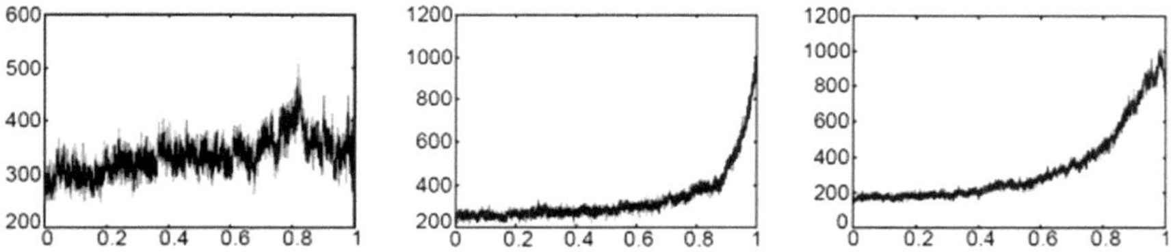

Figure 3.0.28. Histograms of $|\cos(\mathbf{V}^{\mathrm{s}}, \mathbf{u}_{\mathrm{p}})|$ (Kivotides, Barenghi, and Sergeev (2006a)).

The above two examples enable us to conclude that, in the range of parameters typical of PIV measurements, neutrally buoyant particles trace the normal fluid well, and, provided strong interactions of particles with quantized vortices can be neglected, particle velocity fluctuations induced by interactions between particles and quantized vortices are relatively small. However, these conclusion can be invalidated by trapping of solid particles on quantized vortex cores. This phenomenon will be addressed in the next Section.

4 Particle-Vortex Reconnections and Particle Trapping on Quantized Vortices: Self-Consistent "Mutual Interaction" Model

The problem of particle trapping on quantized vortices constitutes a part of a wider problem of reconnections of quantized vortices with the surface of the particle moving in the fluid velocity field. This problem cannot be analyzed based on the "One-Way Interaction" model introduced in Section 3. Instead, a much more elaborate "Mutual Interaction" model must be developed based on dynamically self-consistent calculations (Kivotides, Barenghi, and Sergeev (2007a, 2007b, 2008)) which take into account an influence of the flow field around the sphere on the evolution of the superfluid vortex.

The evolution of the vortex filament, $\mathcal{L}$ represented as a space curve $\mathbf{X}(s,\, t)$ (see Figure 4.0.29), where s is the arclength, is governed by the

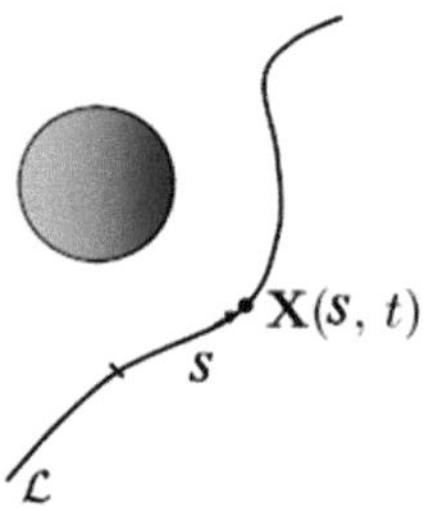

Figure 4.0.29. Superfluid vortex filament in the vicinity of the particle.

equation

$$\frac{\partial \mathbf{X}}{\partial t} = \mathbf{V}^{\mathrm{s}} + \mathbf{V}^{\mathrm{b}} + \mathbf{V}^{\phi} + \mathbf{V}^{\mathrm{f}} . \tag{4.0.1}$$

The contributions, identified by the superscript, are:

s: Biot-Savart integral given by formula (3.0.16).

b: Deformation of vortices due to the presence of a stationary particle:

$$\begin{aligned} &\mathbf{V}^{\mathrm{b}} \equiv \nabla \Phi^{\mathrm{b}} , \quad \nabla^2 \Phi^{\mathrm{b}} = 0 ; \\ &\text{on the particle surface} \quad (\mathbf{V}^{\mathrm{s}} + \mathbf{V}^{\mathrm{b}}) \cdot \hat{\mathbf{n}} = 0 . \end{aligned} \tag{4.0.2}$$

ϕ: potential flow field induced by the motion of the spherical particle:

$$\mathbf{V}^{\phi}(\mathbf{x},\, t|\mathbf{r}_{\mathrm{p}}) = -\frac{1}{2}\left(\frac{a_{\mathrm{p}}}{|\mathbf{x}|}\right)^3 \mathbf{u}_{\mathrm{p}} \cdot \left(\mathbf{I} - 3\frac{(\mathbf{x}-\mathbf{r}_{\mathrm{p}})(\mathbf{x}-\mathbf{r}_{\mathrm{p}})}{|\mathbf{x}-\mathbf{r}_{\mathrm{p}}|^2}\right) , \tag{4.0.3}$$

where $\mathbf{r}_{\mathrm{p}}(t)$ is the current position of the particle centre, and $\mathbf{I}$ is the unit tensor.

f: mutual friction between the superfluid and the normal fluid:

$$\begin{aligned} \mathbf{V}^{\mathrm{f}} &= h_{**}(\mathbf{V}^{\mathrm{s}} + \mathbf{V}^{\mathrm{b}} + \mathbf{V}^{\phi}) \\ &+ h_{*}\frac{\partial \mathbf{X}}{\partial s} \times (\mathbf{v}_{\mathrm{n}} - (\mathbf{V}^{\mathrm{s}} + \mathbf{V}^{\mathrm{b}} + \mathbf{V}^{\phi})) \\ &+ h_{**}\frac{\partial \mathbf{X}}{\partial s} \times \left(\frac{\partial \mathbf{X}}{\partial s} \times \mathbf{v}_{\mathrm{n}}\right) . \end{aligned} \tag{4.0.4}$$

These equations must be considered together with the equations of motion of neutrally buoyant spherical particle; these equations are $d\mathbf{r}_\mathrm{p}/dt = \mathbf{u}_\mathrm{p}(t)$, and

$$m_\mathrm{e}\frac{d\mathbf{u}_\mathrm{p}}{dt} = 6\pi a_\mathrm{p}\mu_\mathrm{n}(\mathbf{v}_\mathrm{n} - \mathbf{u}_\mathrm{p}) + 2\pi\rho_\mathrm{s}a_\mathrm{p}^3\frac{\partial\mathbf{V}^\mathrm{s}(\mathbf{r}_\mathrm{p}, t)}{\partial t} + \frac{1}{2}\rho_\mathrm{s}\int_S dS\,(\mathbf{V}^\mathrm{s} + \mathbf{V}^\mathrm{b})^2\hat{\mathbf{n}}\,, \quad (4.0.5)$$

where $m_\mathrm{e} = \frac{4}{3}\pi a_\mathrm{p}^3(\rho_\mathrm{p} + \frac{1}{2}\rho)$ is the effective mass of the particle, and $\hat{\mathbf{n}}$ is the normal unit vector to the particle surface. At temperatures $1\,\mathrm{K} < T < T_\lambda = 2.17168\,\mathrm{K}$, $\mu_\mathrm{n}(T)$ is the viscosity of the normal fluid. At $T < 1\,\mathrm{K}$, the coefficient μ_n is determined by the drag force due to ballistic scattering of quasiparticles (phonons and rotons) off the particle surface (see the discussion later in this Section).

Numerical method of solution of the system of Equations (3.0.16) and (4.0.2)-(4.0.5) is described in detail by Kivotides, Barenghi, and Sergeev (2006b) (this method is based on generalization of the approach by Schwarz (1974 and 1985), later developed further by Tsubota and Maekawa (1993), to the problem of reconnection of the quantized vortex with the stationary surface or the surface moving with prescribed velocity).

Using this model we may analyze first how incorrect is the "One-Way Interaction" model. Figure 4.0.30[12] shows the distance between the particle and the initially straight vortex vs time found from (a) analytical calculation (Poole *et al.* (2005)) based on the "One-Way Interaction" model (see also Exercise 6), and (b) numerical, dynamically self-consistent model given by Equations (3.0.16), (4.0.2)-(4.0.5). As can be seen, despite the simplicity of the "One-Way Interaction" model, agreement is good until the moment when the particle and the vortex are so close that the vortex reconnects to the particle surface.

The mechanism of particle-vortex collision has many subtle and intriguing features. Unfortunately, limited space does not permit a detailed description, so that we will focus here only on the most important aspects of particle-vortex collision, referring the interested reader to original publications (Kivotides, Barenghi and Sergeev (2006b, 2007a, 2007b, 2008); Barenghi, Kivotides, and Sergeev (2007)).

[12]Reproduced, with kind permission from Springer Science and Business Media, from Fig. 1 of the original publication: C. F. Barenghi, D. Kivotides, and Y. A. Sergeev, Close Approach of a Spherical Particle and a Quantized Vortex in Helium II, *J. Low Temp. Phys.*, **148**:293-297, 2007. Copyright 2007 by Springer.

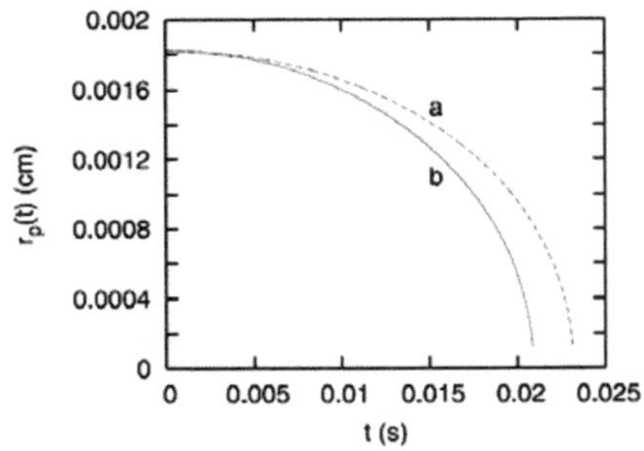

Figure 4.0.30. Distance between the particle and the vortex core vs time: (a) "One-Way Interaction" model; (b) self-consistent "Mutual Interaction" model (Barenghi, Kivotides, and Sergeev (2007)).

All the calculations illustrated below were performed for the neutrally buoyant particle of radius $a_\mathrm{p} = 1\,\mu\mathrm{m}$ located initially at the distance $2a_\mathrm{p}$ from the initially straight vortex filament. To simplify the analysis, in all examples considered below it was assumed that $\mathbf{v}_\mathrm{n} \equiv \mathbf{0}$.

We illustrate first an influence of the initial velocity of solid particle on the particle-vortex collision. Figure 4.0.31[13] shows a sequence of particle-vortex configurations at $T = 1.3\,\mathrm{K}$ for initial velocity $u_\mathrm{p} = 25\,\mathrm{cm/s}$. The particle arrives from the right; *top left*: the vortex is deformed as it tries to avoid the incoming particle; *top middle and top right*: the reconnection of vortex to the particle surface excites Kelvin waves; *bottom left*: the particle drags the vortex, forcing its two strands to come together, thus facilitating a second reconnection; *bottom middle*: following the reconnection, the vortex recoils and the particle breaks free; *bottom right*: the following relaxation creates more Kelvin waves.

The second computation is carried out at the same temperature for the initial velocity $u_\mathrm{p} = 20\,\mathrm{cm/s}$. In this case, illustrated by Figure 4.0.32, the particle is trapped by the vortex: after reconnecting with the vortex the particle slows down and stops since it lacks the kinetic energy to stretch the vortex and to induce the second reconnection.

Calculations illustrated by Figure 4.0.33[14] show the collision with the

---

[13] Figures 4.0.31 and 4.0.32 are reprinted, by permission from the American Physical Society, from D. Kivotides, C. F. Barenghi, and Y. A. Sergeev, *Phys. Rev. B*, **77**:014527, 2008. Copyright 2008 by the American Physical Society.

[14] Reprinted, by permission from the American Physical Society, from D. Kivotides, C. F. Barenghi, and Y. A. Sergeev, *Phys. Rev. B*, **75**:212502, 2007. Copyright 2007 by the American Physical Society.

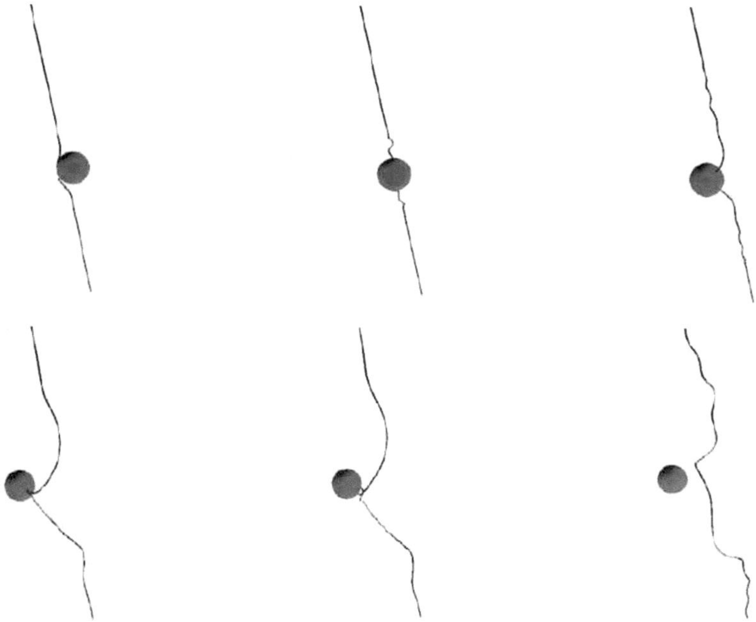

Figure 4.0.31. Particle-vortex collision at $T = 1.3\,\mathrm{K}$ (Kivotides, Barenghi, and Sergeev (2008)). Initial velocity $u_\mathrm{p} = 25\,\mathrm{cm/s}$. Particle moves from the right.

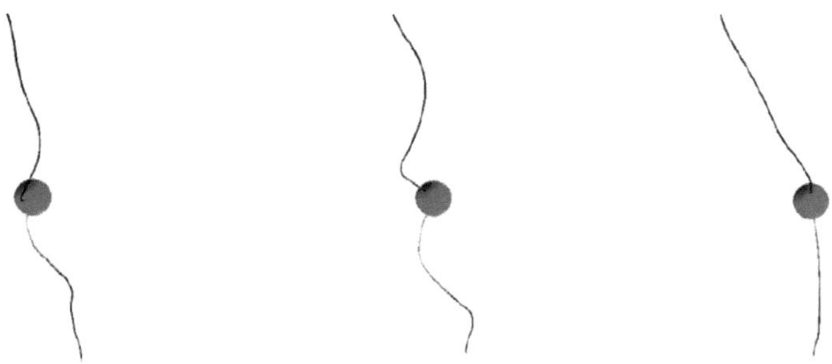

Figure 4.0.32. Particle trapping on the vortex core (Kivotides, Barenghi, and Sergeev (2008)). $T = 1.3\,\mathrm{K}$, initial velocity $u_\mathrm{p} = 20\,\mathrm{cm/s}$.

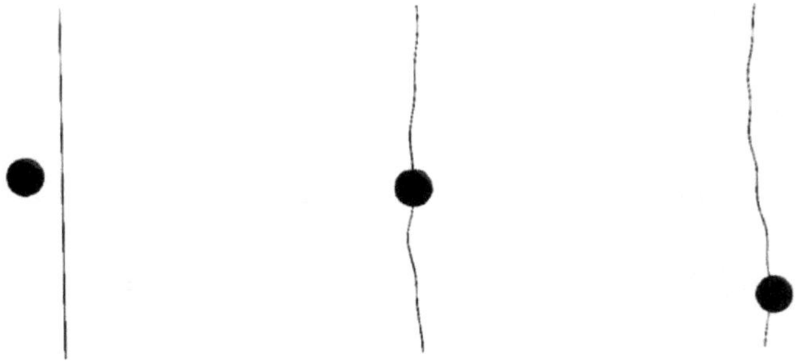

Figure 4.0.33. Particle trapping. $T = 1.3\,\mathrm{K}$, initial velocity $u_{\mathrm{p}} = 0$ (Kivotides, Barenghi, and Sergeev (2007)) After reconnection with the vortex, the particle drifts along the vortex core with the average speed $v_{\mathrm{drift}} \approx 0.5\,\mathrm{cm/s}$.

vortex of the particle initially at rest at the distance $2a_{\mathrm{p}}$ from the vortex. The particle starts moving under the influence of the pressure gradient force induced by the superfluid velocity field of the vortex. In this example the vortex traps the particle which then drifts along the vortex filament. Such a drift is another interesting phenomenon which can be better understood by solving the following exercise.

Exercise 9. Animation of numerical results showed the spherical particle drifting along the superfluid vortex filament with the average velocity $v_{\mathrm{drift}} \approx 0.5\,\mathrm{cm/s}$. Such a drift can be explained by the interaction of the particle with Kelvin waves induced on the vortex by the particle-vortex collision. Model a Kelvin wave of amplitude A as a vortex ring of radius A, and calculate the momentum of the ring. Balancing the momentum of the particle with that of the ring, estimate the amplitude of Kelvin waves.

Influence of temperature on particle-vortex collision and particle trapping. Figure 4.0.33 showed that at temperature $T = 1.3\,\mathrm{K}$ the particle, initially at rest at the distance $2a_{\mathrm{p}}$ from the initially straight filament, will eventually be trapped by the quantized vortex core. The results of calculation for the same initial configuration and particle velocity, but μ_{n} assumed to be only 0.2 of its value at $T = 1.3\,\mathrm{K}$ (we will see later how realistic such a value of μ_{n} is), can be illustrated by a sequence very similar to that shown by Figure 4.0.31: the particle, although undergoing the process of reconnection with the vortex filament, eventually breaks free, so that trapping does not occur. Similar (in fact, almost identical) scenario is typical of $T \to 0$ limit when the (viscous) damping force acting on the particle can be neglected (but beware! This issue is less trivial than it seems – see the discussion below.)

These examples show that the the presence of the damping is crucial for trapping of solid particles by quantized vortices, so that it can be, rather naively, expected that trapping cannot occur at temperatures below 1 K when the normal fluid is absent. However, at $T < 1\,\mathrm{K}$ there still exists a damping force

$$\mathbf{F}^{\mathrm{d}} = -6\pi a_{\mathrm{p}} \mu_{\mathrm{n}} \mathbf{u}_{\mathrm{p}} \tag{4.0.6}$$

caused by ballistic scattering of quasiparticles (phonons and rotons) off the particle surface. Note the Stokesian form of this force (cf. Equation (2.0.20), although $\mu_{\mathrm{n}}(T)$ should now be understood not as a viscosity but as the damping coefficient.

Calculations for various values of the damping coefficient showed that, for the particle initially at rest at the distance $2a_{\mathrm{p}}$ from the initially straight vortex filament, trapping does not occur in the case where the damping coefficient, μ_{n} is smaller than $0.2 \times \mu_{\mathrm{n}}(1.3\,\mathrm{K})$. The value of the damping coefficient was measured experimentally by Jäger, Schuderer, and Schoepe (1995) for the spherical particle of radius $100\,\mu\mathrm{m}$. Figure 4.0.34[15] shows the experimental results of Jäger, Schuderer, and Schoepe (1995) for the

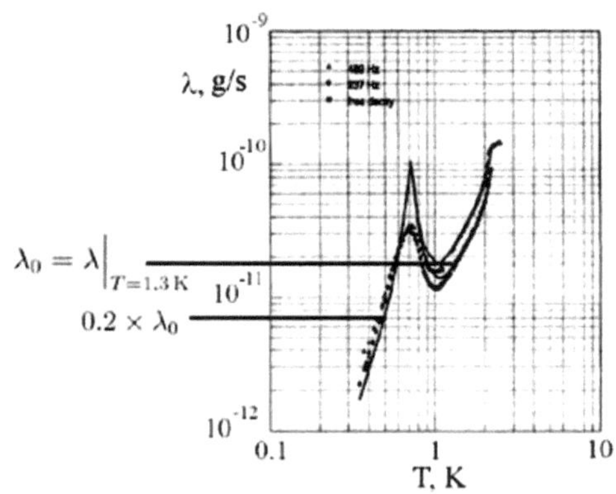

Figure 4.0.34. Damping coefficient $\lambda = 6\pi a_{\mathrm{p}} \mu_{\mathrm{n}}$ as a function of temperature (experimental results of Jäger, Schuderer, and Schoepe (1995) for $a_{\mathrm{p}} = 100\,\mu\mathrm{m}$).

[15]Adapted, by permission from the American Physical Society, from J. Jäger, B. Schuderer, and W. Schoepe, *Phys. Rev. Lett.*, **74**:566-569, 1995. Copyright 1995 by the American Physical Society.

coefficient $\lambda = 6\pi a_{\mathrm{p}}\mu_{\mathrm{n}}$ in the range of temperatures including both the region $T > 1\,\mathrm{K}$ corresponding to the classical viscous dissipation in the normal fluid, and $T < 1\,\mathrm{K}$ corresponding to the regime of ballistic scattering of quasiparticles. It can be seen that in the temperature interval $0.6\,\mathrm{K} < T < 1\,\mathrm{K}$ the value of μ_{n} may even exceed the viscosity at temperatures above $1\,\mathrm{K}$. Only at temperatures below $0.5\,\mathrm{K}$ the damping coefficient becomes smaller than $0.2 \times \mu_{\mathrm{n}}(1.3\,\mathrm{K})$. Therefore, it can be confidently concluded that the trapping of neutrally buoyant, $1\,\mu\mathrm{m}$ particles does not occur at temperatures below $0.5\,\mathrm{K}$.

At these, very low temperatures the motion of solid particle can be modelled by a simpler, "One-Way Interaction" model which is based on the assumption that particles are not trapped on quantized vortex lines. However, at these temperatures the presence and motion of the particles still modifies the vortex filaments, so that a certain modification will be necessary of the "One Way Interaction" model, see the following Section 5.

5 Particle Motion in a Vortex Tangle at Very Low Temperature

In order to understand some features of the particle motion in the vortex tangle at temperature $T \to 0$ when the normal fluid is absent and the damping force exerted on the particle can be ignored, we start with a simple, two-dimensional model of the tangle. In such a model the vortex lines become vortex points, and the Biot-Savart law reduces to a simple statement that each vortex point moves as a fluid point in a flow field created by all other vortices, so that modelling of the motion of vortices becomes much easier than in a real, three-dimensional tangle. (Such a two-dimensional system of vortex points is known as the Onsager's point vortex gas, see Onsager (1949).)

We consider, in the periodic box, the motion of neutrally buoyant particles in the system of vortex points of random polarity. We neglect trapping of particles on quantized vortices as well as any influence of particles on the motion of vortex points.

In this approximation the particle motion is governed by Equations (3.0.10), and the following problem can be immediately identified: in the case where the position of the particle coincides with that of the vortex point, the pressure gradient force, $-\nabla p$ becomes unphysically singular. Based on the mechanism, described in Section 4, of the particle-vortex collision, we will resolve this manifest difficulty by modifying the model (3.0.10) as follows.

The cause of the problem can be easily understood: in the real, three-dimensional tangle at temperature $T < 0.5\,\mathrm{K}$, even when the particle breaks through the vortex, it nevertheless reconnects with the vortex filament when the distance between the particle and the vortex core becomes of the order of particle radius. Since the vortex line attached to the particle is necessarily orthogonal to the particle surface, the reconnection results in a dramatic decrease of the force exerted on the particle; this force is zero when the particle-vortex configuration is symmetric. Then, the "One-Way Interaction" model can be modified by assuming that there exists a force-free region for $r_\mathrm{p} < a_\mathrm{c}$, where $a_\mathrm{c} \sim O(a_\mathrm{p})$ is a cut-off distance.

In the two-dimensional calculation we set $a_\mathrm{c} = a_\mathrm{p}$. We found that the trajectory of solid particle, which starts its motion with the velocity equal that of the fluid particle, looses very quickly any resemblance to the trajectory of the fluid point; moreover, the trajectory of the solid particle soon acquires the ballistic character. The reason for such a behaviour of the solid particle is the instability of its trajectory as discussed in Section 3.

Another rather unexpected feature of particle motion is the saturation of the magnitude of average particle velocity with time: although $|\mathbf{u}_\mathrm{p}(0)| = |\mathbf{v}_\mathrm{s}(0)|$, the magnitude of particle velocity quickly increases above that of the fluid point, $|\mathbf{v}_\mathrm{s}(t)|$ and remains larger than the average value of $|\mathbf{v}_\mathrm{s}|$ at all times. The saturated particle velocity satisfies the scaling

$$\langle u_\mathrm{p} \rangle \sim \sqrt{\frac{\ell}{a_\mathrm{c}}}\, \langle v_\mathrm{s} \rangle\,, \tag{5.0.1}$$

where ℓ is the intervortex spacing, and $\langle v_\mathrm{s} \rangle = \kappa/(2\pi\ell)$ the average superfluid velocity. The following exercise will help to understand why the particle velocity saturates and hence to derive the scaling (5.0.1).

Exercise 10. The reason for increase with time of the mean particle velocity is that, during an interaction with a moving vortex, the particle velocity may change by the value of the order of velocity of the vortex point. All vortex points move as fluid points, so that the mean velocity of vortices is $\langle v_\mathrm{s} \rangle$. Using the cut-off model, estimate the maximum value of particle acceleration during the close particle-vortex encounter. Also, using the mean intervortex distance, ℓ, calculate the magnitude of the mean acceleration along the particle trajectory. Assume that the mean particle acceleration cannot be larger than the acceleration during the close particle-vortex encounter (obviously), and hence derive the above estimate for the saturated particle velocity.

Neglecting any influence of solid particles on the motion and evolution of quantized vortex lines, the three-dimensional calculation of the particle motion in the vortex tangle was performed in the periodic box $0.1\times0.1\times0.1\,\text{cm}^3$. Using Schwarz's method (1985, 1988), the vortex tangle was generated starting with an arbitrary number of superfluid vortices whose evolution is governed by the Biot-Savart law. The numerical technique, described by Samuels (2001) and Kivotides *et al.* (2001), involves discretization along the filament ($2.08\times10^{-3}\,\text{cm}$, in the calculation illustrated below) with time step typically $6.27\times10^{-4}\,\text{s}$. Time sequence illustrating formation of the fully developed tangle is shown in Fig. 8.5.1 of Chapter 1 by C. F. Barenghi. In our calculation the saturated value of the vortex line density in the statistically steady state was $L\approx 18.5\times10^3\,\text{cm}^{-2}$.

The motion, governed by Equations (3.0.8), of micron-size, neutrally buoyant particle was studied numerically in the statistically steady state of the vortex tangle. In the three-dimensional case there is no need to explicitly introduce a force-free region for distances $r_\text{p} < a_\text{c}$ from the vortex core, as we did in the two-dimensional case: in Biot-Savart calculations, the normalization of the velocity when the solid particle approaches too close to a vortex is achieved by the numerical cut-off of the pressure gradient force acting on the particle. A force-free region is automatically provided by the discretization along the vortex filament; it turns out that the size of such a region (i.e. a cut-off distance a_c) is of the order of the particle radius a_p.

In the three-dimensional case, the trajectories of solid particles have the same ballistic character as the two-dimensional trajectories discussed above. Likewise, the phenomenon is observed as well of particle velocity saturation at values of $\langle u_\text{p}\rangle$ much higher than $\langle v_\text{s}\rangle$. In the three-dimensional case the calculated average particle velocity agrees with scaling (5.0.1). Indeed, in this calculation $\ell\approx 7.32\times10^{-3}\,\text{cm}$, $a_\text{c}=a_\text{p}=10^{-4}\,\text{cm}$, $\langle u_\text{p}\rangle\approx 0.2\,\text{cm/s}$, and $\langle v_\text{s}\rangle\approx 0.02\,\text{cm/s}$, so that $\sqrt{\ell/a_\text{c}}\approx 8.6$ and $\langle u_\text{p}\rangle/\langle v_\text{s}\rangle = 10$.

These results clearly show that from the point of view of flow visualization, in the low temperature limit, $T\to 0$ the trajectories of solid particles do not reveal flow patterns of the superfluid.

6 Particle Motion in Turbulent Thermal Counterflow: Experiment and Phenomenological Theory

In Section 3 we have already discussed surprising results of PIV experiment by Zhang and Van Sciver (2005a) who studied the sedimentation of heavy (not neutrally buoyant but of density $\rho_\text{p} = 1.1\,\text{g/cm}^3$) particles in thermal counterflow produced by the heat source situated at the bottom of vertical apparatus, so that the normal fluid flows upwards, see Figure 6.0.35. It seemed natural to expect that the dominating force acting on the particle

Figure 6.0.35. Sedimentation of the solid particle in the thermal counterflow.

will be the viscous drag force and, therefore, the particle velocity will be

$$u_\mathrm{p} = v_\mathrm{n} - v_\mathrm{slip} \,, \tag{6.0.1}$$

where v_slip is the terminal velocity of particle sedimentation given by relation (3.0.14). Were this correct, the experimental data for $v_\mathrm{pa} = u_\mathrm{p} + v_\mathrm{slip}$ plotted against v_n would have collapsed on the straight line shown in Figure 3.0.19. However, as can be seen from this Figure, the results of PIV measurement showed much lower particle velocity. Zhang and Van Sciver (2005a) found that, instead of (3.0.14), the particle velocity can be represented as $u_\mathrm{p} = v_\mathrm{n} - v_\mathrm{slip} - v_\mathrm{add}$, where the additional velocity, v_add can be explained only by strong interactions between sedimenting particles and the vortex tangle. Zhang and Van Sciver also found that $v_\mathrm{pa}/v_\mathrm{n} \approx 0.5$ independently of temperature, and $v_\mathrm{add} \sim q = \rho S T v_\mathrm{n}$.

Exercise 11. Use experimental data of Zhang and Van Sciver (2005a) to show that the additional velocity, v_add can be interpreted only as caused by the body force, but not by the drag force.

Below we will discuss the phenomenological theory, developed by Sergeev, Barenghi, and Kivotides (2006b), of the motion of micron-size particles in thermal counterflow and hence explain the surprising result of Zhang and Van Sciver's experiment.

Imagine that two strands of the quantized vortex are attached to the surface of spherical particle as shown in Figure 3.0.17 (left) (it must be remembered that at the point of reconnection the vortex strand is necessarily orthogonal to the particle surface). The force exerted by a vortex strand attached to the surface is $\mathbf{F} = \int_S p\,\hat{\mathbf{n}}\,dS$ which, as shown by Schwarz (1974), can be written as

$$\mathbf{F} = \frac{\rho_\mathrm{s}}{2}\int\limits_S (\mathbf{V}^\mathrm{s} + \mathbf{V}^\mathrm{b})^2 \hat{\mathbf{n}}\,dS \tag{6.0.2}$$

(we have already encountered this force given by the last term in Equation (4.0.5)). The contribution of $\mathbf{V}^\mathrm{b}$ to this force can be neglected in the case where the radius of curvature of the vortex strand is much larger than the particle radius.

The vortex tangle in the counterflow can be so dense that several vortex strand can be simultaneously attached to the particle, see Figure 6.0.36. Since the quantum of circulation is small, the leading contribution to the

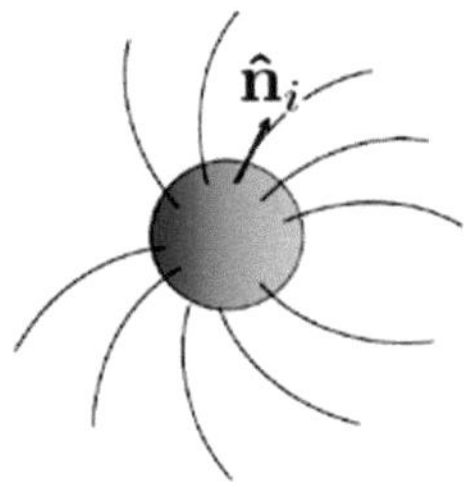

Figure 6.0.36. Spherical particle reconnected with several quantized vortex strands in the dense thermal counterflow.

integral (6.0.2) is provided by a small area around the point where the vortex attaches to the surface. This enables us to find the following analytic approximation for the force exerted on the particle:

$$\mathbf{F} \approx \frac{\rho_\mathrm{s}\kappa^2}{4\pi}\ln\frac{a_\mathrm{p}}{\xi}\sum_{i=1}^{N}\hat{\mathbf{n}}_i\,. \tag{6.0.3}$$

It can be noticed that, in agreement with the results of Zhang and Van Sciver (2005a), $\mathbf{F}$ is indeed a body force.

Several possible particle-vortex configurations shown in Figure 6.0.37 can be imagined. If configuration is symmetric, as in the first two figures, the net force is zero. If one or more vortex loops are asymmetrically attached to the sphere, as in the third and the fourth figures, the contributions from individual vortices will not cancel out resulting in a net body force.

The following scenario seems realistic: the sphere, as it moves between vortex lines, carries along one or more vortex lines or even separate loops

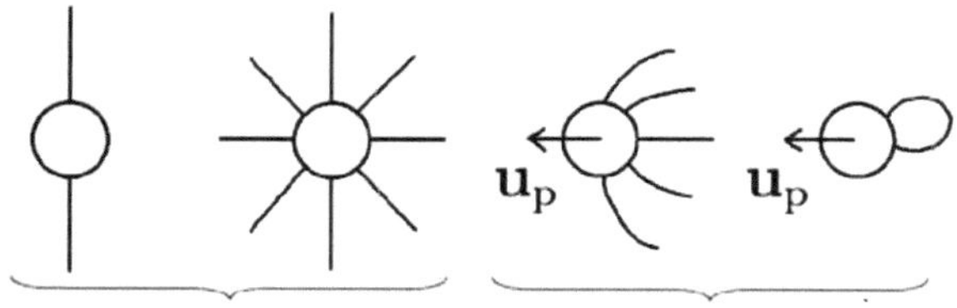

Figure 6.0.37. Possible particle-vortex configurations (Sergeev, Barenghi, and Kivotides (2006b)).

as the result of previous close encounter with vortices. This scenario corresponds to asymmetric particle-vortex configurations, shown in Figure 6.0.37[16], leading to emergence of the force exerted on the particle in the direction opposite to the its motion. In the case of relatively dilute tangle ($a_p \ll \ell$) typical of the experiment of Zhang and Van Sciver (2005a), the average force exerted on the particle by the vortex tangle can be calculated as

$$F \approx \frac{\rho_s \kappa^2}{4\pi} \left(\frac{2\beta a_p}{\ell} \right) \ln \frac{a_p}{\xi} . \tag{6.0.4}$$

In this formula, the only unknown quantity is the parameter β which should be determined by geometrical properties of the vortex tangle in the vicinity of the particle. It can be expected that β is of the order unity.

The additional velocity can now be calculated as $v_{add} = F/(6\pi a_p \mu_n)$, so that for $v_{pa} = u_p + v_{slip}$ we find

$$v_{pa} = v_n - v_{add} \approx \left(1 - \frac{\beta \rho \kappa^2 \gamma \ln(a_p/\xi)}{12\pi^2 \mu_n} \right) v_n , \tag{6.0.5}$$

where $\gamma(T) = L^{1/2}/v_{ns} = \rho_s L^{1/2}/(\rho v_n)$ is the known mutual friction coefficient (see Chapter 1 by C. F. Barenghi), with $v_{ns} = v_n - v_s$. In agreement with experimental results of Zhang and Van Sciver (2005a) we find that v_{pa} is proportional to the normal fluid velocity. Moreover, the temperature dependence of the slope v_{pa}/v_n turns out to be the same as in the cited experiment.

[16] Reprinted, by permission from the American Physical Society, from Y. A. Sergeev, C. F. Barenghi, and D. Kivotides, *Phys. Rev. B*, **75**:019904(E), 2007. Copyright 2007 by the American Physical Society.

Using formula (6.0.5), the ratio of the additional velocity to the heat flux can be calculated as a function of temperature:

$$\frac{v_{\rm add}}{q} = \frac{\beta\kappa^2 \ln(a_{\rm p}/\xi)}{12\pi^2}\left(\frac{\gamma}{\mu_{\rm n} S T}\right), \tag{6.0.6}$$

so that $v_{\rm pa} = u_{\rm p} + v_{\rm slip}$ can now be calculated explicitly. Figure 6.0.38[17] reproduces Figure 3.0.19, shown earlier in Section 3, in which the dashed

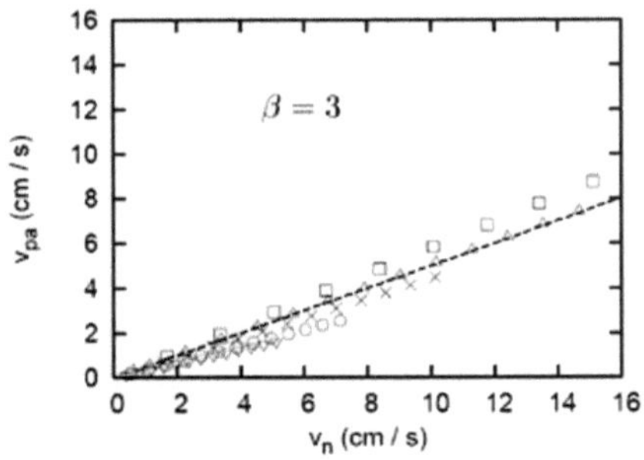

Figure 6.0.38. $v_{\rm pa} = u_{\rm p} + v_{\rm slip}$ vs $v_{\rm n}$ (Sergeev, Barenghi, and Kivotides (2006b)). Dashed line is calculated from Equations (6.0.5) and (6.0.6).

line is added in order to show the calculation of Sergeev, Barenghi, and Kivotides (2006b) with $v_{\rm pa}$ with $\beta = 3$. As can be seen, the developed phenomenological theory agrees well, not only qualitatively but also quantitatively, with PIV measurements of Zhang and Van Sciver and, therefore, explains the mechanism of particle motion in the thermal counterflow.

Calculations similar to those leading to formulae (6.0.4)-(6.0.6) can also be performed in the limit of dense vortex tangle such that $a_{\rm p} \gg \ell$. In this case reconnections between the particle and vortices happen all the time, so that the particle is always attached to vortex filaments. Calculation of the average force exerted on the particle by the vortex tangle yields (Sergeev, Barenghi, and Kivotides (2006b)):

$$F \approx \frac{\rho_{\rm s}\kappa^2}{4\pi}\, 2\beta_{\rm d}\left(\frac{a_{\rm p}}{\ell}\right)^2 \ln\frac{a_{\rm p}}{\xi}\,, \tag{6.0.7}$$

where $\beta_{\rm d} = O(1)$ is again a geometrical factor, and $(a_{\rm p}/\ell)^2$ represents the cross-section of the particle interaction with the network of vortices. Pro-

[17] Reprinted, by permission from the American Physical Society, from Y. A. Sergeev, C. F. Barenghi, and D. Kivotides, *Phys. Rev. B*, **74**:184506, 2006. Copyright 2006 by the American Physical Society.

ceeding as in the case of dilute vortex tangle, we predict a different (cf. Equation (6.0.5)) dependence of $v_{\rm pa}$ on $v_{\rm n}$, T and $a_{\rm p}$:

$$v_{\rm pa} \approx v_{\rm n} \left(1 - \frac{3\pi\mu_{\rm n}}{3\pi\mu_{\rm n} + \beta_{\rm d}\gamma} \frac{\beta_{\rm d} a_{\rm p} \kappa^2 \gamma^2 \rho^2 \ln(a_{\rm p}/\xi)}{12\pi^2 \mu_{\rm n} \rho_{\rm s}} v_{\rm n} \right) . \tag{6.0.8}$$

Finding whether or not this prediction agrees with observations would require a new experiment similar to that of Zhang and Van Sciver (2005a) but for considerably higher values of the vortex line density (i.e. such that $L \gg a_{\rm p}^{-2}$).

Bibliography

T. R. Auton, J. C. R. Hunt, and M. Prud'homme, *J. Fluid Mech.*, **197**:241, 1988.

A. Babiano, J. H. E. Cartwright, O. Piro, and A. Provenzale, *Phys. Rev. Lett.* **84**:5764, 2000.

C. F. Barenghi, D. Kivotides, and Y. A. Sergeev, *J. Low Temp. Phys.*, **148**:293, 2007.

G. K. Batchelor, *An Introduction to Fluid Dynamics*. Cambridge University Press, 1973.

N. G. Berloff and P. H. Roberts, *Phys. Rev. B*, **63**:024510, 2000.

G. P. Bewley, D. P. Lathrop, and K. R. Sreenivasan, *Nature*, **44**:588, 2006.

G. P. Bewley, D. P. Lathrop, K. R. Sreenivasan, and M. S. Paoletti, 2007, in preparation.

D. Celik and S. W. Van Sciver, *Exp. Therm. Fluid Sci.*, **26**:971, 2002

R.J. Donnelly, A. N. Karpetis, J. J. Niemela, K. R Sreenivasan, and W. F. Vinen, *J. Low Temp. Phys.*, **126**:327, 2002.

R. P. Feynman. In C. J. Corter, editor. *Progress in Low Temperature Physics, Vol. 1.* North-Holland, Amsterdam, p. 17, 1955.

H. E. Hall and W. F. Vinen, *Proc. R. Soc. London, Ser. A*, **238**:215, 1956.

O. C. Idowu, D. Kivotides, C. F. Barenghi, and D. C. Samuels, *J. Low Temp. Phys.*. **130**:269, 2000a.

O. C. Idowu, A. Willis, C. F. Barenghi, and D. C. Samuels, *Phys. Rev. B*, **62**:3409, 2000b.

J. Jäger, B. Schuderer, and W. Schoepe, *Phys. Rev. Lett.*, **74**:566, 1995.

I. Kim, S. Elghobashi, and W. A. Sirignano, *J. Fluid Mech.*, **367**:221, 1998.

D. Kivotides, J. C. Vassilicos, D. C. Samuels, and C. F. Barenghi, *Phys. Rev. Lett.*, **86**:3080, 2001.

D. Kivotides, C. F. Barenghi, and Y. A. Sergeev, *Phys. Rev. Lett.*, **95**:215302, 2005.

D. Kivotides, C. F. Barenghi, and Y. A. Sergeev, *Europhys. Lett.*, **73**:733, 2006a.

D. Kivotides, C. F. Barenghi, and Y. A. Sergeev, *J. Low Temp. Phys.*, **144**:121, 2006b.

D. Kivotides, C. F. Barenghi, and Y. A. Sergeev, *Phys. Rev. B*, **75**:212502, 2007.

D. Kivotides, C. F. Barenghi, and Y. A. Sergeev, *Phys. Rev. B*, **77**:014527, 2008.

L. D. Landau and E. M. Lifshitz, *Fluid Mechanics. Course of Theoretical Physics, Vol. 6.* Butterworth-Heinemann, 1987.

M. R. Maxey and J. J. Riley, *Phys. Fluids*, **26**:883, 1983.

R. Mei, *J. Fluid Mech.*, **270**:133, 1994.

D. J. Melotte and C. F. Barenghi, *Phys. Rev. Lett.*, **80**:4181, 1998.

L. Onsager, *Nuovo Cim. Suppl. 2*, **6**:249, 1949.

D. R. Poole, C. F. Barenghi, Y. A. Sergeev, and W. F. Vinen, *Phys. Rev. B*, **71**:064514, 2005.

M. Raffel, C. Willert, and J. Kompenhaus, *Particle Image Velocimetry, a Practical Guide*. Springer, Berlin, 1998.

P.-E. Roche, P. Diribarne, T. Didelot, O. Francais, L. Rousseau, and H. Willaime, *Europhys. Lett.*, **77**:66002, 2007.

D. C. Samuels. In C. F, Barenghi, R. J. Donnelly, and W. F. Vinen, editors. *Quantized Vortex Dynamics and Superfluid Turbulence*. Springer, Berlin, p. 97, 2001.

K. W. Schwarz, *Phys. Rev. A*, **10**:2306, 1974.

K. W. Schwarz, *Phys. Rev. B*, **31**, 5782, 1985.

K. W. Schwarz, *Phys. Rev. B* **38**:2398, 1988.

Y. A. Sergeev, C. F. Barenghi, D. Kivotides, and W. F. Vinen, *Phys. Rev. B*, **73**:052502, 2006a.

Y. A. Sergeev, C. F. Barenghi, and D. Kivotides, *Phys. Rev. B*, **74**:184506, 2006b (for missing figure see Erratum in *Phys. Rev. B*, **75**:019904(E), 2007).

Y. A. Sergeev, S. Wang, E. Meneguz, C. F. Barenghi, *J. Low Temp. Phys.*, **146**:417, 2007.

M. Tsubota and S. Maekawa, *Phys. Rev. B*, **47**:12040, 1993.

T. Zhang and S. W. Van Sciver. In S. Breon, editor. *Advances in Cryogenics Engineering, AIP Conf. Proc. No. 613*. AIP, Melville, NY, p.1372, 2002.

T. Zhang, D. Celik, and S. W. Van Sciver, *J. Low Temp. Phys.*, **134**:985, 2004.

T. Zhang and S. W. Van Sciver, *J. Low Temp. Phys.*, **138**, 865, 2005a.

T. Zhang and S. W. Van Sciver, *Nature Physics*, **1**:36, 2005b.

Counterflow Turbulence in He II and Its Decay

Ladislav Skrbek

Department of Mathematics and Physics,
Charles University, V Holešovičkách 2,
180 00 Prague 8, Czech Republic

Abstract Current studies of quantum turbulence are concerned with a variety of flows, ranging from steady and decaying regimes in thermal counterflow, to turbulence generated by a towed grid, to turbulence near an oscillating boundary, to rotating turbulence in ^{3}He-B in the presence of a stationary normal fluid. This chapter describes the main experiments as well as the theoretical models and the numerical simulations which we use to understand them.

1 Counterflow Turbulence in He II and Its Decay

Thermal counterflow is a form of motion peculiar to two-fluid superfluid hydrodynamics (Landau (1947)) and as such has no direct analog in any ordinary viscous fluid. For this reason it was often stated that counterflow turbulence in He II has nothing to do with classical turbulence, although recent experiments suggest otherwise. Clearly, if one attempted to find a possible connection between thermal counterflow and classical turbulence, the likely candidate would be turbulent thermal convection – this link is yet to be thoroughly investigated. In fact, as noticed by Skrbek *et al.* (2003), the efficiency of the turbulent heat transport in both cases is strikingly similar, in that it displays a power law dependence on the temperature difference with exponent close to 1/3.

Thermal counterflow can be easily set up by applying a voltage to a resistor (heater) located at the closed end of a channel open to the helium bath at the other end. The heat flux is carried away from the heater by the normal fluid alone, and, by conservation of mass, a superfluid current arises in the opposite direction. In this way a relative (counterflow) velocity $\mathbf{v}_{ns} = \mathbf{v}_n - \mathbf{v}_s$ is created along the channel which is proportional to the applied heat flux, $\dot{q} = \dot{Q}/A$, where A is the cross–section of the flow channel, which for simplicity we assume constant. Assuming that the power $\dot{Q}$ applied at the heater is used to convert superfluid into normal fluid, the velocity of the outgoing normal fluid is $v_n = \dot{q}/(ST\rho)$, where S is the specific entropy of He II[1]. The counterflow velocity v_{ns} is then easily established from the condition $\rho_s v_s + \rho_n v_n = 0$. At relatively small values of v_{ns} the heat–flow transfer becomes affected by turbulence due to the appearance of an apparently disordered tangle of superfluid vortex lines. Its intensity is represented by the vortex line density L - total length of vortex lines in a unit volume.

1.1 Historical overview

Experimentally, counterflow turbulence has a long history since the discovery of superfluidity, beginning with the heat transport experiments through capillaries and channels of various sizes. It was observed by Gorter and Mellink (1949) that in wide channels, for not very low flow velocity there is a dissipative process in a form of mutual friction force between the normal fluid and superfluid, of the form $A\rho_s\rho_n(v_s - v_n)^3$ per unit volume. The origin of this process was not known at that time, but later in the early fifties it became clear (thanks to the ideas of Onsager and Feynman) that the mutual friction is the result of interaction between the normal fluid and quantized vortices.

We shall not describe all important experimental results (for a review of early results, we direct the reader to the excellent review by Tough (1982)), but due to space restriction select only a few.

Counterflow turbulence in a wide channel was first investigated and described in detail in an outstanding series of papers by Vinen (1957), who also introduced a phenomenological model for its description based on the concept of a random vortex tangle characterized by the approximately homogeneous vortex line density L. He argued that L obeys the following equation (now called the Vinen's equation):

$$\frac{\partial L}{\partial t} = \frac{\rho_n B}{2\rho}\chi_1 v_{ns} L^{3/2} + g(v_{ns}) - \frac{\kappa}{2\pi}\chi_2 L^2 \,, \tag{1.1}$$

where ρ and ρ_n are the total and normal fluid densities, κ is the circulation quantum, χ_1, χ_2 are undetermined dimensionless constants and B is the mutual friction parameter, tabulated in Donnelly and Barenghi (1998). The arguments leading to Eq. (1.1) are dimensional: The first term on the RHS describes production of turbulence; the last term on the RHS its decay. The so far unspecified function $g(v_{ns})$ had to be included to comply with

[1] Here we limit ourselves by this simple approach, neglecting, for example, the kinetic energy of the flowing normal fluid and superfluid.

the experimental fact that below certain critical v_{ns} there are no quantized vortices and hence no mutual friction in the counterflow.

Eq. (1.1) was later deduced by Schwarz (1988)[2] using the fundamental laws of vortex dynamics in the local induction approximation. Under the assumption that the vortex tangle is homogeneous and the quantity L is enough to characterize it, (from the point of view of turbulence this means that there is only one relevant length scale, $1/\sqrt{L}$ - the average distance between the quantized vortices) this approach accounts for most of the observed phenomena in steady state counterflow turbulence, as described in the review by Tough (1982), whose group contributed over the years a great deal to the experimental investigation of counterflow in He II. Assuming that $g(v_{ns})$ is small, the steady–state solution of Vinen's equation yields

$$L^{1/2} = \frac{\pi \rho_n B \chi_1}{\rho \kappa \chi_2} v_{ns} , \tag{1.2}$$

which means that L is proportional to the square of the counterflow velocity, and the steady–state value determines only the ratio of χ_1 and χ_2.

Besides the powerful technique of second sound attenuation, the relative increase of temperature, ΔT, pressure, Δp, or chemical potential, $\Delta \mu$ over the counterflow channel length as well as the influence of the counterflow on the transmission of positive and negative ions have been measured by various investigators. For details of these techniques, we direct the reader to the review by Tough (1982) and original references therein.

Various techniques allowing for arranging flow of the normal and superfluid He II components in almost arbitrary proportion have been worked out, including a pure superflow, but these results stay beyond the scope of the current lecture.

1.2 Steady-state counterflow

When the heater in the dead end of the counterflow channel is switched on, it does not take long before a steady-state counterflow sets in. When the applied power $\dot{q} = \dot{Q}/A$ is below a critical value, there are no quantized vortices in the flow (except perhaps of the remnant ones). The normal fluid and superfluid velocity fields are not affected by the mutual friction force; therefore the normal fluid behaves as an ordinary viscous fluid of viscosity η, the flow acquires the usual parabolic profile and Δp is observed to agree with the Poiseulle formula for laminar flow:

$$\Delta p_{lam} = G \eta \ell v_n / d^2 , \tag{1.3}$$

[2] see the companion lecture by C.F. Barenghi in this volume

where G is a factor generally known for a particular channel geometry, ℓ is its length and d denotes the small dimension of the channel.

The temperature difference is naturally determined by Δp:

$$\Delta T_{lam} = \Delta p/\rho S = G\eta\ell v_n/\rho S d^2. \tag{1.4}$$

The steady counterflow turbulence becomes stable only when the applied

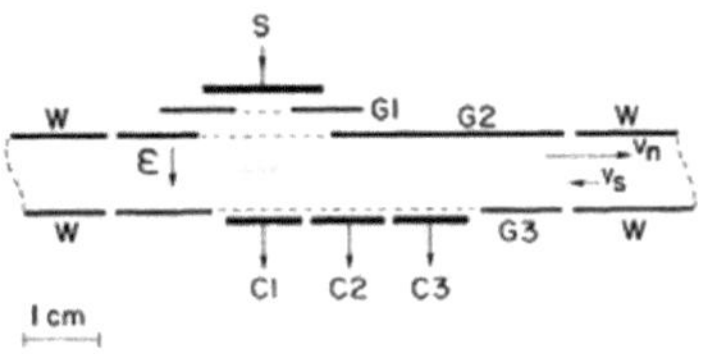

Figure 1. Experimental arrangement of Awschalom, Milliken and Schwarz (1984). S denotes the tritium ion source, G1, G2, G3 are pulsed grids, C1, C2, C3 are colllectors, W denotes the channel wall, v_s and v_n in the counterflow are indicated by arrows. The dotted region indicates a typical ion pulse; the ions are moved by an electric field ε.

power $\dot{q}$ exceeds a critical value. The pressure difference Δp_{turb} as a function of $\dot{q}$ only slightly exceeds the extrapolation of the laminar data - this is so-called Allen and Reekie (1939) rule. On the other hand, the chemical potential and temperature difference increase dramatically in the turbulent state, and are found to exceed their laminar values roughly proportionally to $\dot{q}^3$. The increased dissipation in the turbulent state can be described by a dissipative mutual friction force which also gives rise to an excess second sound attenuation α', which increases approximately as $\dot{q}^2$, as found already by Vinen (1957).

The microscopic origin of the mutual friction force is, of course, scattering of the elementary excitations - phonons a and rotons (i.e., the normal fluid) - off quantized vortex lines. The important question is the form of distribution of these vortex lines. To obtain at least a partial answer to this question, ion currents can be explored. Motion of both ion species[3] is affected by the presence of quantized vortices. Due to the existence of the Magnus force, an attractive interaction between the ions and the vortices

[3]Positive helium ions are effectively solid snowballs made of about 20 helium atoms while negative helium ions represent effectively an electron in otherwise empty bubble of a nm size possessing a rather large hydrodynamic added mass of order 200 bare helium atoms

exists, capable of trapping the ions on the vortex cores. The successful theory of interaction between ions and vortices has been produced by Donnelly and Roberts (1969). It explains, in particular, why the probability to escape from the vortex core, especially at higher temperature, is much smaller for negative ions.

Negative ions have been successfully used by Awschalom, Milliken and Schwarz (1984) to probe the local properties of counterflow turbulence in a large channel of 1 cm$\times$ 2.3 cm rectangular crossection, as schematically shown in Fig. 1. They used pulsed-ion technique based on the tritium source to obtain the spatially resolved measurements of the vortex line density and the normal fluid velocity field in turbulent counterflow. In their measurements, a narrow pulse of negative ions was gated into the channel and allowed to propagate to a particular position under the action of an applied electric field. The field was then switched off, allowing the pulse to remain at this position for as long as several seconds. Some of the ions in the pulse were trapped by quantized vortices; the rest drifted along with the local normal fluid velocity. Later, the electric field was turned back on and the surviving pulse was measured in real time by means of one or more electrometers as it arrived at one of three collectors on the opposite side of the channel, positioned along the channel axis. From the amplitude of the observed pulse and where it fell on the collector, the local values of L and $\mathbf{v}_n$ were determined, with accuracy better than 1 mm - providing what is up to now probably the best information on local structure of counterflow He II turbulence. The remarkable result was that within the experimental uncertainty up to few percent L was found constant over entire crossection of the channel, except within about 1 mm from the channel walls where the technique fails. The conclusion was that the normal fluid profile is approximately flat (except close to the walls, see later) and the vortex line density approximately homogeneous.

TI, TII and TIII states of He II counterflow turbulence The careful extensive measurements of the temperature difference along the counterflow channel of Tough's group (see Tough (1982) for references to early original work) clearly indicate that depending on the geometry, there are more steady counterflow turbulent states. In rather large channels of circular or square crossection, upon increasing the applied power $\dot{q}$, the laminar flow undergoes the first transition at $\dot{q}_{c1}$, above which the measured ΔT exceeds the extrapolated laminar value only slightly, until at $\dot{q}_{c2}$ the second transition occurs, above which the increase of measured ΔT is dramatically bigger. These turbulent states are referred to as TI and TII. It is fair to mention that the existence of the state TI for $\dot{q}_{c1} < \dot{q}_{c1} < \dot{q}_{c2}$ was recognized already by early workers: thanks to its signature in second sound attenuation data by Vinen (1957), using pressure difference data by Brewer and Edwards (1961), and temperature difference by Chase (1962)), but has been experimentally investigated in some detail only later, by Tough's group. Much less is known about the state TI than about state TII (especially about scaling of observed variables with the applied heat flux), as it exists only in the rather narrow temperature dependent region $\dot{q}_{c1} < \dot{q}_{c1} < \dot{q}_{c2}$. Because the agreement between the experimental data and the homogeneous theory of Schwarz (1988) is worse in TI state than in TII state for which this theory works rather well, it was claimed by Tough (1982) that the TI state might not be homogeneous and isotropic.

There is a profound effect of geometry on counterflow turbulence. In channels of high aspect ratio (1:10) rectangular crossection with a small dimension below 100 μm only one transition has been observed at $\dot{q}_{c3}$, into a turbulent steady-state denoted by Tough as as TIII.

No hysteresis associated with the TI – TII transition has been observed, but metastable laminar states can be observed for heat fluxes considerably exceeding $\dot{q}_{c1}$ or $\dot{q}_{c3}$. However, large fluctuations of the chemical potential and temperature difference have been observed in the vicinity of the TI – TII transition.

The existence of the two distinctly different steady turbulent states in He II counterflow turbulence has been a long-standing puzzle. One possibility to understand this problem was considered by Melotte and Barenghi (1998), who addressed the issue of the velocity profile of the normal fluid and its stability. The calculation suggests that in the TI state the superfluid is turbulent but the normal fluid is still laminar. The TI to TII transition thus might correspond to the onset of the normal fluid turbulence.

1.3 Decaying counterflow

Vinen's equation 1.1 has also been tested in the case of time dependent flows. If the power applied to the heater is not constant in time but has a white noise frequency spectrum, the spectral response of L can be derived from Vinen's equation and agrees well with the experiment of Barenghi, Swanson and Donnelly (1982); combining steady–state measurement with measurements of the spectrum, one can determine χ_1 and χ_2 separately. Vinen's equation can also be applied to the transient response of the vortex tangle from one steady state towards another, and even allows an analytical solution, as shown by Schwarz and Rozen (1991). In the case of free temporal decay, setting $v_{ns} = 0$, the solution of Eq. (1.1) reduces to the inverse time dependence $L(t) \propto 1/(t + t_{vo})$, where t_{vo} stands for the virtual

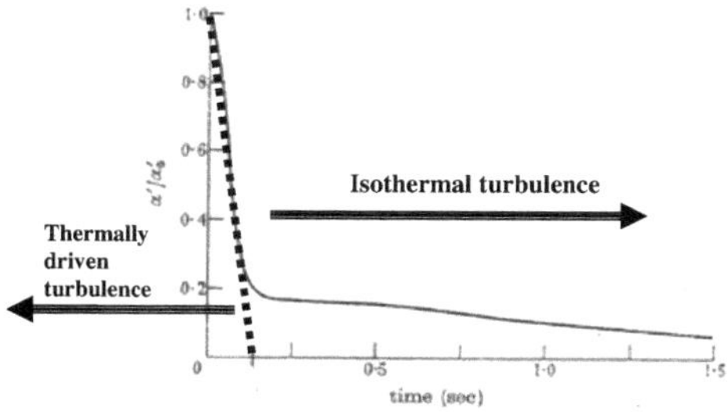

Figure 2. Variation with time of the excess second sound attenuation $\alpha'/\alpha'0$ after a heat current 0.14 W/cm^2 has been switched off (Vinen (1957)). α'_0 is the attenuation in the steady heat current. The dotted line is a plot of Eq. (1.7), with τ estimated to be about 0.14 s, based on the geometry of the apparatus.

origin time. We shall see, however, that this simple prediction in most cases does not hold, as the decaying counterflow turbulence involves additional physical processes closely connected to classical hydrodynamic turbulence.

Thermally–driven decay versus isothermal decay Before we try to analyze the temporal decay of the vortex tangle generated in counterflow along the lines of Barenghi and Skrbek (2007), it is useful to estimate the characteristic time τ for the decay of the temperature difference ΔT across the counterflow channel using simple thermodynamic consideration of Gordeev *et al.* (2005). We assume a channel of a length ℓ and of a constant cross–section A and, for simplicity, with a linear temperature gradient along the channel length, $\nabla T = \Delta T/\ell$. When the heat input $\dot{Q}$ applied to its closed end is switched off, the amount of heat that must be taken away from the channel into the bath kept at the temperature T_0 equals $Q_{ch} = c\varrho A\ell^2 \nabla T/2$ (assuming that ∇T is linear, the average excess temperature over the entire channel is not ΔT but$\Delta T/2$) where $c(T) \cong c(T_0) = c$ is the specific heat of He II. In case there is an additional volume V adjacent to the hot end of the channel in which the heater is placed, we have to add the heat $Q_V = c\varrho V\ell\nabla T$. Thus the total excess heat equals to $Q = Q_{ch} + Q_V = c\varrho\ell(A\ell + 2V)\nabla T/2$. Differentiation with respect of time gives

$$\frac{dQ}{dt} = \dot{Q} = c\varrho\ell(A\ell + 2V)(\dot{\nabla T})/2. \qquad (1.5)$$

On the other hand, as it was established already by Vinen (1957), for high enough counterflow velocity

$$\dot{Q} = -A\zeta(T)(\nabla T)^{1/3}, \tag{1.6}$$

where ζ is a temperature dependent heat efficiency parameter. Eqs. (1.5) and (1.6) provide a differential equation for $\nabla T(t)$, which is easy to solve by separating the variables:

$$[\nabla T(t)]^{2/3} = [\nabla T(0)]^{2/3} - \frac{4\zeta A}{3c\varrho\ell(A\ell + 2V)}\, t, \tag{1.7}$$

where $\nabla T(0)$ is the initial temperature gradient, measured in the steady state before the heater is switched off. Defining τ from Eq. (1.7) as time when $\nabla T(t) \to 0$ yields

$$\tau = \frac{3c\varrho\ell(A\ell + 2V)}{4\zeta A}\, [\nabla T(0)]^{2/3}. \tag{1.8}$$

Fig. 2 applies these simple considerations to the experiments of Vinen (1957). It shows the scan of the original experimental data, with the linear time dependence predicted by Eq. (1.7) superimposed. This procedure is possible, as the excess attenuation of the second sound in the steady state was experimentally found proportional to $\dot{Q}^2$ and hence also to $[\nabla T(t)]^{2/3}$.

The quantitative agreement between the data and our simple approach allows to divide the decay into two distinctly different parts:

(i) $t < \tau$, where the decaying turbulence is still thermally driven by the excess heat accumulated in the hot end of the flow channel,

(ii) $t > \tau$, where the decaying turbulence can be treated as approximately isothermal.

It follows from formula (1.8) that τ is very long if there is a big buffer volume V adjacent to the hot end of the channel and/or if the counterflow channel is very long. Here the decay of counterflow can be understood as a sequence of steady-state pictures, with any dynamics of decaying turbulence suppressed. These simple thermodynamic considerations are important when analyzing the transient behaviour or the free decay of the counterflow turbulence experimentally investigated by many authors.

Isothermal decay of counterflow turbulence The powerful technique of second sound attenuation is capable of direct investigation of rather complex physics of the decaying counterflow turbulence even when it can be considered approximately isothermal. A typical family of decay curves is shown in Fig. 3. In order to compare the decay data with the classical

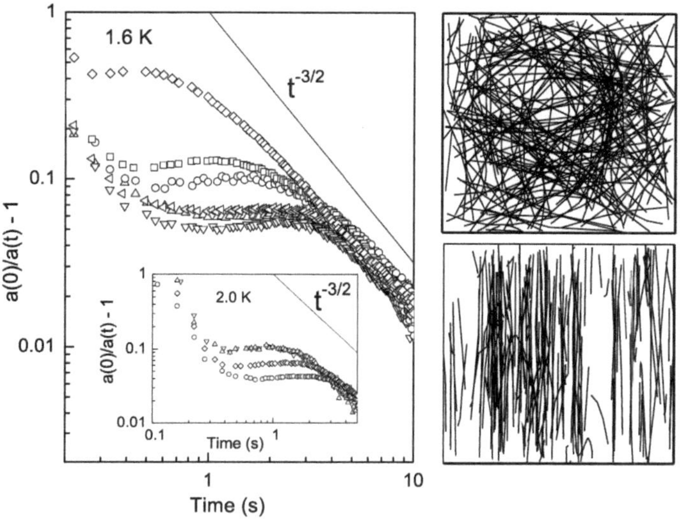

Figure 3. Log–log plot of the quantity $(a_0/a(t) - 1)$, where a_0 and $a(t)$ denote the initial and decaying second sound amplitude, versus time measured at T= 1.6 K. Under the assumption that the tangle is homogeneous and isotropic, this quantity would be proportional to the vortex line density L. For given experimental conditions, the level $a_0/a(t) - 1 = 0.1$ would correspond to the vortex line density $L \approx 10^5$ m/cm^2. The different decay curves correspond to different initial levels of steady–state counterflow turbulence generated at powers 0.5 W ($\diamond$), 0.23 W (squares), 0.18 W ($\circ$) and 0.14 W (various triangles) respectively. Three individual decay curves are shown to appreciate the level of reproducibility for the lowest applied power. The inset shows analogical data measured at T= 2.0 K; applied powers: 0.52 W (various triangles), 0.41 W ($\diamond$) and 0.32 W ($\circ$). Perpendicular and parallel projections of the initial vortex tangle resulting from the computer simulations clearly show that the vortex tangle is strongly polarized.

turbulence data, we display the results in terms of κL, which can be interpreted as root mean square vorticity of the turbulence, assuming for the sake of simplicity (but see the discussion later) that the vortex tangle is isotropic and homogeneous. The most striking feature is that the late stage (past saturation time t_{sat}) of the decay of turbulence displays the classical $t^{-3/2}$ power law, independently of temperature. The same temperature independence was observed in the decay of turbulence generated by a towed grid. These results strongly suggest that for $t > t_{sat}$ the decay of counterflow turbulence and the decay of grid–generated turbulence are essentially similar and follow the pattern predicted by the phenomenological spectral decay model (see Skrbek and Stalp (2000)) of the form

$$\omega(t) = \frac{d(3C)^{3/2}}{2\pi\sqrt{\nu}}(t + t_{vo})^{-3/2} \tag{1.9}$$

where C is the 3D Kolmogorov constant and ν denotes the kinematic viscosity of the fluid. The decay of grid generated turbulence in He II in a channel geometrically similar to ours closely obeys this power law decay up to several orders of magnitude, if superfluid vorticity is defined as $\omega_s = \kappa L$, where κ denotes the circulation quantum and kinematic viscosity is replaced by an effective kinematic viscosity $\nu_{eff}(T)$ of the turbulent He II of order κ - it has been recently shown by Chagovets *et al.* (2007) that values of $\nu_{eff}(T)$ extracted from the decaying grid generated turbulence and decaying counterflow turbulence agree with each other.

The measurement of decaying counterflow turbulence in two channels of square cross-section by Gordeev *et al.* (2005) provided a first direct experimental check that the vortex line density in the late stage of the decay is proportional to the channel size d, in accord with Eq. (1.9). This result confirms the surprisingly close similarity between the late stage of decaying counterflow turbulence in He II and the decay of classical homogeneous and isotropic turbulence.

On the contrary, the early part of the isothermal decay displays complicated features; the decay is slow and sometimes even includes periods during which the second sound signal appears to increase rather than decrease, which would imply a puzzling increase of vortex line density. In fact, as computer simulations of Barenghi *et al.* (2007) show, (see Fig. 3), the vortex tangle induced by the counterflow is polarized by the normal fluid along the channel. When the heater is switched off, this polarizing effect gradually ceases, and vortex reconnections play a more important role. Vortex loops – which were flattened in the plane across the channel in the steady-state – start turning around randomly when reconnecting. Hence the apparent vortex length which is "visible" by second sound can increase according to the sine squared law.

Swanson, Barenghi and Donnelly (1983) showed that polarization of quantum turbulence can also be induced by rotating the counterflow channel around its axis. In this case, according to numerical simulations of Tsubota, Araki and Barenghi (2003), the vortex tangle becomes axially polarized (the vortex lines tend to become aligned along the axis of rotation). It is likely that the spin–down of this system has similarities with the depolarization of the decaying counterflow turbulence.

In the third stage of the decay of counterflow turbulence ($t > t_{sat}$), the vorticity acquires a universal classical $t^{-3/2}$ power law. The classical spectral decay model is based on the existence of a conventional 3D energy spectrum possessing an inertial range of scales of the Kolmogorov type, characterized by a -5/3 roll-off exponent (neglecting intermittency). The so far accepted physical picture of decaying grid generated He II turbulence assumes an existence of a range of normal and superfluid eddies (constituted of bundles of quantized vortex lines) coupled by the action of the mutual friction force. In practice, it is sufficient that a disordered system of vortex lines has enough polarization in the direction of the normal fluid vorticity so that effectively, in a coarse-grained sense, coupled normal fluid and superfluid eddies are formed. Therefore the most plausible interpretation of the experimental results is to say that during the third stage of the decay (for $t > t_{sat}$) we expect superfluid eddies ranging all the way up to the size of the channel. But if there are big superfluid eddies, these will interact via mutual friction with the normal fluid. Therefore the normal fluid is most likely driven into a turbulent state too, so that in this stage the decaying counterflow turbulence should indeed be similar in character to the decaying grid generated turbulence.

On the other hand, the properties of steady-state counterflow turbulence are well described by Vinen's equation, based on the assumption that the vortex tangle is random. Therefore there are most likely no big superfluid (or normal) eddies. This might be understood as a consequence of a large mean counterflow velocity that (in the spanwise direction) tears big normal and superfluid eddies apart on a faster time scale than their turnover time.

Assuming that our outlined physical picture of the structure of the vortex tangle in the first and third stages is correct, we are left to assume that the gradually bigger and bigger superfluid eddies grow during the second stage, until their size becomes saturated by the channel dimension. As they grow, they must in turn generate normal eddies of large size too (due to the action of the mutual friction force – mean counterflow does not exist any more to tear them apart) so that at the beginning of the third stage the decaying counterflow and grid generated He II turbulence appear essentially the same and therefore display the same decay.

1.4 Discussion

Counterflow turbulence in He II as a topic of investigation is celebrating its 50th anniversary since the pioneering experiments of Vinen (1957). Thanks to the work of numerous investigators, a lot is known today about quantum turbulence in helium superfluids in general and about counterflow turbulence in He II in particular. It is perhaps fair to say that over most of this time quantum turbulence was usually viewed as a tangle of quantized vortex lines and its possible relationship to hydrodynamic turbulence in classical fluids was not given much thought. This necessarily resulted in disconnected fields of hydrodynamic and superfluid turbulence, and so it seemed reasonable to think that the two fields were quite different in character.

Recent experiments reveal a deep similarity between the two fields of turbulence, although it breaks down at small scales where the quantization effects of circulation have to be taken into account explicitly. Partly for this reason, partly because thermal counterflow is a form of motion specific to two-fluid superfluid hydrodynamics, a fair level of understanding of steady-state counterflow turbulence in He II was reached without considering the properties of classical turbulent flows. Still, even steady-state counterflow in He II calls for a new level of investigation to finally remove long-lasting puzzles such as the existence of various turbulent stages.

2 Towed Grid Turbulence in He II and Its Decay

Grid generated nearly homogeneous and isotropic turbulence (HIT) in air or water and its decay is one of the most important and extensively explored problems in turbulence. Most of the experimental work is related to wind tunnels, where the turbulence is studied as it decays downstream (e.g., Comte-Bellot and Corrsin (1966); Kistler and Vrebalovich (1966); Bennett and Corrsin (1978)). Turbulence without a mean flow, generated by using an oscillating grid (e.g., by De Silva and Fernando (1994)) and by towing a grid through a stationary sample of fluid (e.g., by van Doorn, White and Sreenivasan (1999)) has also been studied. A general theory describing the decay of turbulence based on first principles has not yet been developed. However, experimental data containing information about the turbulent energy spectra provide a solid foundation for a phenomenological approach first outlined by Comte-Bellot and Corrsin (1966) and Saffman

(1967), as the generally accepted forms of the three-dimensional turbulent energy spectra uniquely determine the temporal decay of turbulence. This approach has been later extended by Skrbek and Stalp (2000), who generalized it by taking into account the finite size of the turbulence box, intermittency and viscosity effects and discussed the final period of decay. Surprisingly, it turned out that the classical model describes very well the experimental data on decaying vortex line density in He II in the Oregon towed grid experiments. In this Lecture, we describe these experiments and give some details of their interpretation.

2.1 Classical grid turbulence

We shall limit our considerations to the simplest form of classical turbulence that can be thought of as homogeneous and isotropic (HIT). Nearly homogeneous and isotropic turbulence can experimentally be generated by a flow of fluid through a grid of bars in a wind tunnel (see Fig. 4), or, alternatively, by towing such a grid through a stationary sample of fluid. Jets of fluid emerging from the grid are unstable and break up into turbulent eddies of various sizes, and at a distance from the grid much greater than the mesh of the grid, M, the turbulent wakes formed behind each element of the grid merge. Such a turbulence is approximately homogeneous in that its statistical properties are locally independent of position. In practice the turbulence is decaying in time, and this imposes some lack of homogeneity. However, the decay turns out as a rule to be sufficiently slow that it has a negligible effect on the local properties of the turbulence.

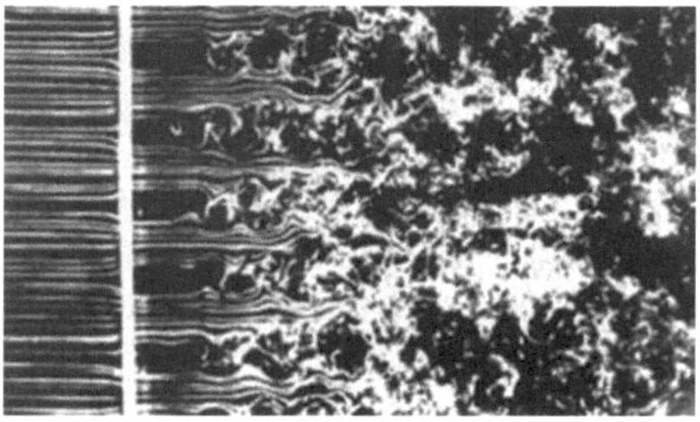

Figure 4. Visualization of grid turbulence in a classical fluid.

As noted already by Leonardo da Vinci, turbulence involves rotational, or eddy, motion on many length scales. Eddies of various sizes are coupled by the nonlinear inertial term in the Navier-Stokes equation in such a way that turbulent energy flows without dissipation (as long as the Reynolds number is large) from the large scale eddies (size D) into smaller eddies (size d). This process, known as Richardson cascade, continues until small eddies with Reynolds numbers of order unity dissipate the turbulent energy by viscosity[4]. The range of length scales over which there is negligible dissipation is called the inertial range.

Let us introduce an important statistical property of HIT: the energy spectrum, $E(k)$, such that the average turbulent energy per unit mass in the range of wavenumbers dk is $E(k)dk$. Following dimensional reasoning of Kolmogorov, inside the fully developed cascade $2\pi/D \ll k \ll 2\pi/d$, that is in the inertial range, the spectrum can depend on only k and the energy decay rate $\varepsilon = -dE/dt$. If there is no dissipation inside the inertial range, ε is also the rate at which energy per unit mass flows down the cascade, to be dissipated ultimately on the Kolmogorov dissipation scale, d. It follows from a dimensional argument that $E(k)$ must have the Kolmogorov form

$$E(k) = C\varepsilon^{2/3}k^{-5/3}(kD)^{\mu} , \tag{2.1}$$

where C is a dimensionless quantity known as the Kolmogorov constant. Using results of many classical experiments confirming this form of the spectrum in the inertial range, Sreenivasan (1995) quotes $C = 1.62 \pm 0.17$. μ is known as the intermittency exponent; it has been rigorously proven by Frisch (1995) that $\mu \neq 0$. Intermittency is associated with the violent, atypical discontinuous nature of turbulence. When a signal from turbulent flow is measured at a single spatial point many times (an Eulerian measurement), the fluctuations in the values appear to be random. Since any random sequence is most naturally explained in terms of its statistical distribution, one typically determines the statistics by constructing a histogram of the signal. The violent nature of turbulence shows itself in the very special shape of the histogram, or the probability distribution function, of the turbulent velocity signal. It is typically wider than the usual Gaussian distributions emerging in the context of equilibrium statistical physics. Although intermittency is a big issue in today's classical turbulence research (and probably will become an important issue in quantum turbulence, too), in this Lecture we shall neglect it.

It follows from Eq. (2.1) that in fully-developed turbulence most of the energy is concentrated in the largest (energy containing) eddies. Examination of the form of the Navier-Stokes equation shows that the non-linear coupling will cause a transfer of energy from these energy-containing eddies in a time of order D/U (the turnover time), where U be the characteristic velocity associated with these eddies. Thus the rate at which the energy

[4]This applies to the 3D turbulence, e.g., in the 2D turbulence the situation is very different

containing eddies lose energy per unit mass $\varepsilon \approx U^3/D$. This energy is being dissipated at large wavenumbers by viscosity, and this dissipation rate can be shown to be $\varepsilon = -dE/dt = \nu\omega^2$, where $\boldsymbol{\omega} = curl\,\mathbf{v}$ is the mean square vorticity in the flow. This mean square vorticity is due largely to the smallest eddies, on the Kolmogorov scale d, which from a dimensional reasoning can be estimated as $d \approx (\varepsilon/\nu^3)^{1/4}$, where ν stands for the kinematic viscosity.

Another important notion for the grid turbulence is so-called Taylor frozen hypothesis, meaning that past the grid advection contributed by turbulent circulations themselves is small and therefore the advection of a field of turbulence past a fixed point can be taken to be entirely due to the mean flow in the wind tunnel. Mathematically it means that the time derivative $\partial/\partial t$ can be replaced by $U_0\partial/\partial x$, where U_0 is the mean velocity and x denotes the downstream distance from the grid in a wind tunnel.

2.2 The classical spectral decay model

Let us introduce a simplified version (for full version, see Skrbek and Stalp (2000)) of a classical spectral decay model of decaying HIT. It assumes that at early times we have a generally accepted form of the 3D energy spectrum

$$E(k) = 0 \; ; \; k < 2\pi/D \tag{2.2}$$
$$E(k) = Ak^m \; ; \; m = 2 \; ; \; 2\pi/D \leq k_1(t) \tag{2.3}$$
$$E(k) = C\varepsilon^{2/3}k^{-5/3} \; ; \; k_2(t) \leq k \leq \gamma(\varepsilon/\nu^3)^{1/4} = 2\pi\eta_{eff} \tag{2.4}$$
$$E(k) = 0 \; ; \; k > \eta_{eff} \tag{2.5}$$

which reflects the fact that eddies larger than the size of the turbulence box cannot exist. The high wave number exponential tail is approximated by a sharp cutoff at the effective Kolmogorov dissipation scale η_{eff} by introducing the dimensionless factor γ of order unity. In the vicinity of the energy containing length scale $\ell_e = 2\pi/k_e(t)$, where $k_1(t) < k_e(t) < k_2(t)$, the spectral energy density displays a broad maximum whose shape does not have to be exactly specified. Evaluating the total turbulent energy by integrating the 3D spectrum over all k leads to a differential equation for decaying turbulent energy and applying $\varepsilon = \nu\omega^2$ leads to a differential equation for decaying vorticity, the quantity we are primarily interested here.

At the early decay the spectrum stays self-similar. For $D \gg \ell_e \gg \eta_{eff}$ the decay is

$$E(t + t_{vo1}) = E(\tau) \propto \tau^{-2\frac{m+1}{m+3}} \; ; \; \ell_e \propto \tau^{\frac{2}{m+3}} \; ; \; \omega \propto \tau^{\frac{3m+5}{2m+6}} \; . \tag{2.6}$$

Here t_{vo1} is the virtual origin time, when the decaying turbulent energy would have been infinitely large. Assuming validity of the Saffman invariant ($m = 2$), we obtain the **first regime** for decaying vorticity, $\omega \propto \tau^{-11/10}$.

As the turbulence decays further and $\ell_e \propto \tau^{2/5}$ grows, the lowest physically significant wave number becomes closer to the broad maximum around $2\pi/\ell_e$. The low wave number part of the spectrum can no longer be approximated as Ak^m with $m = 2$. Instead, it can be characterized by an effective power that decreases as the turbulence decays, such that $0 < m < 2$. Formula (2.6) then shows that the decay rate slows down. As ℓ_e approaches D, m becomes effectively zero and we arrive at the **second regime** of the decay $\omega \propto \tau^{-5/6}$.

At the saturation time, t_{sat}, the vorticity reaches its saturation value, ω_{sat}, and the growth of ℓ_e is completed. Still neglecting the cutoff of the spectrum at η_{eff}, the **universal third regime** is predicted as

$$\omega(t + t_{vo2}) = \omega(\tau) = \frac{\sqrt{27}D}{2\pi}\sqrt{\frac{C^3}{\nu}}\tau^{-3/2} \tag{2.7}$$

with the different virtual origin time t_{vo2}. This regime is universal in that no matter whatthe starting level of turbulence was (providing it was high enough to neglect viscosity corrections), the decaying system must sooner or later reach it.

So far we neglected a role of the high wave number cutoff of the energy spectrum at η_{eff}. As the vorticity decays and the Kolmogorov scale grows, the relative importance of this cutoff increases and a simple power cannot any longer describe the decay of vorticity. It was shown by Skrbek, Niemela and Donnelly (2000) that after saturation of the energy containing length scale by the size of the turbulence box the decay of vorticity is more accurately described as

$$\omega(\tau) = \frac{3\nu}{2^3\gamma}\frac{2\pi}{D}^2\frac{t_B}{\tau}^{3/2}\cos^3\theta \ , \tag{2.8}$$

where $\cos^2(3\theta) = \tau/\tau_B$ and $\tau_B = \frac{16C\gamma^{4/3}}{9\nu}\frac{2\pi}{D}^2$. Naturally, for $\gamma \to \infty$ expression (2.8) reduces to the simple power law (2.7). Formally, as the vorticity decays, η_{eff} becomes the size of the turbulent box D and both the turbulent energy and vorticity vanish. Close to this stage, however, applicability of the spectral decay model is no longer justified.

This decay model has been succesfully applied to a number of wind tunnel experiments as well as to the decaying turbulence generated in water by an oscillating grid by De Silva and Fernando (1994), where the effect of saturation the energy containing length scale is clearly seen. Moreover, the validity of the model was confirmed by computer simulations of decaying turbulence in a bounded domain by Touil, Bertoglio and Shao (2002).

2.3 Oregon towed grid experiments in He II

Experiments on decaying quantum turbulence in He II generated by the towed grid have been performed over years in a group of R.J. Donnelly in Eugene, Oregon. The essential experimental results have been published in a series of papers by Smith, Donnelly, Goldenfeld and Vinen (1993b), Stalp, Skrbek and Donnelly (1999) and Skrbek, Niemela and Donnelly (2000).

The schematic of the experimental apparatus is shown in Fig. 5. The turbulence is generated by towing a grid through a stationary sample of He II. Two different grids have been used over time. The first one was of rather unconventional design – a 65% open brass monoplanar grid of rectangular tines, 1.5 mm thick, with a mesh size, M (tine spacing) of 0.167 cm, the second one was consisting of 28 rectangular tines of width 0.012 cm forming 13 × 13 full meshes across the channel of approximate dimension $M = 0.064$ cm. The grid was attached to a stainless steel pulling rod which exited the cryostat via a pair of tight sliding seals. The space between the seals was continually evacuated to prevent the introduction of impurities inside the cryostat during the experiment. Above the cryostat, the rod was attached to a computer controlled linear servo motor that positioned the grid with 0.01 cm accuracy and provided the towing velocity, v_g, up to 2.5 m/s. This enabled the exploration of a wide range of mesh Reynolds numbers up to $Re_M = v_g M\rho/\mu \leq 2 \times 10^5$, where μ is the dynamic viscosity of the normal fluid and ρ the total density. The channel of square cross section was typically 29 cm long and $D = 1$ cm wide, manufactured by an electroforming process. It was suspended vertically in the helium cryostat and during the measurement totally submersed in superfluid helium, which entered via 16 holes, 500 μm in diameter, placed near the top end. Except for these, the channel was leak-tight.

To probe the quantized vortex line density, L, resulting from pulling the grid, second sound attenuation technique was used (see the companion Lecture). Second sound was excited and detected using vibrating nuclepore membranes 9 mm in diameter mounted flush on opposing walls of the channel. These 6 μm thick polycarbonate membranes have a dense distribution of 0.1 μm holes and on one side is evaporated an approximately 100 Å thick layer of gold, which is then pressed against the channel wall. This gold layer forms one electrode of a capacitor transducer, with the other being a brass electrode pressed towards the opposite side of the membrane from outside the channel. Applying an ac signal ($\sim 0.3 - 1$ V_{PP}) in addition

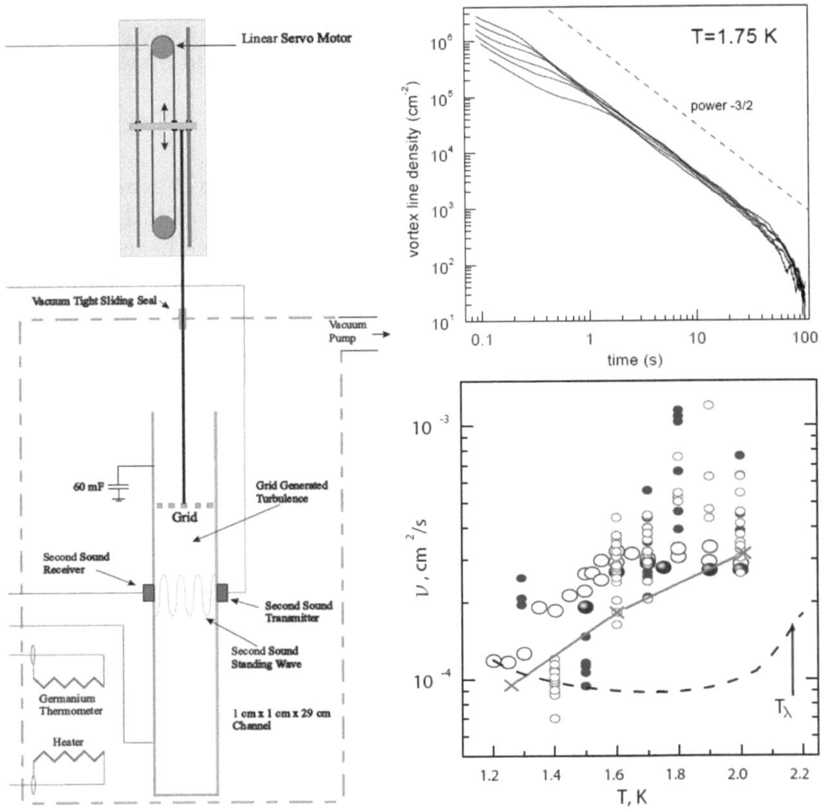

Figure 5. Schematic of the experimental apparatus (left). The log-log plot of the decaying vortex line density versus time after grid passes 2 mm above the measuring volume. Each decay curve represents an average of three identical pulls. The decay curves, in order, correspond to $Re_M = 2 \times 10^5$ (the uppermost one), 1.5×10^5, 10^5, 5×10^4, 2.5×10^4, 10^4, 5×10^3 and 2×10^3. For each Re_M, the decaying vortex density joins and follows thereafter the universal power decay with exponent -3/2 (top right), which is used to deduce the effective effective kinematic viscosity of turbulent He II at various temperatures (bottom right). Big circles represent the corrected values of $\nu_{eff}(T)$ (see text) deduced from the decaying vortex line density in the Oregon experiments with unconventional (open big circles) and conventional (filled big circles) grids. The small open circles are the data deduced from the late stage of decaying counterflow turbulence in a square channel of 1×1 cm^2 crossection; the small filled circles from the smaller channel 0.6×0.6 cm^2. The dotted line is a plot of kinematic viscosity of He II based on the total fluid density tabulated by Donnelly and Barenghi (1998). The crosses connected by the solid line represent a model calculation by Vinen and Niemela (2002) for $\nu_{eff}(T)$.

to a dc bias ($\sim$ 100 V) results in an oscillatory motion of the membrane. In He II, the normal fluid is clamped by viscosity inside the small holes of the membrane, while the superfluid component passes freely, thereby exciting second sound, an entropy wave in He II. Directly across the channel the second sound wave produces a corresponding oscillation of the other membrane (i.e., the receiver) and the induced signal is input to a lock-in amplifier referenced to the transmitter frequency. The channel acts as a second sound resonator. The excitation amplitude is adjusted to be within the upper half of the linear response range and the n-th harmonic- with n about 50- of the fundamental frequency is used: typically $30 - 40$ kHz. A Lorentzian resonance peak is obtained, having a full width at half maximum that is temperature dependent and typically reaches values of $\Delta_0 = 20 - 500$ Hz without quantized vortices in the channel. The data acquisition process can be briefly described as follows. The cryostat was filled with liquid helium, the bath being pumped and the temperature controlled to a desired value. With the grid "parked" towards the top of the channel, the second sound resonance curve was measured and fitted to a Lorentzian, giving the linewidth Δ_0. The grid was then lowered to the bottom and after a necessary waiting time ($\approx$ 2 minutes) pulled through the channel such that its velocity was constant and equal v_g for at least 5 cm below and above the experimental volume. It was then slowed down and smoothly "parked" against the top again. The data acquisition was triggered when the grid passed a predetermined position 2 mm above the measuring volume, as determined by a signal from a photodiode. Typically 100 s of data were recorded at a rate of 100 Hz from the output signal of the lock-in amplifier, representing the recovering second sound standing wave amplitude. Under assumption that the vortex tangle is random it may be converted into decaying vortex line density. Typical family of decay curves is shown in Fig. 5.

2.4 Theoretical description of decaying grid turbulence in He II

In order to compare the decay of classical grid turbulence and the measured decay of L in the Oregon experiment, we have to discuss the role of quantized vortices in He II turbulence. The simplest system to assimilate the role of quantized vortex lines is a rotating bucket of He II. In steady state, the normal fluid is in solid body rotation, with vorticity $\boldsymbol{\omega}_n = \operatorname{curl} \mathbf{v}_n = 2\boldsymbol{\Omega}$, where $\boldsymbol{\Omega}$ is the angular velocity of rotation. The fluid is threaded with a rectilinear array of quantized vortex lines parallel to the axis of rotation. Although the superfluid vorticity is restricted to vortex lines, their areal density, n_0, evolves to match the vorticity of the fluid in solid body rotation. This results in $n_0\kappa = \kappa L = 2\boldsymbol{\Omega}$. Thus on length scales larger than the inter-vortex line distance, the superfluid is also in solid body rotation, with a velocity field essentially that of the normal fluid, i.e. $\langle \boldsymbol{\omega_s} \rangle = \boldsymbol{\omega_n}$.

The role of quantized vortex lines in a turbulent flow becomes more complex. Computer simulations show that similar to how vortex lines align in a rotating bucket, bundles of quantized vortices align on the cores of normal fluid eddies. The superfluid vorticity is comparable to the vorticity of the normal fluid eddy and the effective fluid density is equal to the total fluid density. In a turbulent flow there will be a complex tangle of vortices evolving in such a way as to minimize the difference between the two otherwise independent velocity fields. Turbulent He II flow thus resembles classical flow possessing an effective kinematic viscosity ν_{eff} of order μ/ρ, where μ is the dynamical viscosity and ρ denotes the total density of He II. If so, then the mentioned above usual HIT relationship should apply:

$$\varepsilon = -\frac{dE}{dt} = \nu_{eff}(\kappa L)^2 \ , \tag{2.9}$$

i.e., vorticity would be defined as κL in analogy with the rotating bucket discussed above.

The underlying physics of Eq. (2.9) is discussed in detail by Vinen (2000). It is argued that on length scales significantly greater than both the spacing between the quantized vortex lines in the turbulent superfluid component and the scale on which viscous dissipation occurs in the normal fluid, the two fluids are likely to be coupled together by the mutual friction force and to behave like a conventional fluid. On smaller length scales account must be taken of dissipation, due to viscosity in the normal fluid, frictional interaction between the vortex lines and the normal fluid, and the radiation of sound from the vortex lines. It is shown that Eq. (2.9) is likely to hold but in comparison with classical fluids there are important differences.

If Eq. (2.9) could indeed be applied, then analysis of the experimental data on decaying vortex line density as shown in the top right of Fig. 5 is straightforward, as it allows treating He II as a hypothetical classical fluid possessing an effective kinematic viscosity ν_{eff}. In particular, its temperature dependence can be extracted from comparison of the prefactor associated with the data obeying the third universal decay regime, i.e., with Eq. (2.7), as it was first done by Stalp *et al.* (2002) using the original grid design data and later by Niemela, Sreenivasan and Donnelly (2005) based on the data obtained with the new grid of improved conventional design. We show these data in the bottom right of Fig. 5, however, we have slightly corrected them for the following reason: The original calculations of the measured vortex line density from the Oregon second sound attenuation data did not take into account the "sine square law" (explained in the previous Lecture) which resulted in the fact that the originally published values of ν_{eff} are a factor $(3\pi/8)^2 \approx 1.4$ too low. We show these corrected data together with the analogous data obtained from the Prague second sound experiments of Chagovets *et al.* (2007) on decaying counterflow turbulence (see the companion Lecture on this topic). This comparison is possible, as the late stage of the decaying counterflow He II turbulence clearly obeys the $-3/2$ power law. Despite considerable scatter, the towed grid data (obtained with two different grids) and the counterflow data (from two channels of different sizes) clearly overlap. This strongly suggests that the third universal regime of the decaying He II turbulence can be, perhaps surprisingly, associated with the unique temperature dependent effective kinematic viscosity regardless of the manner by which the starting level of He II turbulence was generated.

Fig. 6 is composed in such a way that it displays all observed regimes of decaying vorticity in the grid generated He II turbulence. We stress that this is possible only if one assumes validity of Eq. (2.9) and, furthermore, the vortex tangle is assumed as random. In the compensated picture on the right the role of finite viscosity in the decay is clearly visible - the decay data follow much better the more accurate Eq. (2.8) rather then the simple $-3/2$ power law predicted by Eq. (2.7). The departure from a single power law illustrates the increasing influence of the growing Kolmogorov length scale in the decay. As the Kolmogorov scale grows, it gradually approaches D and thus ℓ_e which is already saturated by it. Expression (2.8) describes the experimental data down to surprisingly low level of vorticity, of order $0.1\ \mathrm{s}^{-1}$, corresponding to a vortex line density $L = \omega/\kappa \approx 100\ \mathrm{cm}^{-2}$ and to a mean distance between quantized vortex lines $1/\sqrt{L} \approx 1$ mm. Still, the essentially classical description of the decaying vorticity holds.

With no inertial scale left there is no energy transfer towards higher wavenumbers and the only possibility for further decay is the exponential viscous decay. This is the fourth and last regime of decaying vorticity in a finite channel [5], evident from the right inset of Fig. 6. The late decay curves originating from various Re_M display an exponential decay of the form $\omega \propto exp\,(-t/t_0)$ (practically indistinguishable from the spectral model prediction). This last decay regime can be considered in analogy with the decay of the oscillatory motion in viscous fluids, characterized by exponential decay of the energy $E \propto E_0 \exp{(-\beta t)}$, where the decay coef-

[5] Note that it differs from the final period of decay observed in classical wind tunnels, as these can be essentially regarded as infinitely large. As explained, e.g., by Skrbek and Stalp (2000), for such decaying HIT there is a changeover from the third universal regime into final stage of the decay, where each component of energy (not vorticity) decays like $t^{-3/2}$

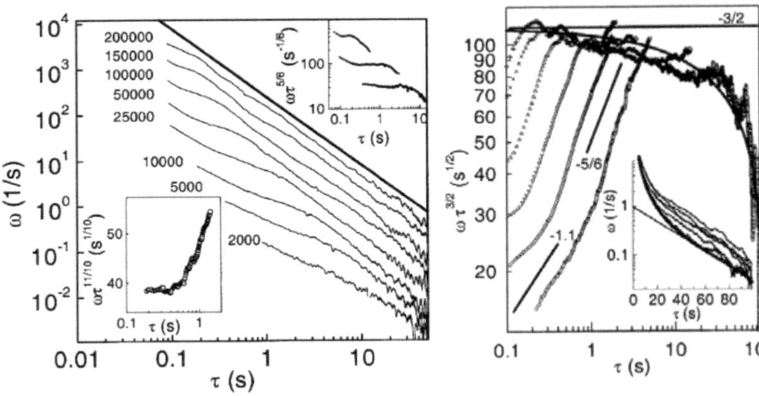

Figure 6. Four regimes of decaying He II grid generated turbulence. Left: The decaying helium II vorticity measured at $T = 1.3$ K for the indicated Re_M. Each curve represents an average of three individual decays. As the decay curves tend to collapse on the universal curve, for clarity they are shifted by a factor of 2 downwards, the uppermost remaining unchanged. The early part of the vorticity decay displays a power law with exponent $-11/10$ (left inset, showing normalized data for $Re_M = 10^4$) and later $-5/6$ (right inset, showing normalized data for $Re_M = 1.5 \times 10^5, 2.5 \times 10^4$, and 5×10^3). After saturation, typically several orders of magnitude of decaying vorticity closely follow the power law with exponent $-3/2$, represented by the thick solid line. Right: The decaying vorticity multiplied by $(t + t_{vo})^{3/2} = \tau^{3/2}$, measured at $T = 1.65$ K. The data sets from left to right correspond to $Re_M = 2 \times 10^5, 1.5 \times 10^5, 10^5, 5 \times 10^4, 2.5 \times 10^4$, and 10^4, each being an average of three individual decays. Except for the last one, the data sets are plotted up to $10t_{sat}$; the late decay for all of them is shown in the inset. The early part of the decay displays power laws with exponents $-11/10$ and later $-5/6$, as indicated by straight solid lines next to the data sets. After saturation, several orders of magnitude of decaying vorticity follow expression (4), using $C = 1.62$, $\nu = 2.28 \times 10^{-4}$ cm^2/s [4] and $\gamma = 0.418$, represented by the curved thick solid line. The influence of growing Kolmogorov length scale is indicated by the difference from the horizontal direction, representing expression (3), i.e., $-3/2$ power law. The inset shows the late decay of the same data sets compared with expression (4). The late decay can be characterized as exponential, of a form $exp(-t/t_0)$ with $t_0 = 29$ s, represented by the straight solid line.

ficient $\beta = 2\nu k^2$. For $k = 2\pi/D$ and ν of order 10^{-4} cm^2/s it suggests a characteristic decay time close to the observed one.

2.5 Discussion

From the point of view of classical fluid dynamics, He II has been used to investigate the decay of grid generated turbulence in a finite channel, displaying four distinctly different regimes over six decades of decaying vorticity. Remarkably, the form of the decay did not vary over the range $1.2 < T < 2$ K, despite ρ_n/ρ changed by more than a factor of 10. The decay can be described classically (except perhaps for a slight departure from the predicted universal curve at later times not discussed in this Lecture). This can be understood as the first case when the quantum working fluid – He II was used in classical turbulence research, thanks to the sensitive second sound attenuation technique that provides a tool to directly probe decaying vorticity. We stress that the classical decay model outlined here is capable of describing most of the decay of grid generated turbulence even in the zero temperature limit where there in no normal fluid (see Lectures by S. Fisher).

Despite a deep similarity between conventional and superfluid grid turbulence as discussed in this Lecture, one has to keep in mind that there are fundamental differences between the two, arising from the two-fluid behaviour and quantization of circulation in the superfluid. Hence applicability of classical description to quantum turbulence and vice versa is generally not possible.

3 Quantum Turbulence in a Stationary Normal Fluid

3.1 Introduction

Most studies on quantum turbulence so far have been concerned with He II. There are, however, other superfluids (such as superfluid ^{3}He-^{4}He mixtures, superfluid ^{3}He phases, neutron stars, Bose-condensed atomic gases) where quantum turbulence is possible and worth studying. In this Lecture we consider quantum turbulence in superfluid phase ^{3}He-B at relatively high temperatures, where its flow can be described by the two-fluid model. The relevant experiments have been carried out by the Helsinki ROTA group using the rotating cryostat that has contributed over many years to our knowledge on rotating superfluid ^{3}He phases. The ^{3}He atom is a fermion with spin 1/2 and transition to superfluidity occurs, similarly as a superconducting transition in metals, thanks to the existence of "superfluid particles"– Cooper pairs with the total spin unity. The non-invasive nuclear magnetic resonance (NMR) technique can therefore be used here to great advantage in investigations of the flow properties and quantum turbulence in ^{3}He-B.

As for the microscopic properties of superfluid phases of ^{3}He and experimental techniques needed to cool the sample down to submillikelvin temperature, we rely on companion Lectures by S.N. Fisher. Here we only summarize some basic features of ^{3}He-B. Quantization of rotational motion in a multiply connected region is quantized here in units $\kappa = h/2m_3 \approx 0.664 \times 10^{-7}$ m^2/s, where m_3 is is the mass of the ^{3}He atom. We consider singly quantized vortices (with about two orders of magnitude larger cores), similar to those in He II. Above the superfluid transition ^{3}He is a Fermi liquid whose viscosity steeply (as $1/T^2$) increases with decreasing temperature and right above the transition its kinematic viscosity is surprisingly high, comparable with that of air or oil. The kinematic viscosity of the normal fluid in ^{3}He-B is therefore so high that in experimental cells of typical size of 1 cm^3 the normal fluid hardly moves. A completely new class of quantum turbulence becomes possible when only the superfluid component is moving with respect to boundaries. The situation is essentially different from He II where the normal fluid possesses very low kinematic viscosity (except perhaps below 1 K, see later) and can easily become turbulent.

3.2 Theoretical description

In conventional liquids the vorticity $\boldsymbol{\omega} = curl\mathbf{v}$ obeys the Navier-Stokes equation, which we write here in the form

$$\frac{\partial \boldsymbol{\omega}}{\partial t} = \nabla \times [\mathbf{v} \times \boldsymbol{\omega}] + \nu \nabla^2 \boldsymbol{\omega} \,. \tag{3.1}$$

The interplay of the two terms on the RHS, the inertial first term and the viscous second term, governs the transition to turbulence (see, e.g., McComb (1990)). It is determined by the external conditions through the Reynolds number, $Re = RU/\nu$, formed by the characteristic size R of the system, the flow velocity U, and the kinematic viscosity ν. At large $Re \gg 1$, the effect of the inertial term is dominating, and laminar flow becomes increasingly unstable. If vorticity is released into a meta-stable laminar state, a sudden transition to a chaotic flow of eddies occurs.

The vorticity of the superfluid component is quantized in terms of the circulation quantum $\kappa = 2\pi\hbar/M$ where $M = 2m_3$ is the mass of a superfluid particle - a Cooper pair consisting of two ^{3}He atoms. The flow of the superfluid component can be characterized by the 'superfluid Reynolds number' $Re_s = RU/\kappa$. The Feynman criterion $Re_s \sim 1$ gives the velocity at which it becomes energetically favorable to form a quantized vortex. It does not mean, however, that on exceeding this criterion the vortex is always created. If a large nucleation barrier exists, the superfluid remains vortex-free, sometimes until much higher values of Re_s. Quantized vortices can then be injected, either by some extrinsic means or by increasing the velocity so that the nucleation barrier is overcome and spontaneous vortex formation may occur.

The limit of large $Re_s \gg 1$ is equivalent to a vanishing Planck's constant, $\kappa \propto \hbar \to 0$, so that the vorticity becomes a continuous variable, like in the classical case. However, the fluid still remains unconventional because of its two-fluid nature: In addition to the superflow, there is the normal component which is at rest in the container frame, $\mathbf{v}_n = 0$. Interactions between the superfluid vorticity and the normal component give rise to a mutual friction force on a unit volume of the superfluid,

$$\mathbf{f}_{mf} = -\alpha' \rho_s [\mathbf{v}_s \times \boldsymbol{\omega}_s] + \alpha \rho_s [\hat{\boldsymbol{\omega}}_s \times [\boldsymbol{\omega}_s \times \mathbf{v}_s]] \, ,$$

which is described by the dimensionless temperature-dependent parameters α and α', experimentally measured by Bevan *et al.* (1997). These correspond to the usual mutual friction parameters, if we are dealing with one vortex or locally polarized vorticity, or they may be renormalized in the general case. Here $\mathbf{v}_s$ and $\boldsymbol{\omega}_s = \nabla \times \mathbf{v}_s$ are the superfluid velocity and vorticity, and $\hat{\boldsymbol{\omega}}_s$ is the unit vector in the direction of $\boldsymbol{\omega}_s$. Inserting the mutual-friction force into the Euler equation one obtains the hydrodynamic equation for $Re_s \gg 1$ (Sonin (1987)):

$$\frac{\partial \boldsymbol{\omega}_s}{\partial t} = (1 - \alpha') \boldsymbol{\nabla} \times [\mathbf{v}_s \times \boldsymbol{\omega}_s] + \alpha \boldsymbol{\nabla} \times [\hat{\boldsymbol{\omega}}_s \times (\boldsymbol{\omega}_s \times \mathbf{v}_s)] \, . \tag{3.2}$$

The energy dissipation is determined by the mutual friction damping α, while the reactive mutual-friction α' renormalizes the first term, the inertial term from Eq. (3.1). Like in classical fluids, the inertial term drives the instability and is responsible for the energy transfer to smaller length scales, whereas the dissipative term acts to stabilize laminar flow. The fundamental difference from conventional hydrodynamics is that the relative importance of these two competing terms is determined by an intrinsic parameter of the superfluid, $q = \alpha/(1 - \alpha')$, in contrast to classical liquids where it is governed via the Reynolds number by the extrinsic quantities R and U.

In the case of a single vortex line, as shown by Barenghi *et al.* (1985), the parameter q marks the crossover from the regime where Kelvin waves can propagate along a vortex ($q \lesssim 1$) to where they are overdamped ($q \gtrsim 1$). As explained by Kopnin (2001), in Fermi superfluids and superconductors with finite energy gap one has $q \approx (\omega_0 \tau)^{-1}$, where ω_0 is the spacing between the bound states of quasiparticles in the vortex core and τ^{-1} is their lifetime

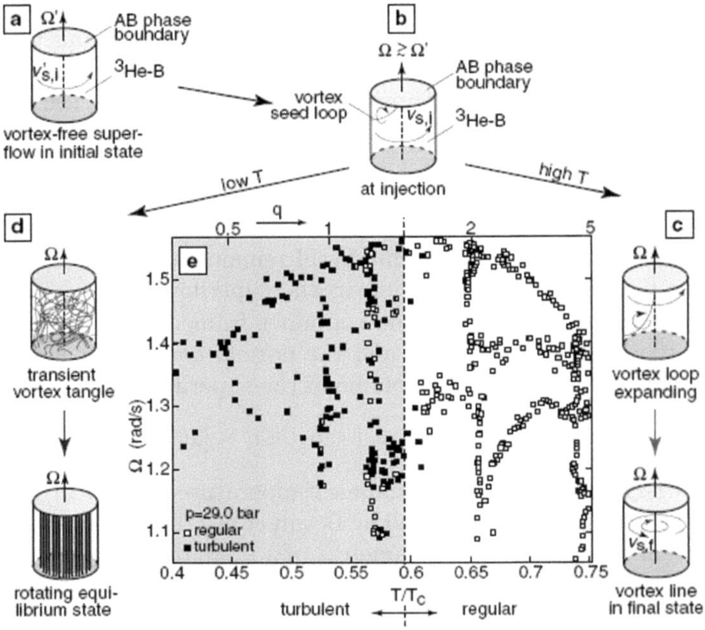

Figure 7. Phase diagram of turbulent superflow in ^{3}He-B found by Finne *et al.* (2003). The principle of the measurements: a) The initial state is vortex-free (Landau state) superflow in rotation at Ω, where the normal component is stationary and the superfluid component flows in the rotating frame. b) A few (ΔN) vortex loops are injected and, after a transient period of loop expansion, the number of rectilinear vortex lines N_f in the final steady state is measured. It is found to fall in one of two categories. c) $N_f = \Delta N$, regular mutual-friction-damped loop expansion. d) $\Delta N \ll N_f \leq N_{eq}$, turbulent loop expansion. e) Phase diagram of measured events.

broadening, owing to scattering from the normal component. Based on measurements of α and α' in ^{3}He-B by Bevan *et al.* (1997), q approaches infinity at T_c and drops roughly exponentially to zero, when cooled down to $T \to 0$. In the following section we describe how such an evolution of q (which in classical fluids would correspond to scanning the Reynolds number from 0 to ∞) influences the superfluid hydrodynamics detected by means of the NMR experiment.

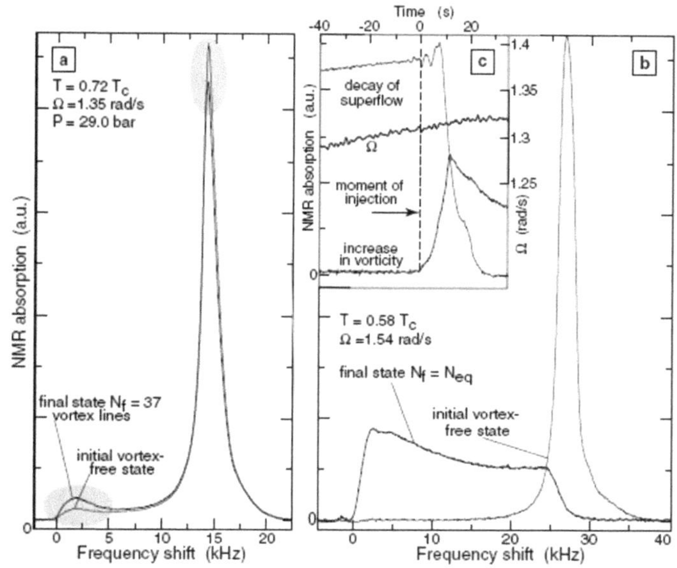

Figure 8. NMR absorption spectra before and after vortex- loop injection. For details, see the text.

3.3 Helsinki NMR experiments on quantum turbulence in ^{3}He-B

The experiment is basically organized as follows: The ^{3}He sample is contained inside a long ($\approx$ 11 cm) quartz tube of $R = 3$ mm in diameter with smooth walls, connected only via a small orifice in the bottom to the heat exchanger volume on the nuclear demagnetization stage of the rotating ROTA cryostat. Three independent superconducting coils produce axially oriented magnetic fields - at both ends (for the NMR measurements) and in the middle, which is called the barrier field and serves for stabilizing the A-phase in this part of the sample. For a wide range of pressures and temperatures this geometry allows stabilizing the AB boundary in the cell.

In rotating ^{3}He-B the critical velocity for intrinsic vortex nucleation is rather high and vortex-free superflow (so called Landau state) can thus be maintained up to high angular velocities. Other metastable states with any intermediate number of rectilinear vortex lines, up to the equilibrium number N_{eq}, are also stable in time. On the other hand, the critical velocity for vortex nucleation in the A-phase is low and for our purpose here can be neglected. Therefore the rotating A-phase can be thought of as always containing (nearly) equilibrium number of vortices. If the cryostat is slowly accelerated from zero up to some angular rotation velocity Ω, the following configuration is possible: below the stable AB boundary the B-phase is in the vortex-free Landau state while above it the A-phase is in solid body rotation. Note that both in the A-phase and in the B-phase the normal fluid always rotates together with the container, providing the unique frame of reference mentioned above.

It has been shown by Blaauwgeers *et al.* (2003) that at some critical angular velocity Ω_{cr} the AB boundary undergoes the shear wave Kelvin-Helmholtz instability and injection of few vortices from the A-phase to the B-phase occurs. This is the mechanism by which in the experiment ΔN "seed" vortex loops are introduced into the B-phase which is in a vortex-free Landau state.

As schematically shown in Fig. 7, the outcome of this injection falls in one of two categories. At high temperature the final number of vortices found after some time in the B-phase, N_f, equal to the number of injected ones, ΔN. It takes some transition time, typically a few seconds, until the injected "seeds" expand and form a rectilinear array in the vicinity of the rotation axis. At low temperature, however, turbulent loop expansion occurs and $\Delta N \ll N_f \leq N_{eq}$. This process leads to a total removal of the macroscopic vortex-free superflow as the superfluid component is forced into solid-body-like rotation (on an average) by the formation of a vortex array with the equilibrium number of rectilinear lines, $N_{eq} \approx \pi R^2 2\Omega/\kappa \sim 10^3$. Fig. 7 also shows the experimentally observed phase diagram of turbulent superflow in ^{3}He-B. Each data point monitors the outcome from the injection event in the final steady state as a function of rotation and temperature. Regular loop expansion is marked with open symbols and turbulent with filled ones. An abrupt transition at 0.60 T_c, independently of Ω, has been found to divide the (Ω, T) plane vertically between the two types of vortex dynamics. Only within a narrow interval $\Delta T \approx 0.05 T_c$ around 0.60 T_c can intermediate behaviour be observed.

Fig. 8 shows how Finne *et al.* (2003) came to these conclusions based on observed NMR spectra. NMR in ^{3}He-B does not directly measure the number of vortices in rotation, but the shape of the NMR spectrum reflects the spatial texture of the sample. The texture problem consists of finding the spatial distribution of the rotation axis $\hat{\boldsymbol{n}}$ of the order parameter. The overall NMR absorption signal from the sample can be thought of as consisting of signal of a number of small volumes (local oscillators) of constant angle β between $\hat{\boldsymbol{n}}$ and magnetic field direction. The normalized frequency shift from the Larmor frequency for each of these oscillators is approximately proportional to $\sin^2 \beta$. Hence the overall shape of the observed NMR spectrum depends on the texture. The type of the texture is a result of interplay of orienting interactions: the magnetic field energy, the flow energy and the surface energy.

It is known that in a long cylindrical sample at rest there is so called flare-out texture: $\hat{\boldsymbol{n}}$ points up at the rotation axis of the container (i.e., $\beta(0) = 0$), off the axis it bends out and reaches an angle $\beta(R) \approx 63$ degrees at the wall. The influence of rotation on the texture in a cylindrical container is rather complicated (there are several textural transitions induced by rotation - see the original work of Korhonen *et al.* (1990)), but the observed NMR spectra shown in Fig. 8 can be interpreted as follows:

The two spectra on the left -(a) represent the vortex-free Landau state and its only minor changes introduced by regular mutual-friction-damped loop expansion above 0.60 T_c. The sharp peak at large frequency shift (from the Larmor frequency) is caused by the large-scale superflow present in the rotating frame. When a few rectilinear vortex lines are formed, some absorption from the large peak is shifted into a tiny new peak at small frequency shift. The heights of both peaks change approximately linearly with N_f (in the highlighted regions).

The two spectra on the right - (b) show a radical restructuring occurring below 0.60 T_c when the large-scale superflow is replaced by an equilibrium array of rectilinear vortices.

The inset - (c) displays peak heights of the two absorption maxima as a function of time during the transient turbulent period: (i) The sharp peak at large frequency shift decreases; this monitors the decay of the large-scale superflow. (ii) The tiny peak at small frequency shift first increases which reflects the build up of the high-density vortex tangle. Subsequently it starts to decrease which corresponds to the rarefaction and polarization of the tangle into an equilibrium array.

The Helsinki data clearly show that the transition between regular high-temperature processes (marked with open symbols in the phase diagram in Fig. 7) and turbulent low temperature processes (filled symbols) observed at 0.60 T_c is surprisingly sharp, with $N_f \sim 1$ above and $N_f \approx 10^3$ below the transition. The main conclusion is that the transition occurs as a function of temperature (or $q = q(T) = \alpha/(1 - \alpha')$), with little or no dependence on the maximum superflow velocity $U\Omega R$ in striking contrast to classical liquids.

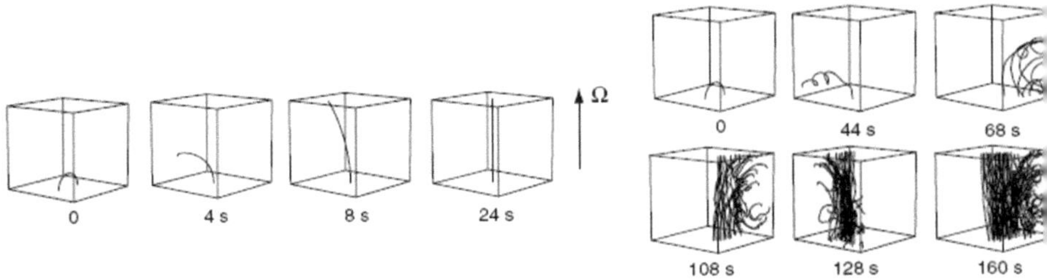

Figure 9. Computer simulations of the Osaka group demonstrating the summary of events observed in the Helsinki experiment at high (left) and low temperatures (right): A vortex half ring is injected into vortex-free superflow, generated by rotation around the vertical axis. The figures represent snapshots in the rotating frame at times indicated.

3.4 Computer simulations

It is of course best if the experiment can progress together with the theory and computer simulations as has been achieved here. Fig. 9 shows the first relevant computer simulations by the Osaka group lead by M.Tsubota that appeared already in the first publication by Finne *et al.* (2003). Although they were performed in a cube (as opposed to the cylindrical geometry of the experiment), they are consistent with the experimental observations and clearly helped a lot in the understanding of the underlying physics. In the simulations, the temperature dependence is contained in the mutual friction parameters α and α'. The situation at high temperature is shown on the left. The loop grows monotonically into a single rectilinear vortex line, pulled by the Magnus force from the superflow. The motion is highly damped since the Magnus force is balanced by the mutual friction force. As shown on the right, at low temperatures the loop moves with the circulating superflow: (44 s) The vortex is unstable with respect to loop formation and develops through reconnections into a tangle at the injection site. (68 s) The Magnus force from the vortex-free superflow extends the tangle along the vertical axis. Experimentally, a flight time τ_F for the tangle to extend across a vertical distance d between the AB boundary and the NMR coil controlled by mutual-friction damping was measured: $\tau_f = d/(\Omega R \alpha)$. (108 160 s) The final step of the observed process is the reconnection of the tangle to rectilinear lines. The replacement of the large-scale superflow by the tangle and its later decay into rectilinear lines are observed as time-dependent frequency shifts in the NMR spectrum, while the spectrum in the final time-independent state yields the total number of rectilinear lines.

3.5 Vortex multiplication – a precursor to turbulence

The described process of generation of quantum turbulence from a seed vortex loop in a rotating container has been further studied by Finne *et al.* (2006), who investigated dynamics of quantized vortices in rotating ^{3}He-B is in the low density (single vortex) regime as a function of temperature. The authors found a surface - mediated process that generates vortices at roughly constat rate - a precursor that precedes a faster form of turbulence described in detail above. This precursor becomes observable when vortex loops are introduced in low velocity rotating flow below about 0.5 Tc. Above this temperature the number of vortex lines remains constant, as they evolve to their equilibrium positions. Below this temperature, however, the number of vortices increases linearly in time until the vortex density has grown sufficiently for turbulence to switch on. The mechanism of this precursor process can be described as a collision of the existing vortex loop with the wall of the container. When a vortex ring approaches the wall, it reconnects with it resulting in two legs connected to the wall. Sharp kinks that are present due to the reconnection process cause Kelvin waves on them. These are able to grow in an applied flow only on one of them (due to so-called Donnelly–Glaberson instabitity; for details we direct the reader to the original work of Glaberson *et al.* (1974) or to the book by R.J. Donnelly (1991)), the largest wave reconnects with the wall again which results in a new separated loop. The conditions for the Kelvin wave to grow are fulfilled only in the underdamped regime, when the parameter q is small, i.e., at low enough temperature.

3.6 Twisted vortex state

The equilibrium state of a superfluid under rotation consists of an array of rectilinear quantized vortices, which are parallel to the rotation axis. Eltsov *et al.* (2006) have investigated twisted vortex states where the vortices have a helical configuration circling a common axis (an example of such a state would be a vortex bundle deformed under torsion) and have shown that the twisted vortex state appears spontaneously when vortex lines expand into vortex-free rotating superfluid, as shown in Fig. 10. The essential features are the vortex-free state in the upper part of the cylinder, the propagating vortex front, and the twisted vortex state that is left behind.The normal fluid is assumed to rotate with the container: $\mathbf{v}_n = \mathbf{\Omega} \times \mathbf{r}$. In the front the azimuthal superflow velocity rapidly increases and therefore

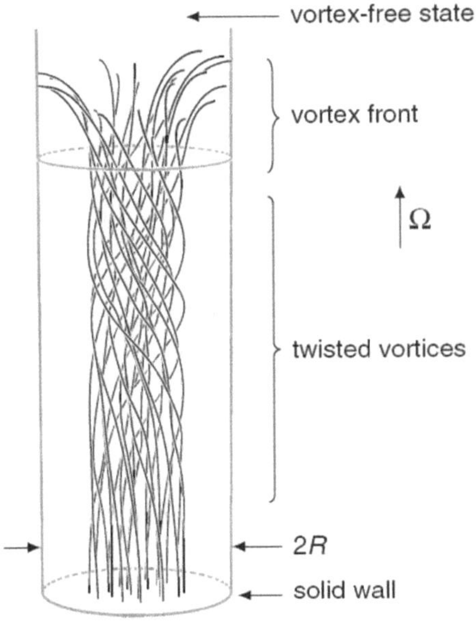

Figure 10. The formation of twisted vortex state. The vortices have their propagating ends bent to the sidewall of the rotating cylinder. As they expand upwards into the vortex-free state, the ends of the vortex lines rotate around the cylinder axis. The twist is nonuniform because boundary conditions allow it to unwind at the bottom solid wall. The figure is a snapshot of a numerical simulation of 23 vortices initially generated near the bottom end.

the azimuthal counterflow is strongly reduced in absolute value. While the vortex front progresses, the vortex ends rotate at a lower speed than the cylinder. This generates the twisted vortex state. A clear signature of the twist is the axial velocity along the vertical direction. It is downwards in the center and upwards in the periphery. At the bottom wall the boundary condition prohibits any axial flow. This implies that the twist vanishes there. Vortex ends can slide with respect to the smooth bottom wall and the winding generated by the front is unwound at the bottom.

The twist implies superflow parallel to the vortex lines. In such a case individual vortices become unstable against Donnelly–Glaberson instabitity when the critical velocity of the parallel flow is reached. The maximum axial velocity therefore remains smaller than this limit – if any tighter twist is created in the front, it is immediately relaxed by instabilities and subsequent vortex reconnections.

3.7 Energy spectrum of quantum turbulence in a stationary normal fluid

Let start with recalling that in case of classical high Reynolds number homogeneous and isotropic turbulence one has the 3D turbulent spectrum of a usual form: For low wavenumber k the spectral energy density is of the form $\Phi(k) = Ak^m$. On increasing k, $\Phi(k)$ around the energy-containing length scale $\ell_e = 2\pi/k_e$ exhibits a broad maximum and the inertial range follows, of the form $\Phi(k) = C\varepsilon^{2/3}k^{-5/3}(k\ell_e)^{\mu}$, where $C = 1.62 \pm 0.17$ is the accepted value of the dimensionless 3D Kolmogorov constant, $\varepsilon = -dE/dt$ is the energy decay rate and μ is the intermittency correction exponent - for simplicity we put $\mu = 0$ and neglect intermittency at this stage. The energy spectrum will be terminated at high $k_{diss} \approx (\varepsilon/\nu^3)^{1/4}$ by viscous dissipation. Here ℓ_{diss} is the Kolmogorov dissipation length scale $\ell_{diss} = 2\pi/k_{diss}$.

In the remaining part of this Lecture we consider possible shape of the energy spectrum of quantum turbulence in a stationary normal fluid. If quantized vortices are present, superfluid velocity field is coupled to stationary normal fluid by the mutual friction force, acting at all length scales. It has been shown by Vinen (2005) that the superfluid eddies of all sizes would decay exponentially, with some characteristic decay time τ_{mf}. To appreciate the role of mutual friction in turbulent flow, τ_{mf} has to be compared with the eddy turnover time, τ_{to}, at any particular length scale. If $\tau_{mf} < \tau_{to}$, superfluid eddies cannot exist. Because the turnover time increases with the size of the eddy, Vinnen's analysis shows that eddies bigger than maximum size of order $R_{max} \approx \left[\varepsilon/(\alpha\kappa L)^3\right]^{1/2}$ cannot exist. This is contrary to classical turbulence, where large eddies (almost) don't decay (so-called permanence of large eddies) and their size is limited only by the boundaries of the container.

In further analysis we shall be interested only in scales that are large in comparison with the mean intervortex distance, in which case Eq. (3.2) can be written in the form similar to the Navier-Stokes equation:

$$\frac{\partial \mathbf{v}_s}{\partial t} + \nabla\mu = \mathbf{v}_s \times \boldsymbol{\omega}_s + q\hat{\boldsymbol{\omega}}_s \times (\boldsymbol{\omega}_s \times \mathbf{v}_s) \tag{3.3}$$

We recall that $\boldsymbol{\omega}_s$ denotes here the course-grained superfluid vorticity and the role of the effective Reynolds number is played by the parameter $Re_{eff} = q^{-1} = (1+\alpha')/\alpha$ which via the mutual friction parameters α and α' depends on temperature but not on geometry.

L'vov, Nazarenko and Volovik (2004) and independently Vinen (2005) analyzed this equation and have shown that owing to the action of mutual friction, there is strong damping of large eddies, with the result that at low wave numbers the energy spectrum rolls off more rapidly, approaching k^{-3}. Beyond a certain critical value k_{cr}, however, the superfluid energy spectrum gradually recovers its usual Kolmogorov form, characterized by the famous roll-off exponent of -5/3.

3.8 Discussion

In superfluids such as ^{3}He-B, where the very viscous normal fluid can be thought of as stationary, the completely new class of quantum turbulence becomes possible when only the superfluid component is moving with respect to boundaries. It is possible only if an intrinsic velocity-independent but temperature dependent criterion is fulfilled and at large scales it is strongly damped by mutual friction, which limits the maximum size of superfluid eddies.

4 Flow due to oscillating objects in cryogenic helium

4.1 Historical overview

Experiments with oscillating structures such as discs, piles of discs, spheres, grids and wires have been widely performed in quantum fluids research since the discovery of superfluidity. The famous Moscow experiment of E.L. Andronikashvili (1946) on torsional oscillations of a pile of closely spaced thin discs that first allowed to directly measure the normal, ρ_n, and superfluid, ρ_s, densities of He II became an integral part of every textbook on low temperature physics. Up to now it is still considered the best demonstration of the validity of the concept of the two fluid model introduced by L.D. Landau (1947).

It was soon realized, however, that the simple physical picture of normal fluid being clamped by finite viscosity and dragged with discs as opposed to stationary superfluid not being involved in flow only holds for sufficiently low oscillation amplitudes. Numerous so called "periodic boundary layer experiments", such as oscillating disc, spheres or U-tube oscillations, capillary flows clearly show "critical conditions" or "critical velocities" above which this simple physical picture does not hold.

A set of results obtained by a Cambridge graduate student, A. C. Hollis-Hallett (1952) showed that the simple two fluid interpretation of the damping of an oscillating disk, and the celebrated Andronikashvili experiment became classical-like at a relatively small, but finite, amplitude of oscillation. Similar results have been obtained using a torsionally oscillating sphere in He II by Benson and Hollis-Hallett (1956).

At the same time, R.J. Donnelly and O. Penrose (1956) had performed an experiment at Yale, showing that oscillations of liquid helium II in a U-tube could be understood as obeying the Landau two fluid equations at low amplitudes, and the Navier-Stokes equations at higher amplitudes.

Subsequently, Donnelly and Hollis-Hallett (1958) wrote a review of the oscillating experiments in liquid helium, and attempted to find a scaling relationship among the various experiments, which was not notably successful. One unintended result of this review was that nearly all experimental work in the field of oscillating systems terminated. Donnelly and Hollis Hallett noted that for most oscillating experiments in liquid helium the motion induced by the moving boundaries is confined to a relatively thin boundary layer near the walls, and they named this class of experiment "periodic boundary layer experiments". These included oscillating discs, oscillating piles of discs, oscillating spheres and oscillating U-tubes. Activity in this area was greatly hampered at that time, by a lack of adequate theories of the classical hydrodynamics of these experiments.

However, it is technically considerably more simple to generate in a cryostat cryogenic helium boundary layer flows due to submerged oscillating objects rather than some well-defined stationary flow, such as pipe flow. Therefore, many investigators later continued with various boundary layer experiments, including those in the zero temperature limit. We shall describe the important, rather intriguing and so far not fully understood results of them later, after considering basic physics of boundary layer flow in classical fluids.

4.2 Mechanical and electrical properties of vibrating sensors

Various vibrating sensors have been used in cryogenic environment. Besides already mentioned discs and spheres, a lot of interesting physics is understood thanks to vibrating wires. A fine conducting wire stretched along the axis of a cylindrical container was used by W.F. Vinen (1961) as a sensor to prove quantization of vorticity in He II. Fine superconducting vibrating wires, nowadays usually in the form of a half-loop few mm in diameter, are probably the most widely used resonators, serving as research tools for investigating the nucleation of quantized vorticity for He II (e.g., Yano *et al.* (2007)) or for ^{3}He-B (e.g.,Bradley (2000)), as well as thermometers, pressure- or viscometers in both ^{4}He and ^{3}He over wide temperature range down to microkelvin temperature. They are driven by the Lorentz force arising from biasing them by small ac current in externally applied magnetic field.

A new addition to this family of oscillating objects is represented by several types of widely commercially available quartz tuning forks, which are mass-produced as frequency standards for digital watches. These piezoelectric oscillators are usually tuned for the frequency $2^{15} = 32768$ Hz at room temperature (resulting in a convenient frequency range when they vibrate in various working fluids) and have a large number of other applications. They are cheap, highly sensitive, robust, for most applications easy to install and use. Only two shielded wires are needed to drive and readout these piezoelectric sensors; no magnetic field (to which they are highly insensitive) is required to drive them. A relatively simple electronic scheme allows detection of their resonant response both in viscous and quantum fluids over up to eight orders of magnitude of the driving force.

Most of a variety of different vibrating sensors at sufficiently small oscillation amplitudes can be approximated as harmonic oscillators subject to a harmonic driving force $F = F_0 \cos(\omega t)$ and a drag force with linear dependence on velocity. The equation of motion is given by

$$\frac{d^2x}{dt^2} + \gamma \frac{dx}{dt} + \frac{k}{m} x = \frac{F}{m} \; . \tag{4.1}$$

We have here four parameters, namely the effective mass m (for the fork of one leg), the drag coefficient γ, the spring constant k, and the amplitude of the driving force F_0. The effective mass and the drag coefficient depend on the medium around the oscillating sensor. The solution of this differential equation is well known: It can be written as $x(t) = x_a(\omega)\sin(\omega t) + x_d(\omega)\cos(\omega t)$, where x_d and x_a are the dispersion and absorption, respectively. The mean absorbed power $\langle F\, dx/dt \rangle = F_0 \omega x_a/2$ is at maximum at the resonant frequency $\omega_0 = \sqrt{k/m}$. It is convenient to introduce the quality factor $Q = \omega_0/\gamma$ as the ratio of the resonant frequency ω_0 to the frequency width $\Delta\omega = \gamma$, where $\Delta\omega$ is the full width of the resonance curve at half of the maximum power.

The following calibration is possible for piezoelectric sensors such as the quartz fork. The high Q sensor is excited with ac voltage $U = U_0 \cos(\omega t)$ while the frequency is slowly swept through resonance. The signal received from the sensor is a current $I(t) = a\, dx(t)/dt$, where a is a numerical constant. In a cryogenic setup, a can be obtained *in situ* by comparing the mechanical oscillator with the equivalent electrical RLC series resonance circuit. The corresponding differential equation for the current reads

$$\frac{d^2I}{dt^2} + \frac{R}{L}\frac{dI}{dt} + \frac{I}{LC} = \frac{1}{L}\frac{dU}{dt} \; . \tag{4.2}$$

We see that $\omega_0^2 = 1/(LC)$, $\gamma = R/L$, and $1/L = (F_0/U_0)\, a/m$. Additionally we have the condition that the dissipated power at resonance has to be equal for both equations: The electrical power $U_0^2/(2R)$ drives the sensor - in case of the fork its two legs which dissipate $2 \cdot F_0^2/(2m\gamma)$. Thus we have a closed set of equations connecting the electrical and mechanical properties of the fork via the fork constant a: $F_0 = (a/2)\, U_0$, $R = 2m\gamma/a^2$, $L = 2m/a^2$, $C = a^2/(2k)$. Experimentally a can be obtained as $a = \sqrt{2m\, \Delta\omega/R}$; here $\Delta\omega$ is determined from the width of the resonance curve while $1/R$ is the linear slope of the experimental $I_0(U_0)$ dependence, where I_0 is the current amplitude at resonance. The only parameter which cannot be directly determined from the experiment is the effective mass m, however, its value is usually independently fairly reliable.

In the experiment the absorption $I_a(\omega)$ and dispersion $I_d(\omega)$ components of the current $I(t) = I_a \cos(\omega t) + I_d \sin(\omega t)$ are usually measured separately with a lock-in amplifier[6]; they can be fit to determine the needed parameters. In particular, the absorption component $I_a(\omega)$ reaches its maximum value I_0 at a frequency which is exactly ω_0 and the full width $\Delta\omega$ of the absorption curve at 1/2 of the maximum height I_0 gives exactly γ. If the fork constant a is known, then $m = a^2 U_0/(2 I_0 \Delta\omega)$ and $k = m\omega_0^2$ can be determined independently. For detailed discussion, see Blaauwgeers *et al.* (2007).

4.3 Inviscid and viscous laminar flow due to an oscillating body

Submerged oscillating objects in an ideal fluid. "Inviscid" hydrodynamic mass enhancement. Because we shall be interested in properties of superfluids, it is useful to start our analysis with the simplest case of incompressible flow of an ideal fluid. Such a flow is always potential[7], there is no energy dissipation. If a body is moving uniformly in an ideal fluid, there is no drag and no lift (so–called d'Alembert's paradox). The equation of motion of a body immersed in an ideal fluid is

$$\frac{dv_k}{dt}\,(M\delta_{ik} + m_{ik}) = f_i, \tag{4.3}$$

where m_{ik} is the induced mass tensor.

It follows that the mass of any submerged body executing oscillations is renormalized. One may define a mass enhancement factor, β, such that

[6] Alternatively, voltage could be measured: e.g., voltage drop across some resistor for the fork or voltage induced across the superconducting wire moving in external magnetic field.

[7] We assume that there is no free surface in the vicinity of the oscillating body, so that the energy cannot be carried away by excited surface waves - ripplons.

the apparent change in mass of the body, Δm, is given by $\Delta m = \beta m_{fl}$, where m_{fl} is the mass of fluid displaced by the submerged body. We may therefore write the "inviscid" mass enhancement in a convenient form $\Delta m = \beta V \rho(p,T)$, where $\rho(p,T)$ is the density of the fluid in question at pressure p and temperature T, and V denotes the volume occupied by the oscillating body. The factor β is primarily determined by the shape of the submerged body and the direction along which it oscillates. It can be shown (see, e.g., Milne-Thomson (1987)) that in an infinite liquid $\beta = \frac{1}{2}$ for a sphere, $\beta = r_\perp / r_\parallel$ for an infinitely long cylinder with elliptic cross section, where $r_\perp$ and $r_\parallel$ are the lengths of the axes which are perpendicular and parallel to the oscillation direction, respectively; for a rectangular beam (Sader (1998)) $\beta = (\pi/4) a_\perp / a_\parallel$, where $a_\perp$ and $a_\parallel$ are the lengths of the sides which are perpendicular and parallel to the oscillation direction, respectively. Note that the above is valid for a body vibrating in an infinite volume of fluid; any walls in the vicinity would affect the value of β. Generally, if walls are close and the displacing liquid must pass through a narrower passage, β ought to be enhanced.

Submerged oscillating objects in a viscous fluid. Laminar regime. We start our consideration with the non-linear Navier-Stokes equation

$$\frac{\partial \mathbf{u}}{\partial t} + (\mathbf{u} \cdot \nabla)\mathbf{u} = \frac{1}{\varrho} \nabla p + \nu \nabla^2 \mathbf{u}, \tag{4.4}$$

where $\mathbf{u}$ is the velocity of the fluid, p its pressure and ν denotes the kinematic viscosity. A classical viscous flow due to a submerged oscillating body (see, e.g., Landau and Lifshitz (1987)) is rotational within a certain layer adjacent to the body, while at larger distances it rapidly changes to potential flow. The depth of penetration of the rotational flow is of order

$$\delta = \sqrt{\frac{2\nu}{\omega}} = \sqrt{\frac{2\eta}{\varrho \omega}}, \tag{4.5}$$

where ω is the angular frequency of oscillation and η and $\nu = \eta/\varrho$ are the dynamic and kinematic viscosity of the fluid of the density ϱ. Subject to oscillatory motion in the fluid, the body experiences a force which has components proportional to the velocity of the body v (drag) and to its acceleration $\dot{v}$ (mass enhancement):

$$F = bv + \tilde{m}\dot{v}. \tag{4.6}$$

To determine the values of b and $\tilde{m}$ generally a full solution of the flow field around the oscillating body is required. Simplifications are possible in two limiting cases, which depend on the relative magnitudes of the characteristic size of the oscillating body ℓ, oscillation amplitude x_0, and viscous penetration depth δ:

1) $\ell \ll \delta$ and $\omega x_0 \ell/\nu \ll 1$: In this case the flow at any given instant can be regarded as steady – as if the body were moving uniformly with its instantaneous velocity. As a rule, this case does not apply to the oscillating cryogenic flows considered here.

2) $\ell \gg \delta$ and $\ell \gg x_0$: In this case the layer of rotational flow around the body is very thin while in the rest of the fluid the flow is potential. This case is directly applicable, for example, to quartz tuning forks (see later) in ^{3}He and ^{4}He liquids: The kinematic viscosity of normal liquid ^{4}He-I above the λ-point is $\nu_4 \approx 2 \cdot 10^{-4}$ cm^2/s and of normal ^{3}He above the superfluid transition $\nu_3 \approx 1\,\mathrm{cm}^2/\mathrm{s}$. For typical frequency of oscillation 32 kHz we get the penetration depths $\delta_4 \approx 0.4\,\mu\mathrm{m}$ and $\delta_3 \approx 30\,\mu\mathrm{m}$ while $\ell \approx 400\,\mu\mathrm{m}$.

Note that other oscillating objects such as spheres or wires may not always be in this flow regime since they usually have smaller characteristic size and smaller oscillation frequency.

When the conditions for case (2) above are valid, a major contribution to both b and $\tilde{m}$ in Eq. (4.6) is found by solving the potential flow field $\mathbf{u}$ around the body (Landau and Lifshitz (1987)). In particular, b is expressed as

$$b = \sqrt{\frac{\rho\eta\omega}{2}} \left[\frac{1}{|v_0|^2} \oint |u_0|^2 dS \right] = \sqrt{\frac{\rho\eta\omega}{2}}\, \mathcal{C}\, S \;, \tag{4.7}$$

where v_0 and u_0 are the amplitudes of the velocities of the body and the flow, the integral is taken over the surface of the oscillating body, S is the surface area of the body, and $\mathcal{C}$ is some numerical constant which depends on the exact geometry of the body. For example for a sphere $\mathcal{C} = 3/2$, while for an infinitely long cylinder oscillating perpendicular to its axis $\mathcal{C} = 2$.

The largest contribution to mass enhancement $\tilde{m}$ comes from the potential flow around the body and can be expressed through the mass ρV of the liquid displaced by the body of volume V. This was considered in a previous paragraph. A smaller contribution is caused by the fact that the viscous drag force experienced by the body is usually phase shifted with respect to the velocity of the body. This can be interpreted such that a volume of order $S\delta$ of the liquid is clamped to comotion with the oscillating body. Thus for $\tilde{m}$ we have to add a term containing another geometry-dependent coefficient B (for a sphere, $B = 3/4$)

$$\tilde{m} = \beta\rho V + B\rho S\delta \;. \tag{4.8}$$

In the case of more complicated geometries (such as a tuning fork) rigorous calculation of the parameters β, B, and $\mathcal{C}$ is difficult. Here β, B, and $\mathcal{C}$ could be considered as fitting parameters, to be determined from the experiment.

Hydrodynamic model of oscillating sensor The addition of the force F to the equation of motion of the sensor (4.1) leads to a reduction in the resonant frequency and an increase in the width of the resonance curve:

$$\omega_0^2 = \omega_{0vac}^2(m_{vac}/m), \tag{4.9}$$
$$\gamma = \gamma_{vac}(m_{vac}/m) + b/m, \tag{4.10}$$

where $m = m_{vac} + \tilde{m}$ is the effective mass of the oscillating body immersed in the fluid. For convenience we redefine the fork parameters β and B from Eq. (4.8) as relative to the effective fork mass in vacuum:

$$\tilde{m} = m_{vac}\,\frac{\rho}{\rho_q}\left[\beta + B\delta(S/V)\right]\ . \tag{4.11}$$

Ignoring the vacuum resonance width Δf_{vac}, we finally obtain the dependence of the resonant frequency f_0 and of the full width of the absorption curve at half height Δf on the fluid density and viscosity:

$$\left(\frac{f_{0vac}}{f_0}\right)^2 = 1 + \frac{\rho}{\rho_q}\left(\beta + B\,\frac{S}{V}\sqrt{\frac{\eta}{\pi\rho f_0}}\right)\ ;\ \Delta f = \frac{1}{2}\sqrt{\frac{\rho\eta f_0}{\pi}}\,\mathcal{C}\,S\,\frac{(f_0/f_{0vac})^2}{m_{vac}}. \tag{4.12}$$

Here V and S denote the volume and surface area of the sensor. Eqs. (4.12) can be used to determine experimentally the hydrodynamic parameters from measurements in a fluid with known ρ and η[8]. Once the parameters are known, the fork can be used for measurements of ρ and η for any other medium and thus as a pressure and temperature sensor, if $\rho(P,T)$ and $\eta(P,T)$ are known.

Laminar He II flow - crossover from hydrodynamic to ballistic regime Within the two-fluid model, assuming there are no quantized vortices in the flow, the drag experienced by the oscillating body is only that due to the viscous normal fluid flow. In He II with decreasing temperature below 1 K the mean free path increases and the hydrodynamic description ceases to be valid. From the hydrodynamic point of view it can be seen as normal fluid penetration depth exceeding all relevant length scales.

[8]This approach does not take into account a steady secondary flow through the action of viscosity in the boundary layer, so called "streaming" (see, e.g.,Schlichting and Gersten (1996)) and β therefore weakly depends on the geometry of the surrounding container.

At low temperature the drag is caused by scattering of quasiparticles. The drag coefficients λ_{lam} for phonons and rotons for He II have been calculated by Baym, Barrera and Pethick (1969)

$$\lambda_{ph} = \varrho_{ph} c \sigma; \qquad \lambda_{rot} = \varrho_{rot} \overline{u} \sigma, \tag{4.13}$$

where ϱ_{ph} and ϱ_{rot} are the densities of phonons and rotons, c and $\overline{u}$ are their mean velocities and σ is the scattering cross section. From Eq. (4.13) we obtain the laminar drag coefficients

$$\lambda_{ph} = \frac{2\pi^2 (kT)^4}{45 \hbar^3 c^4}\, \sigma; \qquad \lambda_{rot} = \frac{\hbar {k_0}^4}{3\pi^2} \exp \frac{-\Delta}{T}\, \sigma \tag{4.14}$$

where k is the Bolzmann constant, c denotes the temperature and pressure dependent velocity of the first sound, k_0 is the roton wavenumber and Δ stands for the roton gap.

A crossover from the hydrodynamic to the ballistic regime has been observed by a number of investigators and is described in detail for vibrating wires by Morishita, Sawada and Satoh (1998) and for a vibrating sphere by Jäger, Schuderer and Schoepe (1995).

4.4 Transition from laminar to turbulent drag regime in flow due to a vibrating object

The laminar drag regime is characterized by the driving force $F \propto v$, where v is the velocity of the submerged body, while in the the turbulent drag regime $F \propto v^2$.

Viscous fluids In a steady classical flow past a submerged object, transition from laminar to turbulent drag regime usually occurs at some critical value of Reynolds number $Re_{cr} = v_{cr}\ell/\nu$, where v_{cr} denotes the critical velocity and ℓ is a characteristic size of the object (see, e.g., Schlichting and Gersten (1996)).

In a viscous flow due to an oscillating submerged body of the angular frequency of oscillation $\omega = 2\pi f$ a new important length scale $\delta = \sqrt{2\nu/\omega} = \sqrt{2\eta/\rho\omega}$ – the viscous penetration depth emerges. If, however, $\ell \ll \delta$ and, additionally, the Reynolds number $U\ell/\nu$ is small, then the flow at any given instant can be regarded as steady - as if the body were moving uniformly with its instantaneous velocity. If, on the other hand, $\ell \gg \delta$ and the amplitude of motion $U/\omega \ll \ell$, then the Reynolds number need not be small in order to neglect the nonlinear term in the Navier-Stokes equation. It is shown in, e.g., Landau and Lifshitz (1987) that in a thin layer near the surface of the body the flow is rotational but in the rest of the fluid it is potential.

For a full description of oscillatory flows, besides the Reynolds number, one needs to define an additional dimensionless number such as Strouhal number $St = v\tau/\ell$, where $\tau = 2\pi/\omega$ is a characteristic time. Note, however, that if one assumes that the characteristic length scale is the penetration depth, the Reynolds and Strouhal numbers become equal (except for a numerical constant of π). Consequently, the crossover from the laminar to turbulent drag would be described by the Reynolds number $Re^{\delta}_{cr} = v\delta/\nu$ alone.

It is instructive to consider an analytically tractable example of a viscous flow due to an oscillating hydrodynamically smooth sphere. In a laminar regime, the drag force acting upon a sphere of radius R is given $F_{lam} = \lambda v = 6\pi\eta R(1 + R/\delta)v$. For the turbulent drag regime we adopt the standard expression $F_{turb} = \gamma v^2 = C_D P \rho v^2/2 = C_D \rho \pi R^2 v^2/2$, where C_D is the turbulent drag coefficient and P denotes the projected area of the body to the direction of motion.

Assuming that the transition occurs when these forces become equal in a limit of high frequency [9] we arrive at:

$$v_{cr} = \frac{\lambda}{\gamma} \cong \frac{6\sqrt{2}}{C_D}\sqrt{\nu\omega} \simeq 21\sqrt{\nu\omega}\ . \qquad (4.15)$$

Here we assumed that for a smooth sphere $C_D \simeq 0.4$. The prefactor for a hydrodynamically smooth sphere is yet to be experimentally confirmed, but using the vibrating quartz fork the scaling $v_{cr} \propto \sqrt{\nu\omega}$ has been recently experimentally verified by Blazkova *et al.* (2007) with gaseous and liquid ^{4}He over more than two orders of magnitude of the kinematic viscosity.

Quantum fluids In quantum fluids such as He II or ^{3}He-B the transition from laminar to turbulent drag regime involves nucleation of quantized vorticity. Intrinsic nucleation mechanism (involves nucleation of a first seed vortex loop in vortex-free superfluid and its subsequent expansion) and/or extrinsic nucleation mechanism (from existing remnant vortices) are possible.

The intrinsic mechanism is more plausible in superfluid ^{3}He-B, possessing low critical velocity of order few mm/s. Bradley (2000) used a thin superconducting vibrating wire in a form of semicircle and obtained interesting

[9] Note that there is a crossover to this $R \gg \delta$ regime, from the "steady-state like" regime $R \ll \delta$, which depends on the actual geometry of the flow in question, may include additional flow phenomena such as instabilities against Taylor vortices and is generally not fully understood.

experimental data, strongly suggesting that such a wire is sensitive enough to detect the presence of a single vortex loop. A current understanding is that upon reaching a critical velocity a seed vortex loop becomes nucleated at some excrescence on the wire where the flow velocity is enhanced. Until this moment, the amplitude of oscillations of the wire grows roughly linearly with the increasing drive. As soon as the loop becomes nucleated, the amplitude stops growing and stays on a plateau, but a closer look shows small oscillations of this amplitude about the plateau value. The nucleated vortex loop grows until it becomes long enough to support a Kelvin wave of the drive frequency, the amplitude of which grows until the loop twists, reconnects and a free vortex ring flies away, carrying with it a part of the energy supplied by the drive. This cycle becomes faster with the distance beyond the threshold, and closely resembles the behaviour of the tank circuit of the rf squid. This process seems reproducible in that the signature of it - the detected voltage across the wire - shows a steady pattern. At large drive the velocity of the wire reaches the pair breaking velocity for ^{3}He Cooper pairs which results in quasiparticle production and considerable heating.

In He II at low temperature (i.e., of order 100 mK or below, where there is hardly any normal fluid) the vortex nucleation marking the transition from laminar to turbulent drag regime is most likely of extrinsic order. Experiments with vibrating wires (e.g., Morishita, Sawada and Satoh (1998)) or spheres (Jäger, Schuderer and Schoepe (1995)) display what could be called a "virgin" behaviour - the very first transition after cooldown is generally not reproducible and usually occurs at much higher drive level than the subsequent ones. In other words, the original laminar drag regime persists up to much higher values of superfluid Reynolds number, defined as a product of the velocity and size of the object upon the circulation quantum. Moreover, intermittent switching between potential flow and turbulence has been observed by Niemetz, Kerscher and Schoepe (2002) (and analyzed in some detail by Schoepe (2004)) with such a levitated oscillating electrostatically driven sphere. Stable laminar flow at low drive and steady turbulence at high drive have been found separated by an intermediate range of driving forces where the flow is unstable, intermittently switching between these two regimes.

Yano *et al.* (2007) found additional information on motion of quantized vortices attached to the boundary of vibrating wires. Their experimental results, similar to those obtained by Charalambous *et al.* (2006) with an oscillating grid, strongly suggest that at low temperature any surface in contact with He II is covered by a "fur" of quantized vortex loops, creating an effective boundary layer that further increases the effective mass of the oscillating structure in question. Yano *et al.* (2007) present a flow phase diagram showing at low temperatures (in a millikelvin range) two critical velocities: $v_{c1} \approx 4-5$ cm/s and $v_{c2} \approx 20$ cm/s between which the intermittent switching occurs. With increasing temperature towards 1 K the upper one decreases and merges with the lower one; above about 1 K this critical velocity smoothly increases up to about 9 cm/s close below the λ - point.

A considerable theoretical effort to understand the nature of the onset of dissipation in pure superfluid in flows due to oscillating objects has recently been made by Hanninen *et al.* (2007). Computer simulations (always starting with a "seed" vortex loop, strongly suggesting that in the experiment the vortex nucleation is most likely of extrinsic origin, relying on existing remnant vortices) of flow due to oscillating sphere using the vortex filament method give qualitative agreement with the experimentally observed facts, most notably with the surprisingly uniform results for the critical flow velocity of order 5 cm/s marking in the zero temperature limit the onset of turbulence. This critical velocity appears to be only weakly dependent on the size and frequency of the oscillating objets, both in experiments and in computer simulations. Despite this progress, a lot more work is needed to understand the nature of the onset of dissipation in He II at low temperature. It is perhaps not surprising, as the long standing corresponding problem of transition to turbulent drag regime in classical boundary layer flows is still far from being fully understood.

Above 1 K the behaviour of the vibrating wire is similar to that of the quartz oscillating tuning fork, for which the critical velocity marking the onset of the turbulent drag regime also smoothly increases with the temperature. The advantage of the tuning fork is that it can be employed as a tool to generate and detect the flow up to high velocities of order few m/s, both in He II and in He I[10]. When crossing the λ - transition, the critical velocity obtained as a crosspoint of best linear fits on a logarithmic velocity versus driving force plots ($F \propto v$ and $F \propto v^2$) seems to change with the temperature smoothly, joining at the λ - point the classical behaviour described earlier. On the other hand, well below the λ - temperature the logarithmic plots of the drag coefficient versus velocity display a peculiar shape: for low velocities the drag coefficient linearly decreases with increasing drive as expected, after a rather sharp minimum displays a broad maximum before adjusting itself to the expected turbulent value of order unity.

The underlying physics is still rather poorly understood and experiments with quartz tuning forks of various sizes and resonant frequencies are under progress. Based on available experimental results and current theoretical model worked out in collaboration of the Prague group with W.F. Vinen,

[10]The velocity of flow generated by the fork can be so high that cavitation in He I as well as in He II can be observed and investigated

we temporarily suggest that transitions to turbulent drag regime occur separately in the superfluid (at lower critical velocity) and in the normal fluid (at higher critical velocity, so that at the λ - point it smoothly joints the classical behaviour of normal liquid He I.

Therefore, much further effort, both experimental and theoretical, is clearly needed in order to fully understand the flow due to oscillating objects in quantum fluids and the nature of its transition from laminar to turbulent drag regime, which is closely connected with the fundamental problem of nucleation of quantized vorticity in both He II and superfluid ^{3}He.

Bibliography

J.F. Allen, J. Reekie, *Proc. Cambr. Phil. Soc.* **35**, 114, 1939.

E.L. Andronikashvili, *J.Phys. U.S.S.R.* **10**, 201, 1946; *ZHETF* **18**, 424, 1948.

D.D. Awschalom, F.P. Milliken and K.W. Schwarz, *Phys. Rev. Lett.* **53**, 1372, 1984.

C.F. Barenghi, A.V. Gordeev, L. Skrbek, *Phys. Rev. E.* **74**, 026309, 2006.

C.F. Barenghi, L. Skrbek, *J. Low Temp. Phys.* **146**, 5, 2007.

C.F. Barenghi *et al.*, *Phys. Fluids* **28**, 498, 1985.

C.F. Barenghi, C.E. Swanson, R.J. Donnelly, *Phys. Rev. Lett.* **48**, 1187, 1982.

G. Baym, R.G Barrera and C.J. Pethick, *Phys. Rev. Lett.*, **22**, 20, 1969.

C. Bennett and S. Corrsin, *Phys. Fluids* **21**, 2129, 1978.

C.B. Benson and A.C. Hollis Hallett, *Can. J. Phys.* **34**, 668, 1956.

T.D.C. Bevan *et al.*, *J. Low Temp. Phys.* **109**, 423, 1997.

R. Blaauwgeers *et al.*, *Phys. Rev. Lett.* **89**, 155301, 2002.

R. Blaauwgeers, M. Blazkova, M. Človečko, V.B. Eltsov, R. de Graaf, J. Hosio, M. Krusius, D. Schmoranzer, W. Schoepe, L. Skrbek, P. Skyba, R.E. Solntsev, D.E. Zmeev, *J. Low Temp. Phys.*, **146**, 537, 2007.

M. Blazkova, D. Schmoranzer, L. Skrbek, *Phys. Rev. E***75**, 025302, 2007.

D.I. Bradley, *Phys. Rev. Lett.* **84**, 1252, 2000.

D.F. Brewer, D.O. Edwards, *Phil. Mag.* **6**, 1173, 1961.

T.V. Chagovets, A.V. Gordeev, L. Skrbek, *Phys. Rev. E* **376**, 027301, 2007.

D. Charalambous, L. Skrbek, P.C. Hendry, P.V.E. McClintock, W.F. Vinen, *Phys. Rev. E***74**, 036307, 2006.

C.E. Chase, *Phys. Rev.* **127**, 361, 1962; **131**, 1898, 1963.

G. Comte-Bellot and S. Corrsin, *J. Fluid Mech.* **25**, 657, 1966.

I.P.D. De Silva and H.J.S. Fernando, *Phys. Fluids* **6**, 2455, 1994.

R.J. Donnelly, *Quantized vortices in He II*, Cambridge University Press, Cambridge, England, 1991.

R.J. Donnelly, C.F. Barenghi, *J. Phys. Chem. Data* **27**, 1217, 1998.

R.J. Donnelly, A.C. Hollis Hallett, *Annals of Physics* **3**, 320, 1958.

R.J. Donnelly, O. Penrose, *Phys Rev.* **103**, 1137, 1956.

R.J. Donnelly, P.H. Roberts, *Proc. Roy. Soc.*, **A312** 519, 1969.

V.B. Eltsov *et al.*, *Phys. Rev. Lett.* **96**, 215302, 2006.

A.P. Finne *et al.*, *Nature* **424**, 1022, 2003.

A.P. Finne *et al.*, *Phys. Rev. Lett.* **96**, 085301, 2006.

U. Frisch, *Turbulence: The Legacy of A. N. Kolmogorov*, Cambridge University Press, Cambridge, UK, 1995.

W.I. Glaberson, W.W. Johnson, R.M. Ostermeier, *Phys. Rev. Lett.* **33**, 1197, 1974; R.M. Ostermeier, W.I. Glaberson, *J. Low Temp. Phys.* **21**, 191, 1975.

A.V. Gordeev, T.V. Chagovets, F. Soukup, L. Skrbek, *J. Low Temp. Phys.* **138**, 549, 2005.

C.J. Gorter, J.H. Mellink, *Physica* **15**, 285, 1949.

R. Hanninen, M. Tsubota, W.F. Vinen, *Phys. Rev. B***75**, 064502, 2007.

L. Kistler and T. Vrebalovich, *J. Fluid Mech.*, **26**, 1966.

J.S. Korhonen *et al.*, *Phys. Rev. Lett.* **65**, 1211, 1990.

L.D. Landau, *J.Phys. U.S.S.R.* **5**, 71, 1941; *J.Phys. U.S.S.R.* **11**, 91, 1947.

L.D. Landau and E.M. Lifshitz, *Fluid Mechanics*, Pergamon Press, Oxford, UK, 1987.

A.C. Hollis Hallett, *Proc. Royal. Soc.* **A210**, 404, 1952.

J. Jäger, B. Schuderer and W. Schoepe, *Phys. Rev. Lett.*, **74**, 566, 1995.

N.B. Kopnin, *Theory of nonequilibrium superconductivity* Clarendon Press, Oxford, 2001.

V.S. L'vov, S.V. Nazarenko, G.E. Volovik, *Pis'ma v ZHETF* **80**, 546, 2004.

W.D. McComb, *The Physics of fluid turbulence*, Clarendon Press, Oxford, 1990.

D.J. Melotte, C.F. Barenghi, *Phys. Rev. Lett.* **80**, 4181, 1998.

L.M. Milne-Thomson, *Theoretical hydrodynamics*, New York: Dover Publications, 1996.

M. Morishita, T. Kuroda, A. Sawada, and T. Satoh, *J. Low Temp. Phys.*, **76**, 387, 1989.

J.J. Niemela, K.R. Sreenivasan, R.J. Donnelly, *J. Low Temp. Phys.* **138**, 537, 2005.

M. Niemetz, H. Kerscher and W. Schoepe, *J. Low Temp. Phys.*, **126**, 287, 2002.

J.E. Sader, *J. Appl. Physics*, **84**, 64, 1998.

P.G. Saffman, *J. Fluid Mech.* **27**, 581, 1967; *Phys. Fluids* **10**, 1349, 1967.

H. Schlichting, K. Gersten, *Boundary-Layer Theory*, Springer, 1996.

W. Schoepe, *Phys. Rev. Lett.*, **92**, 095301, 2004.

K.W. Schwarz, *Phys. Rev. B* **38**, 2398, 1988.

K.W. Schwarz, J.R. Rozen, *Phys. Rev. Lett.* **66**, 1898, 1991; *Phys. Rev. B* **44**, 7563, 1991.

L. Skrbek, A.V. Gordeev, F. Soukup, *Phys. Rev. E* **67**, 047302, 2003.

L. Skrbek, J.J. Niemela, R.J. Donnelly, *Phys. Rev. Lett.* **85**, 2973, 2000.

L. Skrbek, S.R. Stalp, *Phys. Fluids* **12**, 1997, 2000.

M.R. Smith, R.J. Donnelly, N. Goldenfeld, W.F. Vinen, *Phys. Rev. Lett.* **71**, 2583, 1993.

E.B.Sonin, *Rev. Mod. Phys.* **59**, 87, 1987.

K.R. Sreenivasan, *Phys. Fluids* **7**, 2778, 1995.

S.R. Stalp, J.J. Niemela, W.F. Vinen, R.J. Donnelly, *Phys. Fluids* **14**, 1377, 2002.

S.R. Stalp, L. Skrbek, R.J. Donnelly, *Phys. Rev. Lett.* **82**, 4831, 1999.

C.E. Swanson, C.F. Barenghi, R.J. Donnelly, *Phys. Rev. Lett.* **50**, 190, 1983.

J.T. Tough, *Superfluid turbulence*, in Prog. in Low Temp. Phys., Vol.VIII, North-Holland Publ. Co., 1982.

H. Touil, J.P. Bertoglio, L. Shao, *J. Turbulence* **3**, 049, 2002.

M. Tsubota, T. Araki and C.F. Barenghi, *Phys. Rev. Lett.* **90**, 205301, 2003.

E. van Doorn, C. M. White, K. R. Sreenivasan, *Phys. Fluids* **11**, 2387, 1999.

W.F. Vinen, *Proc. Roy. Soc.* **A240**, 114, 1957; **A240**, 128, 1957; **A242**, 493, 1957; **A243**, 400, 1957.

W.F. Vinen, *Proc. R. Soc.* **A 260**, 218, 1961.

W.F. Vinen, *Phys. Rev. B* **61**, 1410, 2000.

W.F. Vinen, J.J. Niemela, *J. Low Temp. Phys.* **128**, 167, 2002.

W. F. Vinen, *Phys. Rev. B* **71**, 024513, 2005.

H. Yano, N. Hashimoto, A. Handa, M. Nakagawa, K. Obara, O. Ishikawa, T. Hata, *Phys. Rev. B* **75**, 012502, 2007.

Bose-Einstein condensation and superfluid turbulence

Natalia G. Berloff *

Department of Applied Mathematics and Theoretical Physics
University of Cambridge, Cambridge, CB3 0WA, United Kingdom

Abstract. We consider superfluid motion that arises from Bose condensation and restricted by quantum conditions. Superfluids have been studied experimentally for many years and have by now become a major focus of cryogenic physics. Applications of the subject are wide-ranging, from engineering to astrophysics. Superfluid turbulence provides insights into classical fluid turbulence, especially at high Reynolds numbers, where the vorticity has an intermittent, fractal character. In this Chapter we consider different modifications of the nonlinear Schrödinger equation to elucidate various aspects of superfluid behaviour: motion of vortices, travelling waves, interactions with normal fluid and superfluid turbulence.

1 Introduction

The processes of self-organization, formation of large-scale coherent localized structures and interactions of these structures with small-scale fluctuations are at the heart of nonlinear sciences, ranging from classical turbulence, superfluids, ultracold gases and Bose–Einstein Condensates (BECs), to the formation of the early Universe. The key to our understanding of turbulence is to elucidate physics of interactions between large scales (eg. large eddies) and small scales (eg. turbulent fluctuations), and to develop mathematical models that account for the effects of small scales without actually solving for them.

Superfluidity is the ability to flow through narrow channels without friction; this phenomenon is characterised by the existence of quantised vortices with the quantum of circulation $\kappa = h/m$, where m is the mass of a boson.

---

*I wish to thank EPSRC-UK for support to my research described in this Chapter. I am grateful to Carlo Barenghi and Yuri Sergeev for organizing the school "Vortices and Turbulence at Very Low Temperature" and for giving me the opportunity to present the lectures at this school.

Superfluidity is closely related to the Bose-Einstein condensation – the state of matter in which a large percentage of bosons collapse into their lowest quantum state, allowing quantum effects to be observed on a macroscopic scale. This phenomenon has been first realised experimentally in dilute atomic gases in 1995 and since that time many more superfluid system has been experimentally obtained: Fermi gases, exciton-polariton condensates, multi-component BEC, etc. The link between superfluidity and BEC is in the existence of the classical field ψ (order parameter, wave function) associated with the macroscopic component of the field operator. The order parameter ψ is the complex function (or several complex functions for multi-component condensates, or 3x3 matrix of complex functions for ^{3}He) where, quantum mechanically, density, ρ_s, is the square of the amplitude and velocity, $\mathbf{v}_s$ is proportional to the gradient of the phase, S:

$$\psi = \sqrt{\rho_s}\exp[\mathrm{i}S], \qquad \mathbf{v}_s = \frac{\hbar}{m}\nabla S. \tag{1.0.1}$$

The existence of quantised vortices immediately follows from this representation. If we draw a closed contour C in a singly-connected superfluid system described by ψ, then the phase S of ψ can only change by a multiple of 2π for ψ to remain single valued as we move around this contour. If the phase S does indeed changes by, say, 2π, then there has to be at least one point (in 2D) or a line (in 3D) inside C where S takes any value between 0 and 2π. To prevent such phase singularity, the density, ρ_s, has to vanish at this point (line). These points or lines are therefore quantised vortices with the unit of circulation $\int_C \mathbf{v}_s \cdot d\mathbf{l} = 2\pi\hbar/m = h/m$, as stated above.

The first successful macroscopic theory for the motion of superfluid helium (HeII) was that of Landau [1], in which the fluid is modelled as an interacting mixture of superfluid of density ρ_s moving with velocity $\mathbf{v}_s$ and a normal fluid of density ρ_n moving with velocity $\mathbf{v}_n$, representing the phonons and rotons that ride on the superfluid. As the temperature T tends to absolute zero, $\rho_n \to 0$ and $\rho_s \to \rho$ (the total density of the mixture). As $T \to T_\lambda$, the temperature of the phase transition to helium I, $\rho_s \to 0$ and $\rho_n \to \rho$. The range $T < 0.6$K is often defined to be the 'low temperature regime', while $0.6\text{K} < T < T_\lambda$ is the 'high temperature regime', in which most experiments on superfluid turbulence have been performed, and in which the normal fluid determined by the turbulence taking place in the normal fluid. This is confirmed by the success of theories of superfluid turbulence, such as that of Barenghi et al. [2], [3], in which the superfluid vorticity is largely tied to that of the normal fluid. In contrast, in the low temperature range ($T \leq 0.6^\circ K$), where ρ_n is smaller than ρ_s, we may expect turbulence in the superfluid largely to determine turbulence in the normal fluid, rather than the reverse.

In the low temperature regime, $\rho_n/\rho_s \ll 1$, and the normal fluid plays a relatively minor role. Even though the superfluid is inviscid, there are significant differences between classical turbulence at large Reynolds number and superfluid turbulence. The most significant is that vorticity is continuously distributed in a classical (Navier-Stokes) fluid, but is quantised in a superfluid in units of κ. Turbulence in the superfluid therefore resembles a tangle of vortex filaments, whose dynamics differs from that of the chaotic but continuous vorticity of classical turbulence; see, for example, [4].

The Landau [1] theory of superfluidity pre-dated the discovery of quantised vortex lines and therefore omitted significant dynamical effects. This was remedied, in the limit in which the mean spacing b between the vortex lines is small compared with any other length scale L of interest, by HVBK theory, so named after Hall and Vinen [5] who first set up such a theory, and Bekharevich and Khalatnikov [6] who generalised it; see also Hills and Roberts [7]. In this limit ($b \ll L$), the superfluid vorticity is treated as a continuum, but the discrete nature of the vorticity gives rise to an extra force $\mathbf{T}$ on the superfluid component, arising from the tension in the vortex lines; this term is absent from the classical Euler equation of motion for an inviscid fluid. The vortex lines also create a force $\mathbf{F}$ of mutual friction between superfluid and normal fluid that is additional to the mutual friction included by Landau in his equations, and represents the effects of collisions of the quasiparticles with the vortex cores. HVBK theory has scored a number of successes. It has given results [8] in quantitative agreement with experiments on the instability of Taylor-Couette flow in HeII [9]. HVBK theory provides the first of three principal ways of studying superfluid turbulence.

The second approach recognises that HVBK theory presupposes a high density of pre-existing superfluid vortex lines, and is incapable of describing processes that create or destroy those lines. For example, the assumption $b \ll L$ implies that the superfluid vorticity, $\boldsymbol{\omega}_s$, varies slowly with position $\mathbf{x}$. This excludes situations favourable to the reconnection of lines, such as when oppositely-directed vortex lines are brought by the flow into juxtaposition. To understand the creation and destruction of lines, it is necessary to study situations in which b is *not* small compared with L, and in which vortex lines are treated as discrete entities. The term "superfluid turbulence" is often used synonymously for the "evolution of superfluid vortices." On the scale much larger than the vortex core radius ($a \approx 10^{-8}\,\mathrm{cm}$) the motion of superfluid vortices is commonly described by the classical theory of vortex filaments. If superfluid vortices are modelled as filaments of zero cross-section with δ–function vorticity, the kinematic statements

$\nabla \cdot \mathbf{v} = 0, \nabla \times \mathbf{v} = \boldsymbol{\omega}$ imply a line integral along the filaments, $\mathbf{s} = \mathbf{s}(\xi)$, where ξ is arc length:

$$\mathbf{v}(\mathbf{x}) = \frac{\kappa}{4\pi} \int \frac{\mathbf{s}' \times (\mathbf{x} - \mathbf{s}) d\xi}{|\mathbf{x} - \mathbf{s}|^3}, \tag{1.0.2}$$

in which $\mathbf{s}'(\xi) = d\mathbf{s}/d\xi$ is the unit tangent vector. The integral (1.0.2) known as the Biot-Savart Law (BSL) diverges however in the limit $\mathbf{x} \to \mathbf{s}_1$, where $\mathbf{s}_1 = \mathbf{s}(\xi_1)$ is any point on a filament; the nature of the divergence is spelled out in detail in §2.3 of [10]. To salvage (1.0.2) the 'cut-off method' was introduced. When computing the velocity $\mathbf{v}(\mathbf{s}_1)$ of the filament at $\mathbf{x} = \mathbf{s}_1$, the segment $|\xi - \xi_1| < \delta$, where $\delta = \mathrm{O}(a)$, is simply deleted from (1.0.2). The best choice of δ has been determined when the radius of curvature, $R(\xi_1)$, at $\mathbf{s}_1$ is large compared with a. When $R \ll L$, the motion of the vortex at $\mathbf{s}_1$ is decided mainly by the distribution of the vorticity $\boldsymbol{\omega}(\mathbf{s})$ in the vicinity of $\mathbf{s}_1$ and, to a first approximation,

$$\mathbf{v}(\mathbf{s}_1) = (\kappa/4\pi)(\mathbf{s}_1' \times \mathbf{s}_1'') \ln(\ell/a), \tag{1.0.3}$$

where $\ell = \mathrm{O}(R)$. This is known as the 'local induction approximation' (LIA), and is the most convenient, but also the crudest, way of determining the motion of vortex lines. It is especially dubious when (as is often the case) the conditions $a \ll R \ll L$ are not met.

It is clear that the application of classical theory is not free of difficulty. Moreover, two mechanisms very relevant to superfluid turbulence are ignored. It has recently been shown [11] that emission of sound by a vortex tangle is very significant in superfluid turbulence. This process is completely removed by the main assumption of classical vortex theory: $\nabla \cdot \mathbf{v}_s = 0$. The dynamics of vortex filaments in a compressible fluid is not as well understood as that for the incompressible case. Secondly, the processes of severance and coalescence of vortex lines are centrally important for the study of superfluid turbulence, but these are explicitly forbidden by the Kelvin-Helmholtz theorem, according to which vortex lines are frozen to an Euler fluid and cannot change their topology. In a normal fluid, the processes have been successfully simulated numerically by restoring viscosity to the fluid. This step is disallowed in a superfluid, and the only way to defeat the theorem is through *ad hoc* procedures. For example, it has been supposed that, whenever one vortex filament comes within a distance Δ of order a of another filament, reconnection will always occur, and that otherwise reconnection will not happen. A precise way of determining Δ is not known, but its value can clearly greatly affect the reconnection rate in a vortex tangle. Moreover, the angle at which the vortex filaments approach one another is undoubtedly an important factor in determining whether they reconnect or not; a clear set of reconnection rules is lacking.

We now move to the third approach, a theory that gives superfluid vortex lines their own unique core structure. At the same time, it provides a mechanism for the severing and coalescence of vortex lines, and includes sound propagation, so that the emission of sound from a vortex tangle can be evaluated. In short, while having the same objectives as the second approach, it evades all of its difficulties (although it has a few difficulties of its own).

In the Hartree approximation, the imperfect Bose condensate is governed by the GP equation [12]. In terms of the single-particle wavefunction $\psi(\mathbf{x}, t)$ for N bosons, the time-dependent self-consistent field equation is

$$\begin{aligned} \mathrm{i}\hbar\psi_t(\mathbf{x}, t) &= -(\hbar^2/2m)\nabla^2\psi(\mathbf{x}, t) + \\ & \psi(\mathbf{x}, t) \int |\psi(\mathbf{x}', t)|^2 V(|\mathbf{x} - \mathbf{x}'|)\, d^3x' - E_v\psi(\mathbf{x}, t) \end{aligned} \tag{1.0.4}$$

where $V(|\mathbf{x} - \mathbf{x}'|)$ is the potential of the two-body interactions between bosons, and E_v is the chemical potential which we explicitly introduce into the system. The normalisation condition is $\int |\psi|^2\, d^3x = N$. For a weakly interacting Bose gas, (1.0.4) is simplified by replacing $V(|\mathbf{x} - \mathbf{x}'|)$ with a δ−function repulsive potential of strength V_0. This does not qualitatively alter the nature of the results since the characteristic length of the weakly interacting Bose gas is larger than the range of the force. Equation (1.0.4) for such a potential is

$$\mathrm{i}\hbar\psi_t(\mathbf{x}, t) = -(\hbar^2/2m)\nabla^2\psi + V_0|\psi|^2\psi - E_v\psi. \tag{1.0.5}$$

Equations (1.0.4)-(1.0.5) define Hamiltonian systems, the following integrals being conserved: mass

$$M = m \int |\psi|^2\, d\mathbf{x}, \tag{1.0.6}$$

momentum density

$$\mathbf{p} = \frac{\hbar}{2\mathrm{i}} \int [\psi^* \nabla\psi - \psi\nabla\psi^*]\, d\mathbf{x}, \tag{1.0.7}$$

and energy; in the case of (1.0.5) this is expressed by

$$E = \frac{\hbar^2}{2m} \int |\nabla\psi|^2\, d\mathbf{x} + \frac{V_0}{2} \int |\psi|^4\, d\mathbf{x}. \tag{1.0.8}$$

It is usually convenient to model phenomena in an infinite domain in which, prior to the onset of a disturbance, $\psi = \psi_\infty$ everywhere, where, by (1.0.5)

$$\psi_\infty = (E_v/V_0)^{1/2}. \tag{1.0.9}$$

We then modify (1.0.7)-(1.0.8) to forms measuring departures from this uniform state

$$\mathbf{p} = \frac{\hbar}{2\mathrm{i}} \int [(\psi^* - \psi_\infty)\nabla\psi - (\psi - \psi_\infty)\nabla\psi^*] \, d\mathbf{x}, \tag{1.0.10}$$

$$E = \frac{\hbar^2}{2m} \int |\nabla\psi|^2 \, d\mathbf{x} + \frac{V_0}{2} \int (|\psi|^2 - \psi_\infty^2)^2 \, d\mathbf{x}. \tag{1.0.11}$$

The Madelung transformation relates the wave function, $\psi = R \exp \mathrm{i} S$, to classical hydrodynamical variables. The mass density ρ, mass flux $\mathbf{j}$ and velocity potential ϕ, are $\rho = mR^2, \mathbf{j} = \rho\mathbf{v} = \rho\boldsymbol{\nabla}\phi, \phi = (\hbar/m)S$. The real and imaginary parts of (1.0.5) then give a continuity equation and an integrated form of the momentum equation:

$$\rho_t + \boldsymbol{\nabla}\cdot(\rho\mathbf{v}) = 0, \tag{1.0.12}$$

$$\frac{\partial\phi}{\partial t} + \frac{1}{2}v^2 + c^2\left(\frac{\rho}{\rho_\infty} - 1\right) = c^2 a^2 \frac{\nabla^2 \rho^{1/2}}{\rho^{1/2}}, \tag{1.0.13}$$

where $\rho_\infty = mE_v/V_0$ is the density at infinity (the density of the fluid when stagnant), c is the speed of sound $c = \sqrt{E_v/m} = (V_0/m^2)^{1/2}\rho_\infty^{1/2}$ and $a = \kappa/2\sqrt{\pi}c$ is the so-called 'healing length'. The healing length is the characteristic length on which fluid heals itself when locally perturbed. It also fixes the vortex core size. The final term in (1.0.13) is sometimes called the 'quantum pressure'. It may be particularly noticed that, while the fluid flows irrotationally, it is not incompressible. According to (1.0.13) and temporarily ignoring its right-hand side, the pressure P is proportional to ρ^2: $P = (c^2/2\rho_\infty)\rho^2$. Motion in the fluid is accompanied by sound waves.

The elementary excitations (small amplitude harmonic modes) above the ground state (the lowest energy state for a given number of particles) $\psi_0 = \sqrt{\rho_0/m} = \sqrt{E_v/V_0}$ are found by inserting $\psi = \psi_0(\mathbf{x}) + \epsilon\psi_1(\mathbf{x}, t)$ into the GP equation (1.0.5) and linearising it to

$$i\hbar\psi_{1t} = -\frac{\hbar^2}{2m}\nabla^2\psi_1 + \frac{\rho_0 V_0}{m}(\psi_1 + \psi_1^*). \tag{1.0.14}$$

We look for plane wave modes: $\psi_1 = A\exp[i(\mathbf{k}\cdot\mathbf{x} - \omega t)] + B^*\exp[-i(\mathbf{k}\cdot\mathbf{x} - \omega t)]$, where A and B are complex constants. The existence of a non-trivial solution for A and B implies the dispersion relation (Bogoliubov law) as

$$\omega^2 = \frac{\rho_0 V_0}{m^2}k^2 + \left(\frac{\hbar k^2}{2m}\right)^2. \tag{1.0.15}$$

For small k (long wavelengths) the dispersion is sound-like $\omega \approx ck$. The corresponding quasiparticles are *phonons*. For large k (short wavelengths) the dispersion is particle-like $\omega \approx \hbar k^2/2m$.

A vortex filament is no more than a curve on which $\psi = 0$, and there is no reason why such curves should not intersect or divide. Using the hydrodynamical interpretation of (1.0.5), these processes correspond to the coalescence or severance of vortex lines. There is no violation of the Kelvin-Helmholtz theorem, which does not apply to integration paths passing through zeros of ρ. In short, the GP equation provides a means of studying these important changes of topology of a vortex tangle without making any *ad hoc* assumptions, and while remaining within the framework of Hamiltonian dynamics, i.e., without introducing viscosity. There is also no need to appeal to the BSL or LIA approximations, which are in any case somewhat misleading, since the fluid is compressible.

In what follows we consider the GP equation and its modifications as models of superfluid turbulence paying a particular attention to vortices and travelling coherent structures that these equations exhibit. We shall often refer to dimensionless GP equation (1.0.5)

$$-2i\psi_t = \nabla^2\psi + (1 - |\psi|^2)\psi, \tag{1.0.16}$$

where time is measured in units of $ma^2/\hbar$ and distance in healing lengths a. The momentum (1.0.10) and energy (1.0.11) become

$$\mathbf{p} = \frac{1}{2i}\int [\nabla\psi(\psi^* - 1) - \nabla\psi^*(\psi - 1)]\, d\mathbf{x}, \tag{1.0.17}$$

measured in $\hbar/a$ and

$$E = \tfrac{1}{2}\int [|\nabla\psi|^2 + \tfrac{1}{2}(1 - |\psi|^2)^2]\, d\mathbf{x} \tag{1.0.18}$$

measured in $\hbar^2\rho_\infty/m^2$.

We discuss a single straight line vortex and the vortex motion on uniform and non-uniform backgrounds in Section II. We modify the GP equation to make it more applicable to superfluid helium in Section III. Travelling coherent structures that exist in the GP equation, the nonlocal GP equation and the coupled GP equations are subject of our study in Section IV. In Section V we consider mechanisms of the formation of these localised structures. We discuss how to introduce the normal component into superfluid in Section VI, where we also study the dissipation of a vortex ring at non-zero temperatures.

2 Vortices

A vortex line is defined by a zero of the wave function $\psi = 0$. In cylindrical coordinates (r, θ, z) the wave function of a straight-line vortex takes form

$$\psi = R(r)\exp[is\theta], \tag{2.0.1}$$

where s is an integer ("winding number", "topological charge"). Fluid rotates around the z-axis with tangential velocity

$$v = \frac{\hbar}{m}\nabla\theta = \frac{\hbar}{m}\frac{s}{r} \tag{2.0.2}$$

and the amplitude satisfies

$$-\frac{\hbar^2}{2m}\frac{1}{r}\frac{d}{dr}\left(r\frac{dR}{dr}\right) + \frac{\hbar^2 s^2}{2mr^2}R + V_0 R^3 - E_v R = 0. \tag{2.0.3}$$

At large distances the density is unperturbed, so $R \to \psi_\infty$. We introduce the dimensionless function $R = \psi_\infty f(\eta)$, where the distance is measured in healing lengths $\eta = r/a$.

$$\frac{1}{\eta}\frac{d}{d\eta}\left(\eta\frac{df}{d\eta}\right) + \left(1 - \frac{s^2}{\eta^2}\right)f - f^3 = 0, \qquad f(\infty) = 1. \tag{2.0.4}$$

This equation was numerically integrated by Pitaevskii [13] and the solution is shown on Fig. 2.0.1 together with a Padé approximation [14].

The simplest Padé approximation of the vortex amplitude that will be used to analyse vortex motion is in the form $f(\eta)^2 = \eta^2/(\eta^2 + 2)$ [15], so the wavefunction of a straight line vortex is approximated by

$$\frac{\psi}{\psi_\infty} = f(\eta)e^{i\theta} = (x + iy)/\sqrt{x^2 + y^2 + 2}, \quad x = \eta\cos\theta, y = \eta\sin\theta. \tag{2.0.5}$$

The energy per unit length of the vortex line is

$$E_\ell = \frac{\kappa^2\rho_\infty}{4\pi}\left(\int_0^\infty \left[\frac{dR}{dr}\right]^2 r\,dr + \int_0^\infty \frac{R^2}{r}\,dr + \frac{1}{2}\int_0^\infty (1 - R^2)^2 r\,dr\right). \tag{2.0.6}$$

The first term can be regarded as a "quantum energy", the second term is the classical vortex kinetic energy that diverges logarithmically unless a cut-off distance L is introduced; the third term in (2.0.6) represents the potential energy. The energy per unit length of the vortex is usually expressed in the form

$$E_\ell = \frac{\kappa^2\rho_\infty}{4\pi}\left(\ln\frac{L}{a} + L_0\right), \tag{2.0.7}$$

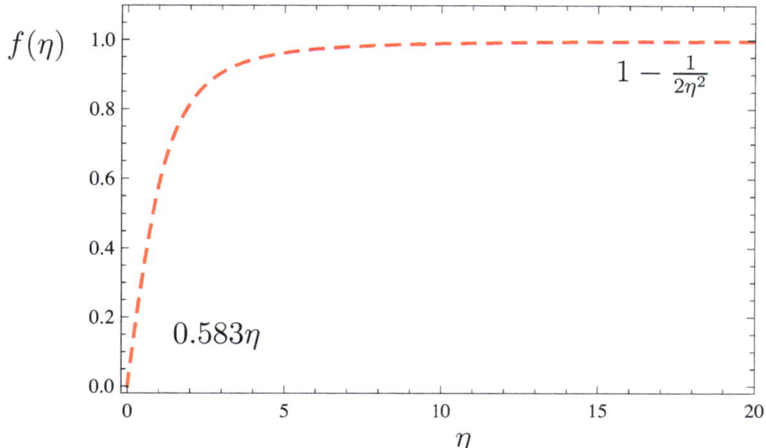

Figure 2.0.1. The amplitude of a straight-line vortex for $s = 1$ as solution of Eq. (2.0.4). The leading order terms of the power series expansions at zero and infinity are shown next to the graph. A Padé approximation of the vortex amplitude $\eta/\sqrt{\eta^2+2}$ is shown by dashed line.

where the constant L_0 is called the "vortex core parameter" and was determined numerically by Pitaevskii [13] as $L_0 = 0.3809$.

Straight-line vortices can transmit energy along their length by Kelvin waves. These have been comprehensively analysed for incompressible Euler fluids; see for example Chapter 11 of Saffman [10]. They have been studied for compressible hollow core and the GP vortices by Roberts [16] who showed that in both situations, the waves may be bound or free, depending on their angular and axial wave numbers, m_θ and k. The free waves radiate energy acoustically to infinity, while the bound states do not. It is found that for the GP vortex one class of $m_\theta = 1$ modes, called 'slow modes' because of their low frequency, consists of bound states for all k. The slow $m_\theta = 2$ modes are also bound when k is sufficiently large, but are free for small k.

2.1 Vortex motion on uniform backgrounds

Here we consider N_v vortices in (x, y)-plane that are separated by distances far exceeding the healing length a and satisfying Eq. (1.0.16).

We introduce a small parameter $\epsilon = a/L$, where L is an average intervortex distance and scales $\nabla \to \epsilon\nabla, \partial_t \to \epsilon^2\partial_t$ and get to the leading order from Eq.(1.0.16)

$$\rho = 1 - O(\epsilon^2), \qquad \nabla^2 S = 0. \tag{2.1.1}$$

This shows that the vortex motion is according to classical inviscid irrotational (except at the vortex positions $\mathbf{x} = \mathbf{x}_i$) incompressible flow dynamics. The velocity potential for a point vortex at origin, with the winding number s, is $s\theta$, so

$$\mathbf{v} = s\nabla\theta = s\frac{\mathbf{e}_z \times \mathbf{x}}{|\mathbf{x}|^2} = sJ\frac{\mathbf{x}}{|\mathbf{x}|^2},$$

where J denotes rotation through $\pi/2$.

Laplace's equation (2.1.1) is linear, so we can linearly superpose a finite number N_v of point vortices with different strengths and positions $\mathbf{x}_i$ $(i = 1, ...N_v)$, thus

$$\mathbf{v} = \sum_i s_i \frac{\mathbf{e}_z \times (\mathbf{x} - \mathbf{x}_i)}{|\mathbf{x} - \mathbf{x}_i|^2}.$$

Classical point vortices move as material points so each vortex is moved by the velocity field due to all the other vortices. The dynamical system of vortex motion takes the form

$$\dot{\mathbf{x}}_i(t) = \sum_{j\neq i} s_j \frac{\mathbf{e}_z \times (\mathbf{x}_i - \mathbf{x}_j)}{|\mathbf{x}_i - \mathbf{x}_j|^2} \qquad (i = 1, ...N_v) .$$

Remarkably, the equations of vortex motion have a Hamiltonian structure:

$$s_i\dot{\mathbf{x}}_i(t) = J\frac{\partial H}{\partial \mathbf{x}_i}; \qquad H = \frac{1}{2}\sum_j \sum_{j\neq i} s_i s_j \ln |\mathbf{x}_i - \mathbf{x}_j|.$$

In general it is quite challenging to understand vortex motion in condensates as vortices move on essentially nonuniform backgrounds, interacting with sound and emitting sound. Next we consider motion of a vortex line near a solid boundary and show that for a vortex parallel to the wall, the motion is essentially equivalent to that generated by an image vortex, but the depleted surface layer induces an effective shift in the position of the image compared to the case of a vortex pair in an otherwise uniform flow.

2.2 Vortex motion near the boundary

Here we consider the problem of vortex motion in an asymptotically homogeneous condensate in the presence of a solid wall where the wave function of the condensate vanishes. Again, the geometry is two dimensional, with the vortex aligned along the z axis, parallel to the surface of the wall. The dynamics of the time-dependent condensate in the presence of the solid wall at $y = 0$ is described by the GP equation (1.0.16) subject to the boundary condition

$$\psi(x, y = 0, t) = 0, \quad |x| < \infty. \tag{2.2.1}$$

In the absence of vortices, the exact solution of (1.0.4) for the stationary state of the semi-infinite condensate is

$$f(y) = \tanh(y/\sqrt{2}). \tag{2.2.2}$$

In classical inviscid fluid dynamics with constant mass density ρ, the relevant kinematic boundary condition at a solid wall with normal vector $\mathbf{n}$ is

$$\rho\,\mathbf{u}\cdot\mathbf{n} = 0, \tag{2.2.3}$$

where ρ is the local density of the fluid and $\mathbf{u}$ is the velocity of the fluid. The corresponding problem of a vortex moving parallel to the wall is solved by placing one or more image vortices in such a way that condition (2.2.3) is identically satisfied.

For the dynamics described by the GP equation (1.0.4), the density $\rho \propto |\psi|^2$ is no longer constant, but rather vanishes at the surface of the wall. Thus condition (2.2.3) is automatically satisfied, and all components of $\mathbf{u}$ can in principle remain finite on the boundary. Therefore, it may seem that image vortices are irrelevant in the case of the GP equation, so that the vortex should remain stationary away from the boundary (where the fluid density is constant apart from exponentially small corrections). It was shown in [17] that this is not true. In fact, the vortex moves parallel to the boundary, and it moves faster than a corresponding pair of vortices of opposite circulation in a uniform condensate in the absence of the depletion caused by the boundary.

The asymptotics of the vortex velocity was obtained using the Hamiltonian group relation

$$U = \frac{\partial E}{\partial p}, \tag{2.2.4}$$

which is satisfied by classical as well as the GP vortex pairs and vortex rings moving on homogeneous backgrounds [18]. The Hamiltonian relation (2.2.4) is derived by considering the variation $\psi \to \psi + \delta\psi$ in (1.0.17) and (1.0.18) that results in

$$\delta p = \mathrm{i}\int\left(\delta\psi\frac{\partial\psi^*}{\partial x} - \delta\psi^*\frac{\partial\psi}{\partial x}\right)dxdy \tag{2.2.5}$$

and

$$\delta E = \tfrac{1}{2}\int\{\delta\psi^*\left[-\nabla^2\psi + \psi\left(|\psi|^2-1\right)\right]+\delta\psi\left[-\nabla^2\psi^* + \psi^*\left(|\psi|^2-1\right)\right]\}\,dxdy\,, \tag{2.2.6}$$

provided $\delta\psi \to 0$ as $y \to \infty$. It follows that $\delta E = U\delta p$ and thus the Hamiltonian group relation of the energy and momentum taken along the sequence of solitary-waves that satisfy Eq. (1.0.16) with $\partial_t \to -U\partial_x$. We compare the motion of a pair of vortices of opposite circulation distance $2y_0$ away from each other with the approximate form of the wavefunction $\psi_1 = \psi_v(y_0)\psi_v^*(-y_0)$ where $\psi_v(y_0) = \psi(x, y - y_0)$ and ψ is given by Eq. (2.0.5) with the motion of a vortex distance $y_0 + l$ away from the solid wall, approximated by $\psi_2 = \tanh(y/\sqrt{2})\psi_v(y_0 + l)\psi_v^*(-y_0 - l)$. If the pair and a vortex move with the same (classical) velocity $1/2y_0$, then

$$l = y_0 - \frac{1}{2}\frac{\partial p/\partial y_0}{\partial E/\partial y_0}, \tag{2.2.7}$$

where E and p are energy and momentum of a vortex next to a solid wall (given by Eqs.(1.0.18)–(1.0.17) but with ψ_∞ replaced by the new ground state $f(y)$) that can be integrated to the leading order in large y_0 to yield

$$l = \int_0^\infty [1 - \tanh(y/\sqrt{2})^2]\, dy = \sqrt{2}. \tag{2.2.8}$$

Therefore, the velocity of the vortex next to the wall is (to the leading order in large y_0)

$$U = 1/(2y_0 - 2\sqrt{2}). \tag{2.2.9}$$

The generalisation of this result were made [25] for inhomogeneous (trapped) condensates. In general, the velocity can be found by using the area of the displaced density at the boundary to determine the shift l in the position of the image vortex.

2.3 Vortex motion in linear potentials

The problem of vortex motion on gradually varying backgrounds, such that the background field does not change considerably at the vortex core scale, can be treated using the method of matched asymptotic expansions [19, 20]. The small parameter of the expansion is associated with the background variation. In particular, it was established that the vortex is expected to move in the direction orthogonal to the gradient of the background density with a (slow) velocity proportional to the magnitude of the gradient of the logarithm of the background density. Unfortunately, there is no universal mobility relation as was emphasised in [21], so the detailed asymptotic expansion procedure has to be carried out for each particular system. Notably, this task has been achieved for the linear density background in a semi-infinite system by Anglin [22] again under the assumption that the vortex core size is much smaller than the distance of the density background variation and assuming that the vortex is far from the surface (boundary) of the condensate cloud.

For an example of vortex motion on inhomogeneous background we consider the case of a linear external potential [25]. Such a potential is a good estimate of the form of a condensate near the Thomas-Fermi (TF) surface of the quadratic trapping potential [23]. A linear trap has also been used experimentally [24]. To consider a linear trap has the advantage that while the dynamics take place on an inhomogeneous density background there is a constant density gradient. Near the surface of a condensate in a harmonic trap $V(r) = m\omega^2 r^2/2$ can be replaced by its linear approximation [23] via

$$V_{tr}(y) = E_v - \frac{1}{2}\gamma y, \tag{2.3.1}$$

where $\gamma = m\omega^2 R$ with Thomas-Fermi (TF) radius R. Thus, this expression for the potential, taken with the full dimensional GP equation (1.0.5) and scaled according to

$$\psi \to \sqrt{\frac{\epsilon\gamma}{V_0}}\psi, \qquad t \to \frac{\hbar}{2\epsilon\gamma}t, \qquad \mathbf{x} \to \epsilon\mathbf{x}, \tag{2.3.2}$$

where

$$\epsilon = \left(\frac{\hbar^2}{2m\gamma}\right)^{1/3}, \tag{2.3.3}$$

produces the non-dimensional GP equation for a condensate in a linear potential trap

$$2i\frac{\partial\psi}{\partial t} = -\nabla^2\psi - (\frac{1}{2}y - |\psi|^2)\psi. \tag{2.3.4}$$

We consider a vortex distance y_0 away from the TF surface. For large y_0, the wave function can be approximated by

$$\psi = \psi_{TF}\psi_{v+}\psi_{v-}, \tag{2.3.5}$$

where the vortex and its image are given by

$$\psi_{v\pm} = \frac{x \pm \mathrm{i}(y \mp y_0)}{\sqrt{x^2 + (y \mp y_0)^2 + 2\xi_0^2}}, \tag{2.3.6}$$

the TF ground state is $\psi_{TF} = y/2$ and the correlation length $\xi_0 = 1/|\psi_{TF}|$. Asymptotics of the integrals for energy and momentum for large y_0 lead to

$$E = \frac{\pi y_0}{4}\left(\frac{1}{2} + 3\ln(y_0)\right), \tag{2.3.7}$$

$$p = \frac{\pi y_0^2}{2}. \tag{2.3.8}$$

The velocity is evaluated according to Eq. (2.2.4) as

$$U_L = \frac{1}{2y_0}\left(\frac{7}{4} + \frac{3}{2}\ln(y_0)\right), \tag{2.3.9}$$

which agrees well with numerical solutions [25] and with boundary layer theory [22].

3 Nonlocal Model

Although the GP equation is a popular model of superfluid helium, several aspects of the local GP model (1.0.5) are qualitatively or quantitatively unrealistic in application to superfluid helium. Superfluid helium is not a weakly interacting Bose gas, instead it is a strongly correlated fluid dominated by many-body effects. The velocity, c, of long wavelength sound waves is proportional to $\rho^{\frac{1}{2}}$ in the GP equaton, but it is known from helium experiments [26] that $c \propto \rho^{2.8}$. Next, the dispersion relation (1.0.15) does not describe a characteristic dip (roton minimum) in the superfluid helium dispersion curve. The omission of the roton branch is not too serious at low temperatures, where only the phonon branch is significantly populated, but it may have some repercussions when studying reconnection processes. The last two of these weaknesses are removed by replacing the local model (1.0.5) by a suitably chosen nonlocal model as in [27, 28, 29]. The minimum requirements on such a potential would be (i) the correct position of the roton minimum and (ii) the correct speed of sound. Actually such a fit can be obtained with a variety of potentials. Pomeau and Rica [27] pioneered the use of nonlocal models for study superfluidity, but their model did not have the correct sound velocity (slope of the dispersion curve at the origin) of superfluid helium. The applicability of (1.0.4) with a potential that adequately represents the dispersion curve has been analysed in [28]. It was shown that for liquid helium having the correct Landau dispersion curve, solutions of equation (1.0.4) develop non-physical mass concentrations. In particular, the "Eulerian part" of the momentum equation (without the quantum stress tensor) may become no longer hyperbolic in some parts of the integration volume. A virial theorem, similar to the one used to establish the catastrophic blow-up in the focusing nonlinear Schrödinger equation, can be used to establish similar catastrophes in bounded volume for (1.0.4). This indicates that the assumptions underlying the derivation of the equation break down and that higher order nonlinearities must be introduced.

A more accurate approach in modelling liquid helium is through density-functional theory done by Dalfovo et al. [30], which attempts to give an adequate microscopic description of interactions. In this approach the total energy is still written as a functional of the one-body density, but it includes short-range correlations. This approach has provided a quantitatively and qualitatively reliable representation of the superfluid properties of free surfaces, helium films, and droplets (see [30] and references therein). At the same time this approach is phenomenological and results in rather complicated forms of the energy functionals with many parameters that are chosen to reproduce liquid helium properties.

The nonlocal model (1.0.4) was modified in [29] in the spirit of a density - functional approach, by introducing only one additional nonlinear term in the expression for the correlation energy. This remedies the nonphysical features of model (1.0.4), while retaining not only an adequate representation of the Landau dispersion relation, but also simplicity in the analytical and numerical studies.

The idea is to replace the potential part of (1.0.8) which we write as $E_{int} = \frac{V_0}{2m^2}\int \rho^2 \, d\mathbf{x}$ by

$$E_{int} = \frac{1}{m^2}\int \left[\frac{1}{2}\int \rho(\mathbf{x}')V(\mathbf{r}' - r)\rho(\mathbf{x})\, d\mathbf{x}' + \frac{W}{4\gamma+1}\rho^{4\gamma+1}\right] d\mathbf{x}, \qquad (3.0.1)$$

where W and γ are phenomenological constants. $V(|\mathbf{x}-\mathbf{x}'|)$ is chosen so that the implied dispersion relation is a good fit to the Landau dispersion curve, for instance by considering a potential of the form

$$V(|\mathbf{x}-\mathbf{x}'|) = V(r) = (\alpha + \beta A^2 r^2 + \delta A^4 r^4)\exp(-A^2 r^2). \qquad (3.0.2)$$

where A, B, α, β, and δ are parameters that can be chosen to give excellent agreement with the experimentally determined dispersion curve.

On adopting (3.0.1), one can see that (1.0.4) is replaced by

$$i\hbar\psi_t = -\frac{\hbar^2}{2m}\nabla^2\psi + \psi\int |\psi(\mathbf{x}',t)|^2 V(|\mathbf{x}-\mathbf{x}'|)\, d\mathbf{x}' + W\psi|\psi|^{4\gamma} - E_v\psi. \quad (3.0.3)$$

This model not only produces the structure and energy per unit length of the straight-line vortex that are very close to those obtained from the Monte-Carlo simulations by Sadd et al. [31], but it also made it possible to bring the vortex core parameter (2.0.7) and the healing length into agreement. Fig.3.0.1 gives the density in the core of the straight line vortex, both for (3.0.3) and for the GP model. This nonlocal modification of the GP equation has been used to elucidate the differences between roton emission and vortex nucleation [32] and to predict what was later identified as a phase transition from a spatially homogeneous state to a layered state characterised by a periodic density modulation [33].

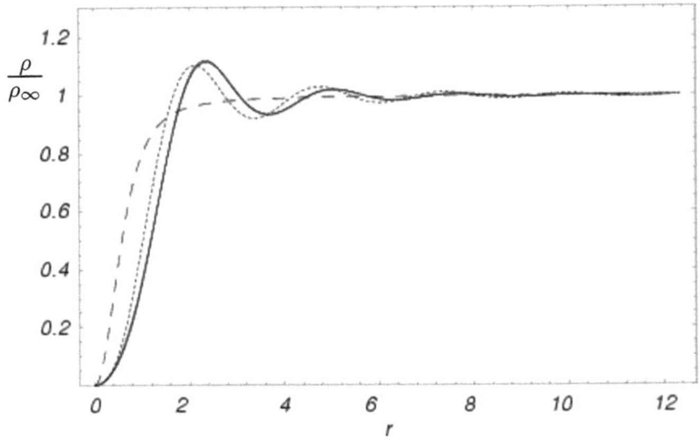

Figure 3.0.1. The amplitude of the vortex line as given by Eq. (3.0.3) with $\gamma = 2.8$ (solid line), $\gamma = 1$ (dotted line) and by the GP equation (1.0.5) (dashed line).

4 Travelling coherent structures in condensates

In this section we address the question of what kinds of travelling waves can exist in superfluid models. In particular we consider three models: the local GP equation (1.0.5), the nonlocal GP model (3.0.3) and the system of coupled GP equations that describe two-component BECs.

4.1 Travelling waves in the local GP equation

Firstly, we consider the travelling coherent structures that exist in the local GP equation (1.0.16). We are interested in localized disturbances moving with constant velocity U in x-direction, so they satisfy

$$2\mathrm{i}U\frac{\partial\psi}{\partial x} = \nabla^2\psi + (1 - |\psi|^2)\psi, \qquad |\psi| \to 1 \quad \text{as} \quad |\mathbf{x}| \to \infty. \tag{4.1.1}$$

Velocity of sound is given by $c = 1/\sqrt{2}$.

Equation (4.1.1) in 1D with $\nabla^2\psi$ replaced by $d^2\psi/dx^2$ can be integrated exactly [34]. We multiply the GP equation (4.1.1) by ψ^*, subtract c.c, and look for solutions with constant $Im(\psi)$. From compatibility of equations for real and imaginary parts we obtain

$$Im(\psi) = \sqrt{2}U = U/c, \qquad \sqrt{2}\frac{Re(\psi)}{dz} = \left(1 - \frac{U^2}{c^2} - Re(\psi)^2\right). \tag{4.1.2}$$

The solutions are found by integrating the second of Eqs. (4.1.2) and using the boundary conditions. Dark soliton (in dimensional units) has form

$$\psi(x-Ut)=\sqrt{\frac{\rho_\infty}{m}}\left(i\frac{U}{c}+\sqrt{1-\frac{U^2}{c^2}}\tanh\left[\frac{x-Ut}{\sqrt{2}a}\sqrt{1-\frac{U^2}{c^2}}\right]\right). \quad (4.1.3)$$

Note that the minimum of the density is at $\rho(0)=\rho_\infty U^2/c^2$, so that for $U=0$, $\psi=0$. This dark soliton is often referred to as black soliton and the solitons with $U\neq 0$ are gray solitons. The width of the solutions is $\sqrt{2}a/\sqrt{1-U^2/c^2}$ and the phase of the wavefunction experiences a jump $\Delta S=2\arccos(U/c)$ between $-\infty$ and $+\infty$. Finally, as $U\to c$, the energy E decreases: therefore, the dissipative effects result in an acceleration of soliton and its disappearance.

Large vortex rings and vortex pairs were investigated in [35] where the following expressions for the energy per unit length, momentum, and velocity, v, were obtained:

$$E=\tfrac{1}{2}\rho\kappa^2R\left(\ln\frac{8R}{a}+L_0-2\right),\quad p=\rho\kappa\pi R^2, \quad (4.1.4)$$

$$v=(\kappa/4\pi R)\left(\ln\frac{8R}{a}+L_0-1\right). \quad (4.1.5)$$

As expected, $E\approx 2\pi RE_\ell$, but also rings obey Hamilton group relation

$$v=\partial E/\partial p. \quad (4.1.6)$$

Jones and Roberts [18] determined the entire sequence of vortex rings numerically for the GP model (1.0.16). They calculated the energy E and momentum p and found two branches meeting at a cusp where p and E assume their minimum values, p_m and E_m. As $p\to\infty$ on each branch, $E\to\infty$. On the lower branch the solutions are asymptotic to the large vortex rings (4.1.4)-(4.1.5).

As E and p decrease from infinity along the lower branch, the solutions begin to lose their similarity to large vortex rings, and (4.1.4) - (4.1.5) determine E, p, and v less and less accurately, although (4.1.6) still holds. Eventually, for a momentum p_0 slightly greater than p_m, the rings lose their vorticity (ψ loses its zero), and thereafter (for $U>0.625$) the solitary solutions may better be described as 'rarefaction waves'. The upper branch consists entirely of these and, as $p\to\infty$ on this branch, the solutions asymptotically approach the rational soliton solution of the Kadomtsev-Petviashvili (KP) equation and are unstable. The pE dispersion curve of the solutions in 3D is shown on Fig. 4.1.1(a). The stability of 2D and 3D solutions was analysed in [36]. The lower branch is linearly stable and the upper branch is linearly unstable to axisymmetric infinitesimal perturbations, but the growth rates are small.

In two-dimensions, there is only one branch of the solutions that terminates at $p = 0$ and $E = 0$. Large p and E correspond to vortex pairs of opposite circulation. As both p and E decrease the velocity increases, the distance between vortices decreases until at some non-zero p and E the zeros of ψ coincide and for even smaller values of p and E the solutions become rarefaction pulses. Figure 4.1.1(b) shows the pE dispersion curve in 2D. The solutions in 2D are linearly stable [36].

4.2 Travelling waves in the nonlocal GP model

Equation (3.0.3) was integrated numerically in [29] to elucidate the behaviour of vortex rings. These calculations indicate that when the velocity of the vortex ring reaches the Landau critical velocity the ring becomes unstable and evanesces into sound waves. For any ring travelling with speed greater than the Landau critical velocity, the amplitude of the far-field solution will not decay exponentially at infinity, which makes the existence of such a ring impossible. One of the goals of these calculations was to clarify Onsager's concept of the roton as "the ghost of a vanished vortex ring." One can hope that the transition from the vortex ring to the sound pulse and the concomitant loss of vorticity would occur close to the roton minimum in energy-momentum space, or (more probable) close to the point where the group velocity and the phase velocity are equal (the Landau critical velocity u_L). Their calculations show that indeed there is a point on the $pE-$ plane where the ring ceases to exist and where $u_L = \partial E/\partial p$, but this point lies far from the roton minimum [29]. It remains to be seen whether the idea of the roton as a ghostly vortex ring will ever be vindicated. As one has a great variety of potentials that lead to the Landau dispersion curve one can tune the parameters so that the line $E = u_L p$, meets the $pE-$ curve for the family of the vortex rings, to allow this sequence of vortex rings to be terminated at a lower energy and momentum level. Whether this process will lead to coalescence with the roton minimum is not yet clear.

4.3 Travelling waves in the coupled Gross-Pitaevskii system

The simplest example of a multi-component system is a mixture of two different species of bosons, for instance, ^{41}K-^{87}Rb [37]. Since alkali atoms have spin, it is also possible to make mixtures of the same isotope, but in different internal spin states, for instance, for ^{87}Rb [38]. The multi-component

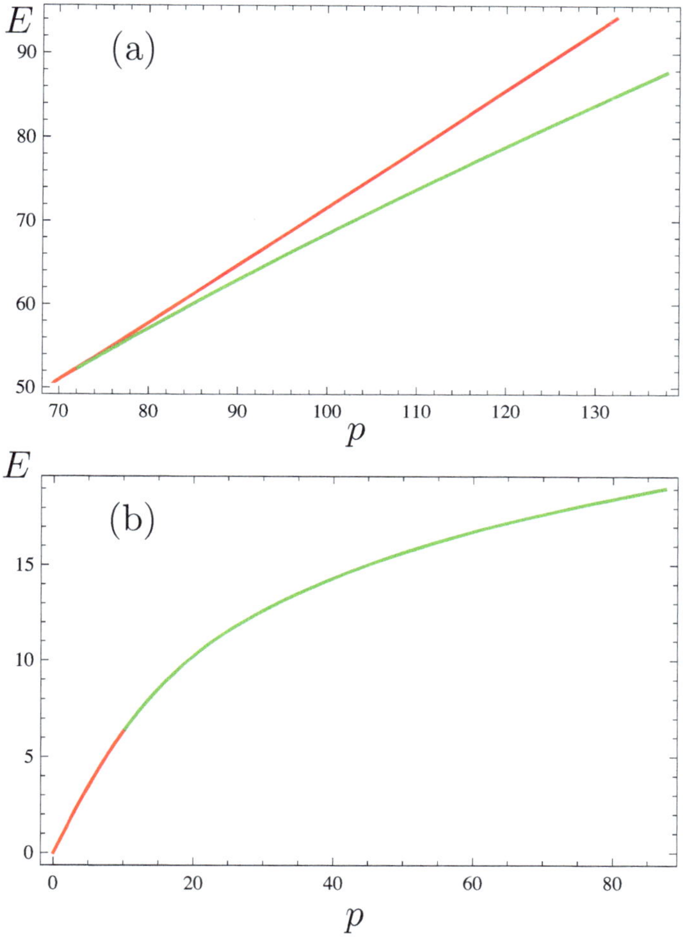

Figure 4.1.1. Energy E as a function of momentum p for three-dimensional (a) and two-dimensional (b) travelling coherent solutions of the GP equation (1.0.16). Rarefaction pulses are shown as dark lines, vortex rings (pairs) are shown as light lines.

BECs are far from being a trivial extension of a one-component BEC and present novel and fundamentally different scenarios for their excitations and ground state. For two components, described by the wave functions ψ_1 and ψ_2, with N_1 and N_2 particles respectively, the GP equations become

$$\begin{aligned} i\hbar\frac{\partial\psi_1}{\partial t} &= \left[-\frac{\hbar^2}{2m_1}\nabla^2 + V_{11}|\psi_1|^2 + V_{12}|\psi_2|^2\right]\psi_1, \\ i\hbar\frac{\partial\psi_2}{\partial t} &= \left[-\frac{\hbar^2}{2m_2}\nabla^2 + V_{12}|\psi_1|^2 + V_{22}|\psi_2|^2\right]\psi_2, \end{aligned} \tag{4.3.1}$$

where m_i is the mass of the atom of the ith condensate, and the coupling constants V_{ij} are proportional to scattering lengths a_{ij} via $V_{ij} = 2\pi\hbar^2 a_{ij}/m_{ij}$, where $m_{ij} = m_i m_j/(m_i + m_j)$ is the reduced mass. The energy functional of the system (e.g. [39]) is

$$E = \int\left[\sum_{i=1}^{2}\{\frac{\hbar^2}{2m_i}|\nabla\psi_i|^2 + \frac{1}{2}V_{ii}|\psi_i|^4\} + |\psi_1|^2 V_{12}|\psi_2|^2\right]d\mathbf{x}. \tag{4.3.2}$$

In this section we study the solitary wave solutions of the coupled GP model (4.3.1). Many solitary wave structures have been recently identified in two-component one-dimensional BECs [40] such as bound dark-dark, dark-bright, dark-antidark, dark-grey, etc. complexes. In higher dimensions, domain walls [41] and skyrmions (vortons) [42] have been found. To study the equilibrium properties the energy functional (4.3.2) has to be minimized subject to the constraints on the conservation of particles leading to introduction of two chemical potentials $\mu_1 = V_{11}n_1 + V_{12}n_2, \mu_2 = V_{12}n_1 + V_{22}n_2$, where $n_i = |\psi_i|^2$ is the number density in equilibrium. The dispersion relation between the frequency ω and the wave number k of the linear perturbations ($\propto \exp[i\mathbf{k}\cdot\mathbf{x} - i\omega t]$) around homogeneous states is obtained as

$$(\omega^2 - \omega_1^2)(\omega^2 - \omega_2^2) = \omega_{12}^4, \tag{4.3.3}$$

where $\omega_i^2(k) = c_i^2 k^2 + \hbar^2 k^4/4m_i^2$ coincides with a one-component Bogoliubov spectrum with the customary defined sound velocity $c_i^2 = n_i V_{ii}/m_i$ and $\omega_{12}^2 = c_{12}^2 k^2$ where $c_{12}^2 = n_1 n_2 V_{12}^2/m_1 m_2$. The system is dynamically stable if the spectrum (4.3.3) is real and positive which implies that $V_{11}V_{22} > V_{12}^2$, $V_{ii} > 0$ for stability [39]. The acoustic branches of Eq. (4.3.3) are $\omega_\pm \approx c_\pm k$ with the corresponding sound velocities

$$2c_\pm^2 = c_1^2 + c_2^2 \pm \sqrt{(c_1^2 - c_2^2)^2 + 4c_{12}^4}. \tag{4.3.4}$$

The solitary waves we seek below are all subsonic, so their velocity U is less than c_-.

A dimensionless form of (4.3.1) is obtained by introducing the chemical potentials $\psi_i \to \psi_i \exp[-i\mu_i t/\hbar]$ and using dimensionless units

$$\mathbf{x} \to \frac{\hbar}{(2m_1\mu_1)^{1/2}}\mathbf{x}, \qquad t \to \frac{\hbar}{2\mu_1}t, \qquad \psi_i \to \sqrt{\frac{\mu_1}{V_{11}n_i}}\psi_i. \tag{4.3.5}$$

This leads to the system of nonlinear Schrödinger equations

$$\begin{aligned} -2i\frac{\partial\psi_1}{\partial t} &= \nabla^2\psi_1 + (1 - |\psi_1|^2 - \alpha_1|\psi_2|^2)\psi_1 \\ -2i\frac{\partial\psi_2}{\partial t} &= \gamma\nabla^2\psi_2 + (1 - \alpha_1|\psi_1|^2 - \frac{\alpha_1}{\alpha_2}|\psi_2|^2 - \Lambda^2)\psi_2, \\ &\psi_1 \to \psi_{1\infty}, \quad \psi_2 \to \psi_{2\infty}, \quad as \quad |\mathbf{x}| \to \infty, \end{aligned} \tag{4.3.6}$$

where $\alpha_i = V_{12}/V_{ii}$, $\gamma = m_1/m_2$ and $\Lambda^2 = (\mu_1 - \mu_2)/\mu_1$ is the measure of asymmetry between chemical potentials (where we assume that $\mu_1 > \mu_2$). Note that the use of Feshbach resonances to vary the interactions between atoms makes this entire range of parameters experimentally accessible. The condition of dynamic stability becomes $\alpha_1\alpha_2 < 1$. To also ensure stability against collapse when only the density of one component is varied, the final stability criterion becomes

$$0 < \alpha_1\alpha_2 < 1. \tag{4.3.7}$$

The acoustic branches (4.3.4) are

$$c_\pm^2 = \frac{1}{2}\left(\psi_{1\infty}^2 + \frac{\alpha_1}{\alpha_2}\gamma\psi_{2\infty}^2 \pm \sqrt{(\psi_{1\infty}^2 - \frac{\alpha_1}{\alpha_2}\gamma\psi_{2\infty}^2)^2 + 4\alpha_1^2\gamma\psi_{1\infty}^2\psi_{2\infty}^2}\right). \tag{4.3.8}$$

To find solitary wave solutions moving with velocity U in positive z-direction, the following equations were solved [43]

$$\begin{aligned} 2iU\frac{\partial\psi_1}{\partial z} &= \nabla^2\psi_1 + (1 - |\psi_1|^2 - \alpha_1|\psi_2|^2)\psi_1 \\ 2iU\frac{\partial\psi_2}{\partial z} &= \gamma\nabla^2\psi_2 + (1 - \alpha_1|\psi_1|^2 - \frac{\alpha_1}{\alpha_2}|\psi_2|^2 - \Lambda^2)\psi_2, \\ &\psi_1 \to \psi_{1\infty}, \quad \psi_2 \to \psi_{2\infty}, \quad as \quad |\mathbf{x}| \to \infty. \end{aligned} \tag{4.3.9}$$

We are interested in the case when both condensates have nonzero uniform states. The values of the wave-functions of the solitary waves at infinity in (4.3.6) are given by

$$\psi_{1\infty}^2 = (1 - \alpha_2 + \alpha_2\Lambda^2)/(1 - \alpha_1\alpha_2), \tag{4.3.10}$$

$$\psi_{2\infty}^2 = \frac{\alpha_2}{\alpha_1}(1 - \alpha_1 - \Lambda^2)/(1 - \alpha_1\alpha_2), \tag{4.3.11}$$

so that we will assume that

$$\alpha_1 < 1 - \Lambda^2, \qquad \alpha_2 < \frac{1}{1-\Lambda^2}. \tag{4.3.12}$$

In a compact form the equations (4.3.9) can be written as

$$2iU\frac{\partial \psi_i}{\partial z} = \frac{m_1}{m_i}\nabla^2\psi_i + f_i(\psi_1,\psi_2)\psi_i, \quad i = 1,2 \tag{4.3.13}$$

where we used the notation

$$\begin{aligned} f_1(\psi_1,\psi_2) &= 1 - |\psi_1|^2 - \alpha_1|\psi_2|^2, \\ f_2(\psi_1,\psi_2) &= 1 - \alpha_1|\psi_1|^2 - \frac{\alpha_1}{\alpha_2}|\psi_2|^2 - \Lambda^2. \end{aligned} \tag{4.3.14}$$

Each solitary wave complex that belongs to a family of the solitary wave solutions for a chosen set of $(\alpha_1, \alpha_2, \gamma, \Lambda^2)$ is characterised by its velocity, U, vortex radii b_i, momenta $\mathbf{p}_i = (0, 0, p_i)$, and energy. The momentum of the i–th component [43] is

$$\mathbf{p}_i = \frac{1}{2i}\int [(\psi_i^* - \psi_{i\infty})\nabla\psi_i - (\psi_i - \psi_{i\infty})\nabla\psi_i^*]\, d\mathbf{x} \tag{4.3.15}$$

We form the energy, E, by subtracting the energy of an undisturbed system of the same mass for which $\psi_i = const$ everywhere, from the energy of the system with a solitary wave, so that the energy of the system becomes

$$\begin{aligned} E &= \frac{1}{2}\int \{|\nabla\psi_1|^2 + \gamma|\nabla\psi_2|^2 + \frac{1}{2}(\psi_{1\infty}^2 - |\psi_1|^2)^2 \\ &+ \frac{\alpha_1}{2\alpha_2}(\psi_{2\infty}^2 - |\psi_2|^2)^2\}\, d\mathbf{x} + \frac{\alpha_1}{2}\int \prod_{i=1}^{2}(\psi_{i\infty}^2 - |\psi_i|^2)\, d\mathbf{x}. \end{aligned} \tag{4.3.16}$$

Alternatively, we can write

$$E = \frac{1}{2}\int \sum \left\{\frac{m_1}{m_i}|\nabla\psi_i|^2 + \frac{1}{2}f_i(\psi_1,\psi_2)(\psi_{i\infty}^2 - |\psi_i|^2)\right\} d\mathbf{x}. \tag{4.3.17}$$

Several families of solitary wave complexes were found: (1) vortex rings of various radii in each of the components (VR-VR complexes), (2) a vortex ring in one component coupled to a rarefaction solitary wave of the other component (VR-RP and RP-VR complexes), (3) two coupled rarefaction waves (RP-RP complexes), (4) either a vortex ring or a rarefaction pulse coupled to a localised disturbance of a very low momentum (slaved wave)

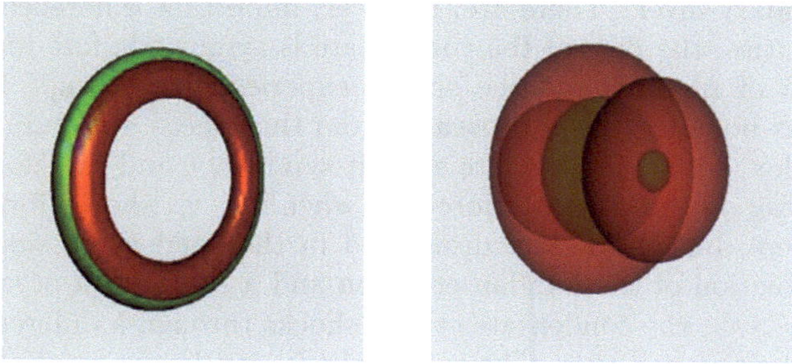

Figure 4.3.1. Isosurfaces of a vortex ring–vortex ring solution (left panel) and a rarefaction pulse – slaved wave solution (right panel).

(VR-SW, SW-VR, RP-SW, SW-RP complexes). Fig.4.3.1 shows the isosurfaces of VR-VR and RP-SW solutions.

5 Mechanisms of formation of travelling coherent structures

In the previous section we showed that various modifications of the GP equation have quite different travelling wave sequences. The goal of this section to discuss the ways in which these structures can be formed in superfluids. In spite of differences in the solutions the mechanisms of their formation are quite similar.

5.1 Critical superflow velocities

Vortex nucleation by an impurity such as the positive ion $^4He_2^+$ moving in superfluid helium at low temperature with velocity v has been studied experimentally and theoretically (see, e.g. [4]), and has uncovered some interesting physics. The flow round an ion that is moving with a sufficiently small velocity, v, is well represented by one of the classical solutions of fluid mechanics, namely the flow of an inviscid incompressible fluid around a sphere. In this solution, the maximum flow velocity, $\mathbf{u}$, relative to the sphere is $3v/2$, and occurs on the equator of the sphere (defined with respect to the direction of motion of the sphere as polar axis). Above some critical velocity, v_c, the ideal superflow around the ion breaks down, leading to the creation of a vortex ring. There is some similarity between the flow of the condensate past the ion and the motion of a viscous fluid past a sphere at large Reynolds numbers, the healing layer being the counterpart of the viscous boundary layer. There are, however, important differences. At subcritical velocities, the flow of the condensate is symmetric fore and aft of the direction of motion, and the sphere experiences no drag. In contrast, the viscous boundary layer separates from the sphere, so evading D'Alembert's paradox, destroying the fore and aft symmetry, and therefore bringing about a drag on the sphere. Moreover, when $v > v_c$, shocks form at or near the sphere, but shocks are disallowed in the condensate since they represent a violation of the Landau criterion and a breakdown of superfluidity. When $v > v_c$, the condensate evades shocks through a different mode of boundary layer separation. The sphere sheds circular vortex rings that move more slowly than the sphere and form a vortex street that trails behind it, maintained by vortices that the sphere sheds. As the velocity of the ion increases such a shedding becomes more and more irregular. Each ring is born at one particular latitude within the healing layer on the sphere. As it breaks away into the mainstream, it at first contributes a flow that depresses the mainstream velocity on the sphere below critical. As it moves further downstream however, its influence on the surface flow diminishes. The surface flow increases until it again reaches criticality, when a new ring is nucleated and the whole sequence is repeated. The vortex street trailing behind the ion creates a drag on the ion that decreases as the nearest vortex moves downstream, but which is refreshed when a new vortex is born.

Frisch et al. [44] and Winiecki et al. [45] have solved numerically the GP equation (1.0.5) for flow past a circular cylinder and Berloff and Roberts [46] for the flow around a positive ion and have confirmed the main features of the scenario just described. Below the critical velocity steady solutions of the GP model exist.

The formation of vortex rings and vortex pairs can be also understood mathematically by considering when the hydrodynamical equations change their type [47]. From the Madelung transformations $\psi = R\exp[iS]$ of the GP equation one gets the following hydrodynamical equations for the number density $n = R^2$ and the phase $\phi = \hbar S/m$ for the superflow with $\phi = u_\infty x$ as $x^2 + y^2 \to \infty$

$$\frac{\partial n}{\partial t} + \nabla \cdot (n \nabla \phi) = 0\,,$$
$$\frac{\partial \phi}{\partial t} + \frac{1}{2}|\nabla \phi|^2 - \frac{1}{2}u_\infty^2 + c^2(\frac{n}{n_\infty} - 1) = c^2 a^2 \frac{\nabla^2 n^{1/2}}{n^{1/2}}. \qquad (5.1.1)$$

Consider a stationary flow and neglect the quantum pressure term. We fix a point outside of the disk at which the components move with velocities $\mathbf{u}$, introduce the local orthogonal coordinates such that the $x-$axis is tangent to the flow and expand ϕ in the neighbourhood of this point as $\phi \approx ux + \tilde{\phi}$.

Linearising for small $\tilde{\phi}$ gives

$$\partial_u(n(u)u)\partial_{xx}\tilde{\phi} + n(u)\partial_{yy}\tilde{\phi} = 0.$$

At low velocities this equation is elliptic and becomes hyperbolic beyond a critical velocity. This happens when $\partial_u(n(u)u) = 0$ on the equator of the disk.

For two-component condensates [48] the nucleation condition was obtained in a similar fashion and becomes

$$\frac{\partial(n_1 u_1)}{\partial u_1}\frac{\partial(n_2 u_2)}{\partial u_2} < u_1 u_2 \frac{\partial n_1}{\partial u_2}\frac{\partial n_2}{\partial u_1}. \tag{5.1.2}$$

Let us consider the motion around a disk in 2D. At subcritical velocity ($U < 0.225$), the flows of the condensates are symmetric fore and aft of the direction of motion, and the disk experiences no drag. When the condition (5.1.2) is satisfied, which happens first on the disk equator where the velocities are maximal, the condensates evade shocks through a boundary layer separation. Fig. 5.1.1 shows the emission of various complexes for the disk of the radius 10 healing lengths that moves with supercritical velocity $U = 0.28$. The disk sheds SW-VP, VP-SW and VP-VP complexes in the order and frequency that depends on the value of the disk's velocity. These complexes move more slowly than the disk and form a vortex wave street that trails behind it, maintained by other complexes that the disk sheds. As the velocity of the disk increases such a shedding becomes more and more irregular. Each complex is born at one particular latitude within the healing layer on the disk. As it breaks away into the mainstream, it at first contributes a flow that depresses the mainstream velocities on the disk below critical. For larger values of the disk velocity ($U > 0.265$), more energy is required for this depression and VP-VP complex that has larger energy than SW-VP complexes is born first. At low supercritical velocities SW-VP complex is born first. As it moves further downstream however, its influence on the surface flow diminishes. The surface flow increases until it again reaches criticality, when a new complex is nucleated and the whole sequence is repeated. The vortex and slaved wave street trailing behind the disk creates a drag on the disk that decreases as the nearest complex moves downstream, but which is refreshed when a new complex is born. The complexes downstream of the disk move with different velocities and interact among themselves. These interactions may lead to a transformation from one type of the complex to another; Fig. 5.1.1 shows the splitting of the VP-VP complex into SW-VP and VP-SW complexes. The mechanism in which solitary waves transfer energy from one to another is discussed in Sec. 5.3 below.

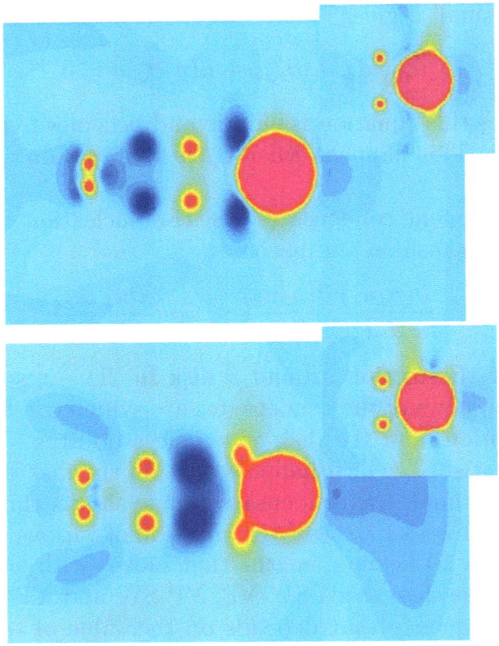

Figure 5.1.1. The time snapshots of $|\psi_1|^2$ (top) and $|\psi_2|^2$ (bottom) of the solution of Eq. (4.3.9) with $-2i\psi_{it}$ added to the left hand sides with $\alpha = 0.5$, $\Lambda^2 = 0.1$ for the flow around a disk of radius 10 moving to the right with velocity $U = 0.28$. The solitary wave street is seen in the wake of the disk. The complexes were emitted in the order VP-VP, SW-VP, VP-SW, and SW-VP complex has just got emitted from the disk boundary. For this large supercritical velocity the VP-VP complex is first nucleated from the surface of the disk and the insets show this moment at an earlier time. On the main panels this complex is in the process of splitting into the VP-SW and SW-VP complexes. Only parts of the computational box are shown.

5.2 Transverse "snake" instability

The velocity of dark solitons in the GP equation in 1D depends on the density at the minimum as we previously discussed. This implies [49] that if a dark soliton is placed in dimension greater than one it breaks due to transverse perturbation along the front and evolves into vortex ring/pairs or other travelling wave structures.

5.3 Energy transfer

The vortex pairs and vortex rings can appear as a result of an interaction among the solitary wave solutions of the GP equation as was shown in [14]. The interactions between various, even vortex-free, solitary waves result in energy and momentum transfer that can lead to vortex nucleation. Rarefaction pulses on the lower branch of the dispersion curve have lower energy and momentum than vortex rings, therefore, such rarefaction pulse may evolve into a vortex ring if interactions with other solutions add enough energy and momentum to the rarefaction pulse. This is illustrated by placing two rarefaction pulses a distance $10a$ apart that move in the same direction. The effect two solitary waves have on each other is non-symmetric. As a result, the rarefaction pulse moving behind transfers part of its energy and momentum to the pulse moving at front, so that the latter transforms into a vortex ring and slows down, whereas the former spreads out and speeds up. This process leads to an even closer interaction of the two solitary waves and an even more rapid transfer of energy from the solitary wave that moves behind to the one moving at front. Eventually almost all of the energy and momentum of the former is transferred to the latter, which becomes a vortex ring of energy and momentum that are only slightly less than twice the energy and momentum of each of the initial rarefaction pulses. The remaining small energy is emitted as sound waves. Figure 5.3.1 gives the graphical illustration of this process through the snapshots of the density cross-sections.

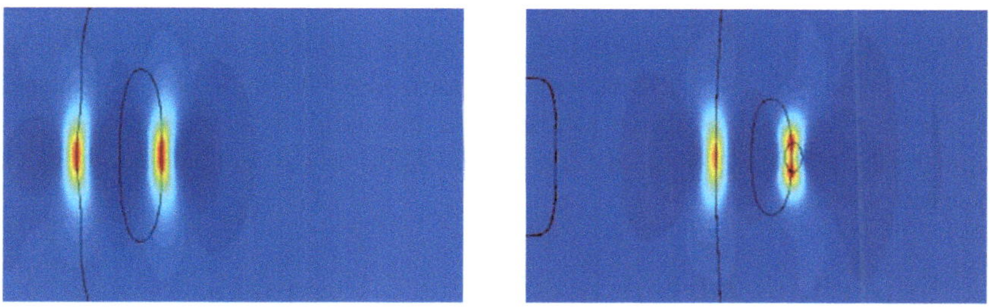

Figure 5.3.1. The snapshots of the contour plots of the density cross-section of a condensate obtained by numerically integrating the GP model (1.0.16) at $t = 6$ (left panel) and at $t = 26$ (right panel). Initial condition consists of two rarefaction pulses on the lower branch of the Ep cusp. Black solid lines show zeros of real and imaginary parts of ψ, therefore, their intersection shows the position of topological zeros. Both low and high density regions are shown in darker shades to emphasise intermediate density regions. Only a portion of an actual computational box is shown.

5.4 Collapsing ultrasound bubbles

Experiments in superfluid helium have demonstrated the production of quantised vortices and turbulence [50] by the collapse of cavitated bubbles [51] generated by ultrasound in the megahertz frequency range.

The time-dependent evolution of the condensate during and after collapse of the bubble involves several stages that were studied in [52]. During the first stage dispersive and nonlinear wave trains are generated at the surface of the collapsed bubble. The Fourier components propagate at different velocities generating wave packets moving in opposite directions. This stage of the evolution is characterised by a flux of particles towards the centre of the cavity, while oscillations of growing amplitude are being formed on the real and imaginary parts of the wave function and the slope of the steep density front is getting smaller. There is a moment of time when the density at the centre of what used to be a cavity reaches the maximum value. This moment indicates the start of a qualitatively new stage of the evolution in which there is an outward flux of particles as the condensate that overfilled the cavity begins to expand. The radial density depletions propagate away from the centre and this is when an instability sets in leading to formation of vortex rings and rarefaction pulses, as Fig. 5.4.1 illustrates. Note the growth of the vortex rings as they move away from the centre. This is due to the mechanism of energy transfer that we discussed in the previous section.

5.5 BEC formation

Finally, the localized disturbances, vortex rings, rarefaction pulses are formed during the BEC formation from a strongly degenerate gas of weakly interacting bosons. As this topic bring us immediately to the discussion of superfluid turbulence we shall discuss the BEC formation in some detail in the next section.

6 Superfluid turbulence in the GP equation

An important and often overlooked feature of the GP equation is that it gives an accurate microscopic description of the formation of BEC from the strongly degenerate gas of weakly interacting bosons [53, 54]. In terms of the coherent-state formalism, it was demonstrated that if the occupation numbers are large and somewhat uncertain, then the system evolves as an ensemble of classical fields with corresponding classical-field action leading, in our case to the GP equation. It is important to emphasise that in the context of the strongly non-equilibrium BEC formation kinetics the con-

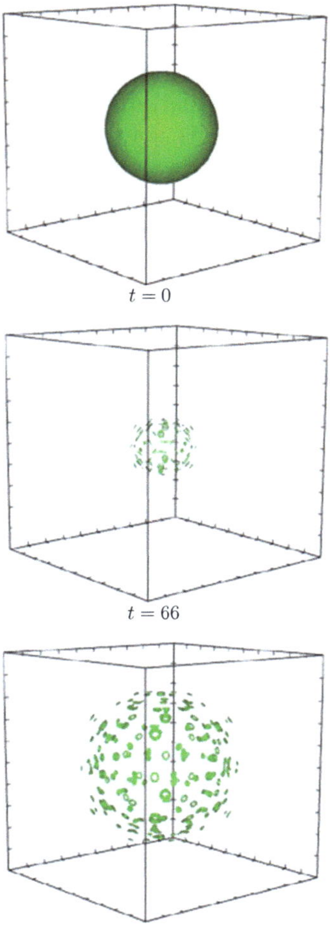

Figure 5.4.1. Time snapshots of the density isoplots $\rho = 0.2$ after the collapse of the bubble of radius $a = 50$ centred at the origin. The side of the computational box is 300 healing lengths and the distance between ticks on the side corresponds to 30 healing lengths. The vortex rings nucleate at about $t \sim 50$.

dition of large occupation numbers is self-consistent: the evolution leads to an explosive increase of occupation numbers in the low-energy region of wavenumber space [55] where the ordering process takes place. Even if the occupation numbers are of order unity in the initial state, so that the classical matter field description is not yet applicable, the evolution that can be described at this stage by the standard Boltzmann quantum kinetic equation inevitably results in the appearance of large occupation numbers in the low-energy region of the particle distribution. The blow-up scenario [55] indicates that only low-energy part of the field is initially involved in the process. Therefore, one can switch from the kinetic equation to the matter field description for the long-wavelength component of the field at a certain moment in the evolution when the occupation numbers become appropriately large.

The formation of the large-scale coherent localized ground state (condensate) from a non-equilibrium initial state has been studied in a number of papers addressing different stages of the formation: weak turbulence [56, 55], strong turbulence in the long-wavelength region of energy space [57], and finally, the formation of a genuine condensate [58, 59]. The superfluid turbulence and evolution of vortex tangle has been extensively studied in [60, 61]. The related question about the effect of finite temperature on the BEC dynamics has also been addressed recently [62].

Unambiguous demonstration of the formation of the state of superfluid turbulence in the course of self-evolution of weakly interacting Bose gas was performed in [57]. The formation of a tangle of well-separated vortices and the decay of superfluid turbulence as well as the formation of a vortex ring are clearly seen on Fig. 6.0.1.

When the turbulent tangle of vortices decays the system reaches a state of thermodynamical equilibrium. Number of particles in condensate depends on the total number density $\rho = N/V$ and the energy density of the system These relations were found in [59] by adapting the Bogoliubov theory of a weakly interacting Bose gas [63] to the classical nonlinear Schrödinger equation

$$i\partial_t\psi = -\nabla^2\psi + |\psi|^2\psi. \tag{6.0.1}$$

The dynamics conserves the total number of particles $N = \int |\psi|^2 d\mathbf{x}$, and the total energy $E = \int \left(|\nabla\psi|^2 + \frac{1}{2}|\psi|^4\right) d\mathbf{x}$. In terms of the Fourier amplitudes $a_{\mathbf{k}}(t) = \int \psi(\mathbf{x},t)e^{-i\mathbf{k}\cdot\mathbf{x}}\, d\mathbf{x}$ the total energy is written as

$$E = \sum_k k^2 a_{\mathbf{k}}^* a_{\mathbf{k}} + \frac{V_0}{2V}\sum_{1234} a_{\mathbf{k}_1}^* a_{\mathbf{k}_2}^* a_{\mathbf{k}_3} a_{\mathbf{k}_4} \delta_{\mathbf{k}_1+\mathbf{k}_2-\mathbf{k}_3-\mathbf{k}_4}. \tag{6.0.2}$$

The Bogoliubov transformation [63]

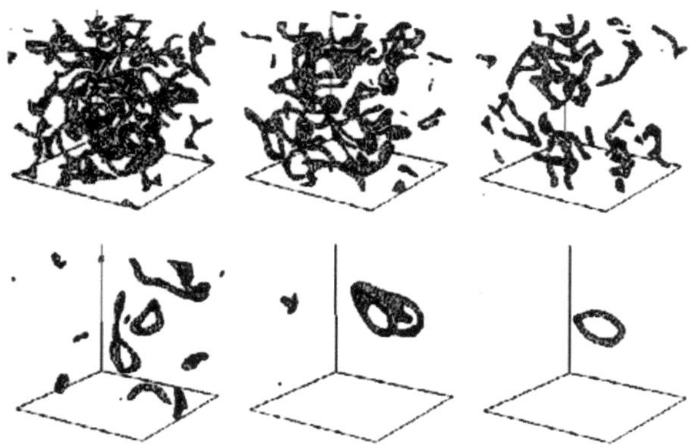

Figure 6.0.1. Evolution of topological defects in the phase of the long-wavelength part of the field ψ in the computational box 128^3.

$$\begin{pmatrix} a_{\mathbf{k}} \\ a^*_{-\mathbf{k}} \end{pmatrix} = \begin{pmatrix} u_+ & u_- \\ u_- & u_+ \end{pmatrix} \begin{pmatrix} b_{\mathbf{k}} \\ b^*_{-\mathbf{k}} \end{pmatrix}, \qquad u_\pm = (1 \pm \Gamma_k^2)/2\Gamma_k, \qquad \Gamma_k = \sqrt{\frac{k^2}{\omega(k)}}$$

is used to diagonalise the quadratic (in $k \neq 0$) term to

$$\sum_{k\neq 0} \omega(k) b^*_{\mathbf{k}} b_{\mathbf{k}}, \qquad \omega(k) = \sqrt{k^4 + 2V_0\rho_0 k^2}, \qquad \rho_0 = n_0/V,$$

where $\omega(k)$ is the Bogoliubov dispersion relation that takes into account nonlinear interactions and $n_0 \equiv |a_0|^2$ is the number of particles in condensate. The particle number density is

$$\rho = N/V = \sum_{\mathbf{k}} \langle a^*_{\mathbf{k}} a_{\mathbf{k}} \rangle = \rho_0 + \sum_{k\neq 0} (u_+^2 + u_-^2) \langle b^*_{\mathbf{k}} b_{\mathbf{k}} \rangle.$$

The equilibrium distribution of non-condensed particles is found from the kinetic equation modified by the presence of condensate: $n_k^{eq} = T/\omega(k)$ where T is a temperature. In the new basis, the uncondensed mass and energy density become

$$\rho - \rho_0 = \frac{T}{V} \sum_{k\neq 0} \frac{k^2 + V_0\rho_0}{\omega^2(k)}.$$

$$\frac{E}{V} = \frac{1}{2} V_0 \left[\rho^2 + (\rho - \rho_0)^2\right] + \frac{T}{V} \sum_{k\neq 0} 1.$$

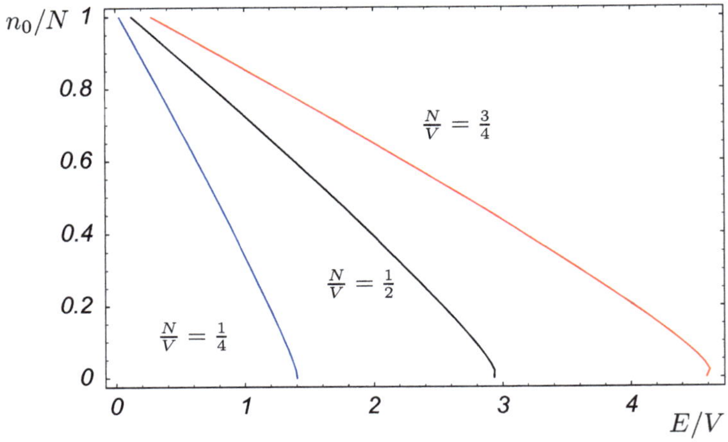

Figure 6.0.2. The proportion of condensed particles to the total number of particles in the system as a function of the energy density for various total number densities.

T/V can be eliminated to yield the expression for the density of the condensed particles as function of the energy density for given total number density.

In Sec.1 we discussed that the vortex lines create a force of mutual friction between superfluid and normal fluid in addition to the mutual friction included by Landau in his equations. We can use the GP equations to represents the effects of collisions of the quasiparticles with the vortex cores. In particular, the vortex line decay law at non-zero temperature was found in [64] in the context of the defocusing NLS equation by inserting a vortex ring into a state of thermal equilibrium and following its decay due to the interactions with the non-condensed particles. The result agrees with predictions of the HVBK theory for superfluid helium [5] according to which the fundamental equation of the motion of a vortex line, $\mathbf{v}_L$, is given by (see also page 90, Eq. (3.17) of [4])

$$\mathbf{v}_L = \mathbf{v}_{sl} + \alpha \mathbf{s}' \times (\mathbf{v}_n - \mathbf{v}_{sl}) - \alpha' \mathbf{s}' \times [\mathbf{s}' \times (\mathbf{v}_n - \mathbf{v}_{sl})], \tag{6.0.3}$$

where $\mathbf{v}_{sl}$ is the local superfluid velocity that consists of the ambient superfluid flow velocity and the self-induced vortex velocity $\mathbf{u}_i$, $\mathbf{v}_n$ is the normal fluid velocity, $\mathbf{s}$ is a position vector of a point on the vortex and $\mathbf{s}'$ is the unit tangent at that point. Mutual friction parameters α and α' are *ad hoc* coefficients in the HVBK theory that are functions of ρ_n, ρ, and T only and are determined numerically from the GP equation as $\alpha \sim \rho_n T/\rho_s T_\lambda$ and $\alpha' \approx 0$.

Self-evolution of two Bose gases from a strongly non - equilibrium initial state has been studied in [65]. It is shown that the large wavelength part of the fields evolves into coherent structures identified as travelling coherent complexes of the coupled nonlinear Schrödinger equations discussed in Sec. 4.3. The evolution of the system is reminiscent of the Kibble-Zurek scenario [66] of a formation of the topological defects when the system is quickly quenched below the point of the second-order transition. This process would correspond to the formation of the cosmological *vortons* and *springs* that are analogous to the vortex ring-slaved wave and vortex ring-vortex ring complexes correspondingly [43]. Figure 6.0.3 shows the evolution fo the long wavelength part of the field and formation of non-axisymmetric VR-VR complex. Collisions with non-condensed quasiparticles lead to the disssipation of the complex and its complete disappearance as the system reaches thermal equilibrium.

Bibliography

[1] L. Landau, *J. Phys. USSR* **5**, 71 (1941).

[2] C.F. Barenghi, D.C. Samuels, G.H. Bauer, and R.J. Donnelly, *Phys. Fluids* **9**, 2631 (1997).

[3] C.F. Barenghi, C.J. Swanson, and R.J. Donnelly, *J. Low. Temp. Phys.* **100**, 385 (1995).

[4] R.J. Donnelly, *Quantized Vortices in Helium II.* Cambridge University Press, 1991.

[5] H.E. Hall and W.F.Vinen, *Proc. R. Soc. Lond. A* **238**, 215 (1956).

[6] I.L. Bekharevich and I.M. Khalatnikov, *Sov. Phys., JETP*, **13**, 643 (1961).

[7] R.N. Hills and P.H. Roberts, *Archiv. Rat. Mech. & Anal.* **66**, 43 (1977); *Int. J. Eng. Sci.* **15**, 305 (1977); *J. Low Temp. Phys.* **30**, 709 (1978); *J. Phys. C* **11**, 4485 (1978).

[8] C.F. Barenghi and C.A. Jones, *Phys. Lett. A* **122**, 425 (1987); *J. Fluid Mech.* **197**, 551 (1988).

[9] C.E. Swanson and R.J. Donnelly *J. Low Temp. Phys.* **67**, 185 (1987).

[10] P.G. Saffman, *Vortex Dynamics*, Cambridge University Press, 1992.

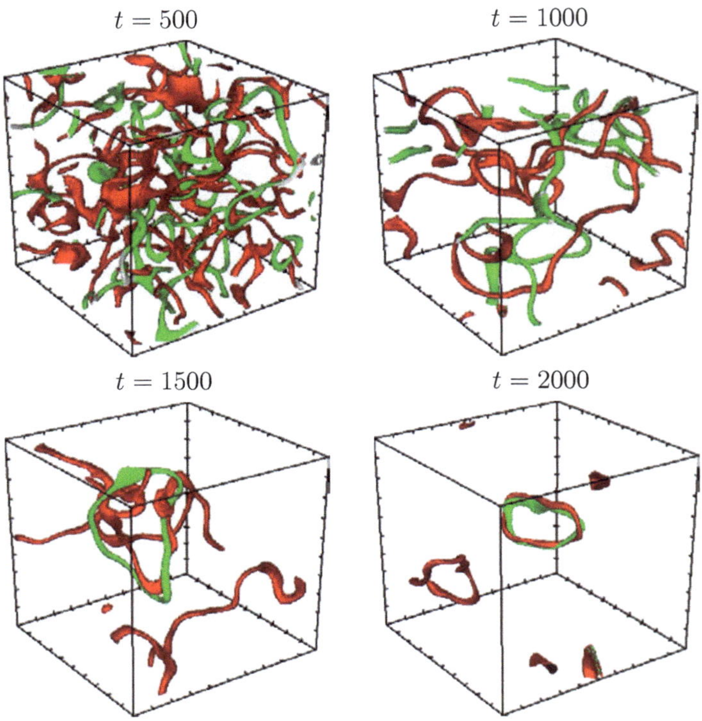

Figure 6.0.3. Evolution of topological defects in the long-wavelength parts $\tilde{\psi}_i$, with $i = 1$ (light grey) and $i = 2$ (dark grey) of the fields ψ_i in the computational box 128^3. The defects are visualised by isosurfaces $|\tilde{\psi}_i|^2 = 0.04\langle|\tilde{\psi}_i|^2\rangle$. High-frequency spatial waves are suppressed by the factor $\max\{1 - k^2/k_c^2, 0\}$, where the cut-off wave number is chosen according to the phenomenological formula $k_c = 9 - t/1000$.

[11] M. Leadbeater M., T. Winiecki, D.C. Samuels, C.F. Barenghi, and C.S. Adams, *Phys. Rev. Letts.* **86**, 1410 (2001); N.G. Berloff, *Phys. Rev. A* **69**, 053601 (2004).

[12] V.L. Ginzburg and L.P. Pitaevskii, *Zh. Eksp.Teor. Fiz.* **34**, 1240 (1958); [*Sov. Phys. JETP* **7**, 858 (1958)]; E.P. Gross, *Nuovo Cimento* **20**, 454 (1961); L.P. Pitaevskii, *Soviet Physics JETP*, **13**, 451 (1961).

[13] L.P. Pitaevskii, *Sov. Phys. JETP* **13**, 451 (1961).

[14] N.G. Berloff, *J. Phys. A: Math. Gen.* **37**, 1617 (2004).

[15] A.L. Fetter, in *Lectures in Theoretical Physics*, ed. by K.T. Mahanthappa and W.E. Brittin (Gordon and Breach, New York, 1969), Vol. XIB, p.351.

[16] P.H. Roberts, *Proc. R. Soc. Lond. A* **459**, 597 (2003).

[17] P. Mason, N.G. Berloff, and A.L.Fetter, *Phys. Rev. A* **74** , 043611 (2006).

[18] C.A. Jones and P.H. Roberts, *J. Phys. A: Gen. Phys.* **15**, 2599 (1982).

[19] B.Y. Rubinstein and L.M. Pismen, *Physica D* **78**, 1 (1994).

[20] Y.S.Kivshar, J. Christou, V. Tikhonenko, B. Luther-Davies, and L.M. Pismen, *Optics Comm.* **152**, 198 (1998).

[21] L.M. Pismen, *Vortices in Nonlinear Fields.* Clarendon Press, Oxford, 1999.

[22] J.R. Anglin, *Phys. Rev. A* **65**, 063611 (2002).

[23] E. Lundh, C.J. Pethick, and H. Smith, *Phys. Rev. A* **55**, 2126 (1997).

[24] D.S. Naik, S.R. Muniz, and C. Raman, *Phys. Rev. A* **72**, 051606(R) (2005).

[25] P. Mason and N.G. Berloff, submitted, *Phys. Rev. A* (2007).

[26] J.B. Brooks and R.J. Donnelly, *J. Phys. Chem. Ref. Data* **6**, 51 (1977).

[27] Y. Pomeau and S. Rica *Phys. Rev. Lett.* **71**, 247 (1993).

[28] N.G. Berloff, *J. Low Temp. Phys.* **116**, 359 (1999).

[29] N.G. Berloff and P.H. Roberts, *J. Phys. A* **32**, 5611 (1999).

[30] F. Dalfovo, A. Lastri, L. Pricaupenko, S. Stringari, and J. Treiner, *Phys. Rev. B* **52**, 1193 (1995).

[31] M. Sadd, G.V. Chester, and L. Reatto, *Phys. Rev. Lett.* **79**, 2490 (1997).

[32] N.G. Berloff and P.H. Roberts, *Phys. Lett. A* **274**, 69 (2000).

[33] F. Ancilotto, F. Dalfovo, L.P. Pitaevskii, and F. Toigo, *Phys. Rev. B* **71**, 104530 (2005).

[34] T. Tsuzuki, *J. Low Temp. Phys.* **4** 441 (1971).

[35] P.H. Roberts and J. Grant, *J. Phys. A: Gen. Phys.* **4**, 55 (1971).

[36] N.G. Berloff and P.H. Roberts, *J. Phys. A: Math. Gen.* **37**, 11333 (2004).

[37] C.J. Myatt *et al.*, *Phys. Rev. Lett.* **78**, 586 (1997); D.S.Hall *et al.*, *Phys. Rev. Lett.* **81**, 1539 (1998); D.M. Stamper-Kurn *et al.*, *Phys. Rev. Lett.* **80**, 2027 (1998); J. Stenger, *et al.*, *Nature* **396**, 345 (1998).

[38] G.Modugno *et al.*, *Phys. Rev. Lett.* **89**, 190404 (2002).

[39] C.J. Pethick and H. Smith, *Bose-Einstein Condensation in Dilute Gases.* Cambridge University Press, 2002.

[40] P. Öhberg and L. Santos, *Phys. Rev. Lett.* **86**, 2918 (2001); Th. Busch and J.R. Anglin, *Phys. Rev. Lett.* **87**, 010401 (2001); P.G. Kevrekidis *et al.*, *Europhys. J. D* **28**, 181 (2004).

[41] S. Coen and M. Haelterman, *Phys. Rev. Lett.* **87**, 140401 (2001).

[42] J. Ruostekoski and J.R. Anglin, *Phys. Rev. Lett.* **86**, 3934 (2001); C.M. Savage and J. Ruostekoski, *Phys. Rev. Lett.* **91**, 010403 (2003); R.A. Battye, N.R. Cooper, and P.M. Sutcliffe, *Phys. Rev. Lett.* **88**, 080401 (2002).

[43] N.G. Berloff, *Phys. Rev. Lett.* **94** , 120401 (2005).

[44] T. Frisch, Y. Pomeau, and S. Rica, *Phys. Rev. Lett.* **69** 1644 (1992).

[45] T. Winiecki, J. F. McCann, and C. S. Adams, *Phys. Rev. Letts* **82**, 5186 (1999).

[46] N.G. Berloff, and P. H. Roberts, *J. Phys.: Math. Gen.* **33**, 4025 (2000).

[47] C. Josserand and Y. Pomeau, *Nonlinearity* **14**, R25 (2001).

[48] N.G.Berloff, Arxiv preprint cond-mat/0412743 (2004).

[49] E.A. Kuznetsov and J. J. Rasmussen, *Phys. Rev. E* **51** 4479 (1995).

[50] R.F. Carey, J.A. Rooney, and C.W. Smith, *Physics Letters A* **65**, 311 (1978); K.W. Schwarz and C.W. Smith, *Physics Letters A* **82** 251 (1981).

[51] R.D. Finch, R. Kagiwada, M. Barmatz, and I. Rudnick, *Phys. Rev. A* **134** 1425 (1964).

[52] N.G. Berloff and C.F.Barenghi, *Phys. Rev. Letts.* **93** 090401(2004).

[53] E. Levich and V. Yakhot, *J. Phys. A: Math. Gen.* **11**, 2237 (1978).

[54] Yu. Kagan and B.V. Svistunov, *Phys. Rev. Lett.* **79**, 3331 (1997).

[55] B. V. Svistunov, *Phys. Rev. B* **52**, 3647 (1995).

[56] V.E.Zakharov, S.L. Musher, and A.M.Rubenchik, *Phys. Rep.* **129**, 285 (1985); S. Dyachenko, A.C. Newell, A. Pushkarev, and V.E.Zakharov, *Physica D* **57**, 96 (1992).

[57] N.G. Berloff and B.V. Svistunov, *Phys. Rev. A* **66**, 013603 (2002)

[58] M.J.Davis, S.A. Morgan, and K. Burnett, *Phys. Rev. Lett* **87**, 160402 (2001); *Phys. Rev. A* **66**, 053618 (2002).

[59] C. Connaughton *et al.*, *Phys. Rev. Lett.* **95**, 263901 (2005).

[60] C. Nore, M. Abid, and M. E. Brachet, *Phys. Fluids* **9**, 2644 (1997)

[61] M. Kobayashi and M. Tsubota, *Phys. Rev. Lett.* **94**, 065302 (2005).

[62] M. Brewczyk *et al.*, *J. Phys. B: At. Mol. Opt. Phys.* **40**, R1 (2007) and reference therein.

[63] N.N. Bogoliubov, *J. Physics* **11**, 23 (1947).

[64] N.G. Berloff and A.J. Youd, *Phys. Rev. Lett.* **99**, 145301 (2007).

[65] N.G. Berloff and C. Yin, *J. Low Temp. Phys.* **145**, 187 (2006).

[66] T.W.B. Kibble, *J. Phys. A* **9**, 1387 (1976); W. H. Zurek, *Nature* **317**, 505 (1985).

Turbulence Experiments in Superfluid ^{3}He at Very Low Temperatures

Shaun N. Fisher

Department of Physics, Lancaster University, Lancaster LA1 4YB, U.K.

Abstract Over the past few years, the experimental study of superfluid turbulence has made very significant advances following the discovery of turbulence in the B-phase of superfluid ^{3}He. Experiments in superfluid ^{3}He are far more cryogenically challenging than in ^{4}He since the temperatures required are of order a thousand times lower. However, ^{3}He has some advantages. The main advantages come from the ease of vortex detection at low temperatures. In particular, vortices in ^{3}He have a large cross-section for Andreev Scattering quasiparticle excitations. This property along with ultra sensitive quasiparticle detection techniques, allows superfluid turbulence experiments even in the very low temperature limit. In this chapter, we first give a brief background to the cryogenic techniques used and the basic physics underlying superfluid ^{3}He. We then describe in more detail vibrating wire techniques which are used for all the turbulence experiments in the low temperature limit. Finally we discuss the low temperature superfluid ^{3}He experiments which have been made to date and compare the findings with observations in superfluid ^{4}He and with the ideas of classical turbulence.

1 Refrigeration and Nuclear Cooling

1.1 Low temperature laboratories

The general layout of a laboratory for Ultra Low Temperature (ULT) measurements, such as those at Lancaster are shown in figure 1.1.1. At Lancaster, the refrigerators are purpose-built for the nuclear demagnetisation of small quantities of copper, as described in references (1; 2; 3). The main part of the refrigeration process is provided by a so-called dilution refrigerator (described below). This is housed in a large 'dewar' (figure 1.1.2) which has vacuum layers for thermal isolation, often encompassing a liquid nitrogen layer acting as a thermal shield. The dilution refrigerator is also in a Inner Vacuum Can (the 'IVC') suspended within the dewar and immersed in liquid helium at 4.2 K.

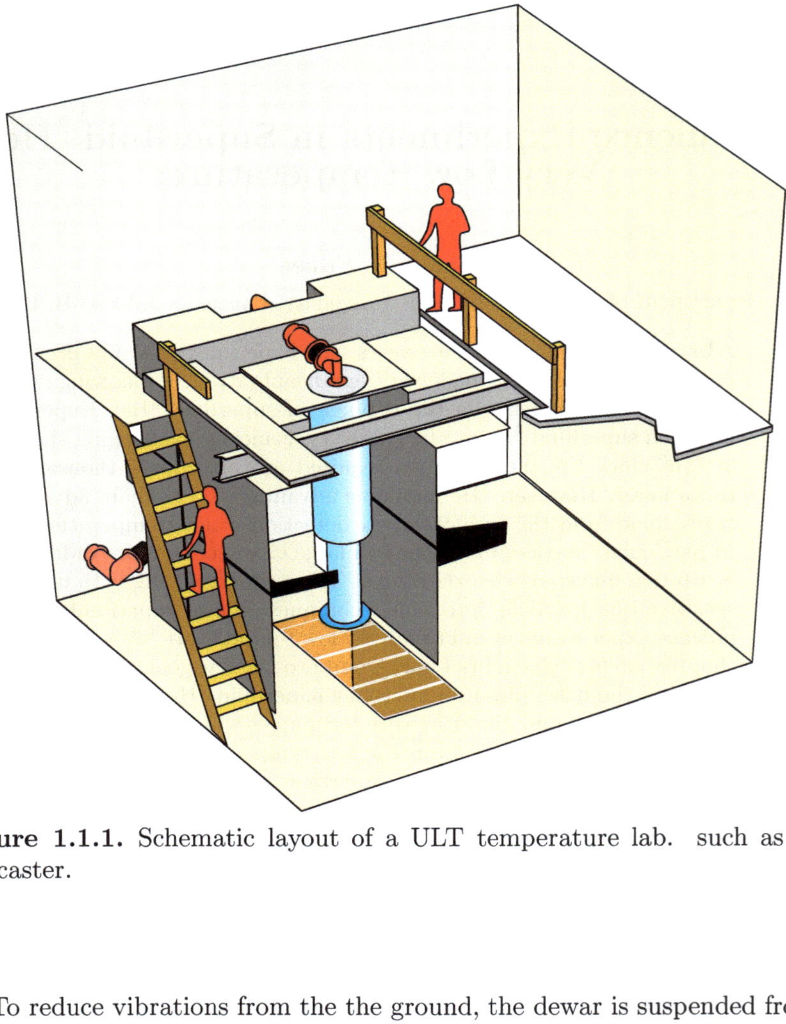

Figure 1.1.1. Schematic layout of a ULT temperature lab. such as in Lancaster.

To reduce vibrations from the the ground, the dewar is suspended from a large block of concrete (~ 50 tons) which is floated off the floor by air springs. Below this lies another block of concrete ~ 20 tons) embedded in the ground. All this is housed in a large screened room to prevent unwanted heating and interference from external EM (radio) signals.

The pumps which operate the refrigerator are located in a separate room. All of the adjoining pumping lines are firmly anchored through a concrete wall outside the screened room. Rubber sections in the pumping lines reduce vibrations transmitted along the pipes to the cryostat.

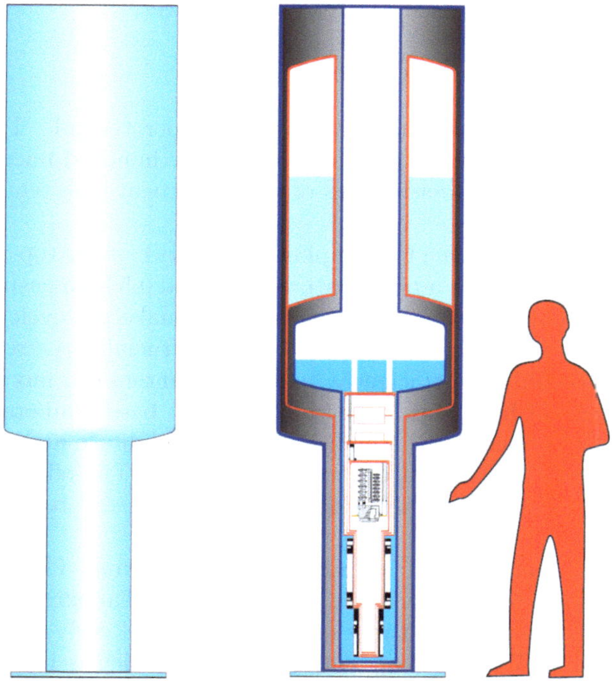

Figure 1.1.2. Schematic diagram of the dewar which houses a nuclear cooling machine as used in the newest Lancaster refrigerator

1.2 The Dilution Refrigerator

Figure 1.2.1 shows a schematic diagram of a dilution refrigerator. It operates by circulating ^{3}He across a phase boundary formed in a liquid mixture of ^{3}He and ^{4}He located in the 'mixing chamber' at the bottom of the fridge. At low temperatures a ^{3}He /^{4}He mixture phase separates into a concentrated ($\sim 100\,\%$) ^{3}He phase which floats on top of a heavier dilute ($\sim 6\,\%$) ^{3}He phase. The entropy of the dilute phase is lower, so if ^{3}He atoms are forced to flow across the boundary into the dilute phase then they require a Latent heat (analogous to evaporation) resulting in cooling. The latent heat per mole is simply $L = T\Delta S$ where $\Delta S = S_{dil} - S_{conc}$ is the molar entropy difference of the ^{3}He in the dilute and concentrated phases. These are both normal degenerate Fermi liquids which have $S \sim R(T/T_F)$ where T_F is the Fermi temperature. Inserting the relevant numbers then gives the following expression for the cooling power:

$$\dot{Q} = \dot{n}_3 L = 84 \dot{n}_3 T^2 \tag{1.2.1}$$

where $\dot{n}_3$ is the ^{3}He circulation rate in moles per second. Typically the circulation rate of a dilution refrigerator is a few hundred micro-moles per second giving cooling powers of tens of micro-watts at base temperatures of typically 10 mK.

The circulation is driven by pumping on the 'still' at the top of the fridge. The still is maintained at a temperature of around 0.5 K to ensure that only ^{3}He is evaporated from the dilute mixture. To make the cycle continuous, the ^{3}He is returned to the concentrated phase. Crucially, the returning ^{3}He must be pre-cooled very efficiently before it re-enters the mixing chamber (or else there will be a high heat leak and poor base temperature). This is achieved by various heat exchanges between the dilute and concentrated phases between the still and the mixing chamber. The efficiency of these heat exchangers ultimately determines the heat leak and base temperature of the fridge.

The newest Lancaster refrigerator is a record 1.7 mK. The low temperatures are achieved by having a large number of step heat exchangers(3). The mixing chamber is made entirely of plastic to avoid eddy-current heating from the main $\sim$ 7 T superconducting solenoid, which is immersed in the main bath of liquid ^{4}He at 4 K. The plastic experimental cells are attached to the bottom of the mixing chamber such that the top of the cell assembly forms the base of the mixing chamber. A super-leak tight seal is made with a soap/glycerene solution on a cone joint, which forms a successful seal on about 9 out of every 10 occasions. Sintered-silver plates on the top of the cell assembly (see figure 1.4.1) make thermal contact directly to the ^{3}He/^{4}He mixture inside the mixing chamber. This dilution refrigerator may pre-cool a Lancaster style nuclear cooling cell (see below) to below 10 mK in less than 1 day. Typically cells are pre-cooled for a period of one week to temperatures of around 5 mK.

1.3 Adiabatic Nuclear Demagnetisation

The standard technique for producing temperatures below that of dilution refrigerators is by adiabatic nuclear demagnetisation of copper. Copper is usually chosen due to its relatively high nuclear heat capacity, its good thermal conductivity, its good coupling between the nuclei and conduction electrons, and very low nuclear ordering temperatures (which means that the nuclei almost always approximate well to an ideal weakly interacting paramagnetic system).

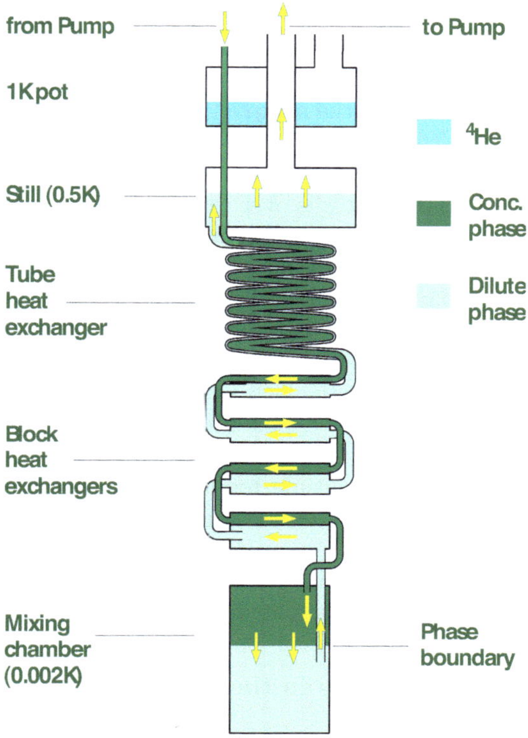

Figure 1.2.1. Schematic diagram of a dilution refrigerator.

The principles of the cooling technique are very simple. An applied magnetic field B Zeeman splits the nuclear spin energy levels by $\sim \mu B$, where μ is the nuclear magnetic moment. As thermal energy is removed during precooling the spins occupy more the lowest energy levels. The level occupation and hence entropy of the system is governed by the factor $\mu B/kT$. After precooling to some initial temperature T_i, the system is thermally isolated by a metallic superconducting heat switch (which is a very thermal good conductor when in the normal state, produced by an independently applied field, but a very poor thermal conductor when superconducting, in zero field). The magnetic field is then slowly reduced producing a reversible adiabatic change in which the entropy must therefore be conserved, i.e. $\mu B/kT$ remains constant. Consequently, when demagnetised to some final field B_f, the final temperature will have fallen proportionately, $T_f = T_i(B_f/Bi)$. So, for instance, demaging from $T_i = 7\,\mathrm{mK}$ and $B_i = 7\,\mathrm{T}$ to a final field of

$B_f = 70\,\mathrm{mT}$ will result in a nuclear temperature of $T_f = 70\,\mu\mathrm{K}$.

The difficulty comes in using the cold nuclei to cool your experimental sample. The minimum sample temperature is limited by the combination of the thermal resistance from the sample to the nuclei and the heat leak into the sample. Making thermal contact to fluids is particularly difficult at low temperatures due to the enormous thermal (Kapitza) boundary resistance between solids an liquids. This is overcome by coating the copper with fine silver sinter, an open silver sponge-like material which has a very large surface area. Conventional techniques for cooling ^{3}He to very low temperatures consist of having a large piece of copper to act as the nuclear refrigerant, with a flange containing silver sintered at one end, onto which is bolted the experimental chamber. However, the problem of cooling superfluid ^{3}He to very low temperatures is particularly acute, since the thermal excitations which carry the heat become exponentially dilute on decreasing temperature. consequently even a small reduction in temperature below $150\,\mu\mathrm{K}$ (say) requires orders of magnitude reduction in the heat leak. The 'Lancaster style' nuclear cooling stages, described below, minimise heat leaks and maximise thermal contact, allowing the very lowest temperatures. The downside of the technique, is that the refrigerant is an intimate part of the cell, making them relatively complex and time consuming to construct.

1.4 Cells for Experiments in the B-phase

A typical experimental cell used for B-phase measurements is shown in figure 1.4.1. Unlike conventional nuclear stages, each cell is cooled by its own copper refrigerant. This design enables a good thermal contact between the ^{3}He sample and the copper refrigerant. The amount of copper which can be fitted into the inner cell is restricted to a mole or two, which in turn limits the useful experimental cold time to a maximum of a couple of weeks after demagnetisation. This usually gives ample time for experiments.

The refrigerant is in the form of roughly one hundred high-purity annealed copper plates with typical dimensions $0.1 \times 30 \times 50\,\mathrm{mm}$. Each plate is silver plated and then silver sintered on both sides with about 0.4 g of 70 nm silver powder. Thicker, 0.5 mm copper plates, sintered with about 2 g of silver powder, are spot welded to lengths of 1 mm annealed silver wire to form the precooling thermal link to the mixing chamber. The plates are cut to leave a free volume for the experiments (see figure 1.4.1) and stacked into the inner cell separated by cigarette paper to reduce eddy-current heating. The inner cell walls are made from paper, impregnated with stycast 1266, giving a wall thickness of 0.3 mm or less. The inner cell is strengthened with turns of cotton on the outside. ^{3}He enters the inner cell from the outer cell

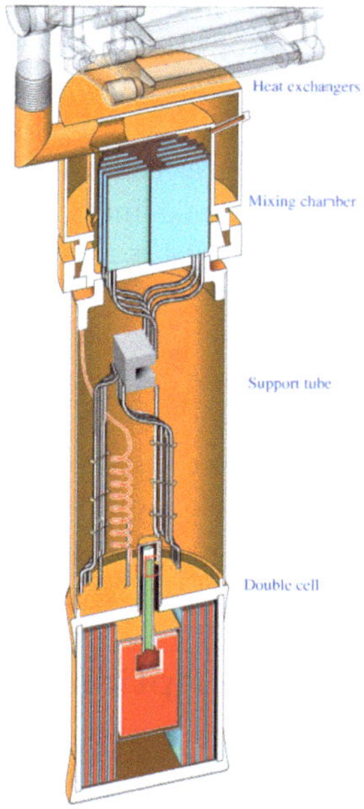

Figure 1.4.1. Schematic diagram of Lancaster style cell used for experiments in superfluid ^{3}He-B.

via a $\sim$ 1 m long, high purity copper filling line ($\sim$ 0.2 mm I.D.). The filling line is coiled around the outside of the inner cell before entering above the copper plates, a large thermal distance away from the experimental volume. Paper filters at the ends of the filling line prevent blockages.

The inner cell is placed inside a machined Araldite tube ($\sim$ 1 mm thick, $\sim$ 70 mm diameter) which forms the outer cell inside wall. The silver precooling wires to the inner cell pass through the top end plate of the outer cell, together with all the electrical leads to the experimental volume. Two more copper fill lines connect the outer cell to sample fill lines on the dilution refrigerator. The outer cell is cooled by several $\sim$ 1 mm thick silver sintered copper plates. These are connected to silver wires to form a precooling link to the mixing chamber.

The outer cell is placed inside another Araldite tube ($\sim$ 100 mm diameter) and the interspace filled with stycast 1266. This ensures that the cell is made leak-tight and also provides enough strength to withstand pressures up to 20 - 30 bar which might be desired for the experiments. The precooling silver wires are spot welded to a high purity single crystal aluminium heat switch which links them to the silver sinters in the mixing chamber. The silver wires from the inner and outer cells are spot welded to different regions of the heat switch to ensure that the two cells are isolated from each other when the switch is superconducting.

The use of an outer cell serves two purposes. It acts as a thermal guard and to reduce heat leaks into the inner cell from external sources (electrical leads, filling lines etc). It also allows the inner cell wall to be constructed with very thin material (since it does not have to be mechanically strong) which greatly reduces heat leaks from the cell walls. Typical heat leaks from the stycast-paper inner cell walls are 1 pW/cm^2. Copper inner cell walls can also be used which give heat leaks at least a hundred times smaller, but these are more difficult to construct. For most applications the paper inner cell walls are sufficient to produce the lowest temperatures that can be measured ($\sim 0.1\,T_c$ or $100\,\mu$K at 0 bar) using the most sensitive thermometry available (vibrating wire resonators, to be discussed later).

2 Simple description of superfluid phases of 3He

2.1 Introduction

Here we give a brief review of the main concepts required to understand the superfluid phases of ^{3}He, discovered in 1972 by Osheroff, Richardson, and Lee(4) who subsequently won a recent Nobel prize for the work. The BCS theory(5) of Fermi superfluids will not be reviewed in any detail since its principles and much of its predictions are, by now, almost common knowledge and can be found in almost any text book on low temperature physics.

The weakly-interacting quasiparticles described by Landau's Fermi Liquid theory(6) which govern the low temperature behaviour of normal ^{3}He, also form the Cooper pairs which are responsible for the superfluid phases. Fermi liquid effects often take a different form in the superfluid phases owing to interactions between paired and unpaired quasiparticles (superfluid and normal components) and also owing to the correlation of paired states in k-space.

In the extreme low temperature limit the number of quasiparticle excitations becomes very small. The mechanical properties of superfluid ^{3}He may then be described in terms of a dilute gas of quasiparticles moving in a potential determined by the condensate. Thus, as far as the mechanical properties of the superfluid are concerned, Fermi liquid effects become trivial in the low temperature limit (they are accounted for by the effective mass of the quasiparticle). On the other hand, many of the magnetic properties are dominated by Fermi liquid interactions over the whole temperature range. A brief discussion of Landau's theory is given in the next section. More comprehensive reviews have been given by Leggett(7) and by Vollhardt and Wölfle(8).

A brief description of the structure of the order parameter and the notation used to describe spin triplet superfluids is given in section 3. Vectors are often used to describe both the individual pairs and the net orientations of the pairs as a whole.

The remaining sections describe the structure of the model states which are believed to correspond to the observed A- and B-phases of superfluid ^{3}He.

2.2 Normal ^{3}He and Landau's Fermi Liquid Theory

Below 3.3 K, weak Van der Waals forces bind ^{3}He atoms into the liquid state. At temperatures somewhat below the Fermi temperature ($T_F \simeq 1\,\mathrm{K}$), the properties of the normal liquid are qualitatively very similar to that of a weakly-interacting degenerate Fermi gas, even though the interatomic spacing is comparable to the size of the atoms. Landau's Fermi liquid theory(6) accounts for this by describing the sea of strongly-interacting atoms in terms of weakly-interacting quasiparticles having an effective mass $m^\star$.

In the absence of external fields Landau's Fermi liquid is then equivalent to an ideal Fermi gas of quasiparticles. The energy of a quasiparticle with momentum $\mathbf{p}$ is

$$E(\mathbf{p}) = \mathbf{p}^2/2m^\star \tag{2.2.1}$$

The specific heat is given by the ideal Fermi gas expression,

$$C_v = (\pi^2/3)k_B^2 G(E_F)T\,, \tag{2.2.2}$$

where $G(E_F){=}3Nm^\star/p_F^2$ is the density of states at the Fermi energy E_F (with corresponding Fermi momentum p_F), and N is the number density of quasiparticles which is equal to the number density of atoms.

A knowledge of the heat capacity and the molar volume (which yields a value for the Fermi momentum) allows the effective mass to be determined. It is found(9) that the effective mass varies from about 3 to 6 times the bare atomic mass between 0 and 34 bar, the entire available pressure range. It is clear that the interactions between the atoms are indeed very significant.

To describe the effect of external fields (pressure gradients, magnetic fields etc.) the interactions between the quasiparticles must be taken into account. This is done on a microscopic level by writing the quasiparticle energy as

$$E(\mathbf{p},\sigma) = p^2/2m^\star + \sum_{\mathbf{p}'} f(\mathbf{p},\mathbf{p}')\delta n(\mathbf{p}') + \sigma\cdot\sum_{\mathbf{p}'}\zeta(\mathbf{p},\mathbf{p}')\sigma(\mathbf{p}') \tag{2.2.3}$$

where $\delta n(\mathbf{p})$ is the deviation of the occupancy of the quasiparticle state $\mathbf{p}$ from its groundstate value and $\sigma(\mathbf{p})$ is the expectation value of the quasiparticle spin in this state in units of $\hbar$. Since the theory can only be applied to degenerate Fermi systems all the relevant quasiparticle states have a momentum close to the Fermi momentum. The functions f and ζ can therefore be expanded in terms of Legendre polynomials $P_l(\cos\theta)$ at the Fermi surface. Multiplication of f and ζ by the density of states at the Fermi surface then yields a set of dimensionless quantities F_l and Z_l as

$$\begin{aligned} G(E_F)f(\mathbf{p},\mathbf{p}') &= \sum_l F_l P_l(\cos\theta) \\ G(E_F)\zeta(\mathbf{p},\mathbf{p}') &= \sum_l Z_l P_l(\cos\theta) \end{aligned} \tag{2.2.4}$$

where θ is the angle between $\mathbf{p}$ and $\mathbf{p}'$. Most of the properties of liquid ^{3}He can be accounted for by the first few Landau parameters F_0, F_1 and Z_0.

The effects of these parameters can be conveniently described in terms of molecular fields. If only the first Landau parameter describing spin interactions Z_0 is retained, then the quasiparticle energy becomes

$$\begin{aligned} E(\mathbf{p},\sigma) &= p^2/2m^\star + \sigma\cdot\sum_{\mathbf{p}'}(Z_0/G(E_F))\sigma(\mathbf{p}') \\ &= p^2/2m^\star + (Z_0/G(E_F))\sigma\cdot\mathbf{S} \end{aligned} \tag{2.2.5}$$

where $\mathbf{S}$ is the net spin of the system in units of $\hbar$. Rewriting this in terms of a molecular field $\mathbf{H}_{mol}$ gives

$$\begin{aligned} E(\mathbf{p}) &= p^2/2m^\star - \gamma\hbar\sigma\cdot\mathbf{H}_{mol} \\ \mathbf{H}_{mol} &= -\frac{Z_0}{\gamma\hbar G(E_F)}\mathbf{S} \end{aligned} \tag{2.2.6}$$

where $\gamma\hbar\sigma$ is the magnetic moment associated with the occupancy of the state $\mathbf{p}$. The static magnetisation $\mathbf{M}$ in an external field $\mathbf{H}$ is then given by the ideal Fermi gas expression with the external field replaced by $\mathbf{H}+\mathbf{H}_{mol}$:

$$\mathbf{M} = \chi_0(\mathbf{H}+\mathbf{H}_{mol}) = \chi\mathbf{H} \tag{2.2.7}$$

where $\chi_0{=}\gamma^2\hbar^2G(E_F)/4$ is the ideal Fermi gas susceptibility. The susceptibility corrected for Fermi liquid effects χ is then found to be

$$\chi = \frac{\chi_0}{1+Z_0/4} \tag{2.2.8}$$

Measurements of χ yield a rather pressure-insensitive value for Z_0 of around -2.8, which indicates a ferromagnetic tendency of the normal state liquid.

The term involving F_0 generates a Hartree-type field proportional to the number density of particles and thus relates to the compressibility and the speed of hydrodynamic sound. Its value varies between about 10 and 100, over the available pressure range. Liquid ^{3}He is thus seen to be much stiffer than the corresponding ideal Fermi gas.

The F_1 term is more interesting. Since $P_1(\cos\theta)=\mathbf{p}\cdot\mathbf{p}'/p_F^2$, it generates a molecular field $\mathbf{A}_{mol}$ which couples the quasiparticle momentum to flow. If only this term is retained, one finds:

$$\begin{aligned} E(\mathbf{p}) &= p^2/2m^\star - \mathbf{p}\cdot\mathbf{A}_{mol} \\ \mathbf{A}_{mol} &= -\frac{F_1}{G(E_F)p_F^2}\mathbf{P} \end{aligned} \tag{2.2.9}$$

where $\mathbf{P}=\sum_{\mathbf{p}}\mathbf{p}$ is the net fluid momentum. The quasiparticle energy must however obey Galilean relativity. Consider a stationary volume of the liquid made up of quasiparticles with velocity $\mathbf{v}_g(\mathbf{p})=\mathbf{p}/m^\star$. An observer moving with velocity $\mathbf{v}$ will see each quasiparticle moving with velocity $\mathbf{v}_g - \mathbf{v}$. Hence the quasiparticle energies will be observed as

$$\begin{aligned} E(\mathbf{p}) &= m^\star(\mathbf{v}_g-\mathbf{v})^2/2 \\ &= p^2/2m^\star - \mathbf{p}.\mathbf{v} \end{aligned} \tag{2.2.10}$$

to first order in v. Now consider a stationary observer viewing the liquid moving with velocity $-\mathbf{v}$. Each quasiparticle transports a mass m rather than $m^\star$, because of the one-to-one correspondence with the bare atoms. So each has an added momentum of $-m\mathbf{v}$ corresponding to the net flow which carries a total momentum $\mathbf{P}=Nm\mathbf{v}$ per unit volume. Therefore, a quasiparticle with momentum $\mathbf{p}$ in the fluid frame of reference has a momentum $\mathbf{p}-m\mathbf{v}$ as viewed by the observer and an energy

$$\begin{aligned} E(\mathbf{p}) &= (\mathbf{p}-m\mathbf{v})^2/2m^\star - (\mathbf{p}-m\mathbf{v})\cdot\mathbf{A}_{mol} \\ \text{where}\quad \mathbf{A}_{mol} &= -\frac{F_1}{G(E_F)p_F^2}Nm\mathbf{v} \end{aligned} \tag{2.2.11}$$

Substituting for the density of states and retaining terms up to first order in v gives

$$E(\mathbf{p}) = p^2/2m^\star - \mathbf{p}\cdot\mathbf{v}(1+F_1/3)m/m^\star\,. \tag{2.2.12}$$

Obviously a quasiparticle in a uniform flow with velocity $-\mathbf{v}$ as viewed by an observer at rest must have the same energy as in a stationary fluid observed by an observer moving with velocity $\mathbf{v}$. Therefore, equating the two expressions for the energy, we find

$$m^\star/m = 1 + F_1/3 \tag{2.2.13}$$

which is the well-known Landau effective mass relation. It appears that the effective mass $m^\star$ and the Landau parameter F_1 are equivalent and any uniform flow is equivalent to a simple Galilean transformation. This is generally not the case for the superfluid phases, since the condensate has its own reference frame but still interacts via the term involving F_1 with the normal component which may be moving at a different velocity. However, at very low temperatures (in the ballistic regime), the normal fluid fraction is almost zero and therefore the effect of F_1 again reduces to a simple replacement of m by $m^\star$, whereas the quasiparticle energies are now defined by the reference frame of the condensate.

2.3 Spin Triplet Pairing: The Order Parameter

The binding energies of the Cooper pairs which make up Fermi superfluids are conveniently described in terms of a gap parameter. Each paired state may be described as $(\mathbf{k}\,\alpha, -\mathbf{k}\,\beta)$, which is a state in which a particle with wavevector $\mathbf{k}$ and spin α=$\pm 1/2$ is paired with one with wavevector $-\mathbf{k}$ and spin β=$\pm 1/2$. The gap parameter is expressed as a complex 2×2 matrix with components $\Delta(\mathbf{k})_{\alpha\beta}$ and the quasiparticle excitation energies are given by the matrix(7)

$$\mathbf{E}(\mathbf{k}) = \sqrt{\xi^2 + \mathbf{\Delta}(\mathbf{k})\mathbf{\Delta}^\dagger(\mathbf{k})} \tag{2.3.1}$$

where ξ=$(p^2$-$p_F^2)/2m^*$ is now defined as the quasiparticle kinetic energy relative to the Fermi energy.

For a simple superconductor, all the pairs have zero spin (α=$-\beta$), zero angular momentum and no anisotropy. Thus, only one independent element is required to describe the gap parameter, which is expressed as a antisymmetric matrix to show its spin singlet character:

$$\mathbf{\Delta}(\mathbf{k}) = \begin{pmatrix} 0 & \Delta \\ -\Delta & 0 \end{pmatrix} \tag{2.3.2}$$

The quasiparticle excitation energy for the isotropic spin singlet state is thus given by

$$E(\mathbf{k}) = \sqrt{\xi^2 + \Delta^2}\,. \tag{2.3.3}$$

This is shown by the solid curve in figure 2.3.1(d). Figures 2.3.1 (a)-(d) show a schematic representation of how the excitation states are constructed from

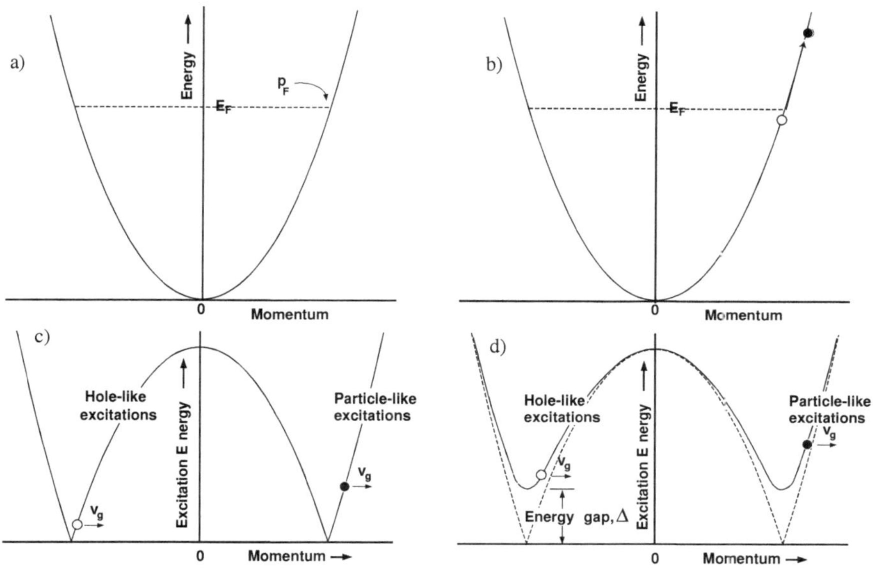

Figure 2.3.1. a) The dispersion curve for fermions in a Fermi gas/Liquid, at zero temperature all states are filled up to the Fermi Energy E_F. b) To generate an exctitation (thermal or otherwise) an atom/quasiparticle below the Fermi energy is excited to a state above the Fermi energy leaving an empty state below (a hole). c) The corresponding excitation dispersion curve showing the hole and particle produced in b). Note the hole has negative momentum (since it corresponds to the removal of a particle), and that the energies are measured relative to the Fermi energy. This is produced as follows. Starting from the curve shown in b), the states with energies below E_F are reflected about E_F ($E_F - E \to E_F - E$) and spun around ($\mathbf{p} \to -\mathbf{p}$), then E_F becomes the new base line for the excitation energies ($\xi = E - E_F$). d) Superfluidity (or superconductivity) lifts the low energy excitations states forming an energy gap Δ corresponding to half the binding energy of a Cooper pair (a broken pair produces two excitations requiring a minimum energy of 2Δ). Note that the higher energy states are relatively unaffected by the pairing.

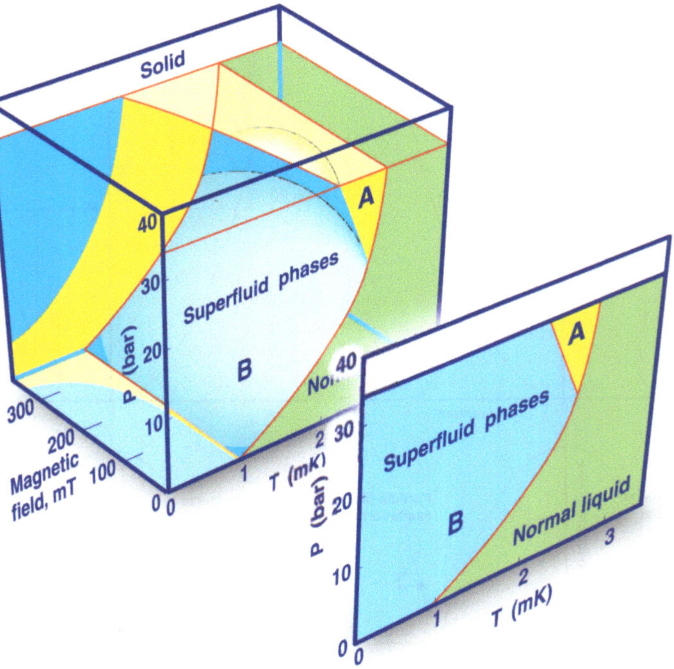

Figure 2.3.2. The superfluid ^{3}He phase diagram. Front panel shows the pressure - temperature plane in zero magnetic field. Note the A-phase is stablised by a magnetic field (horizontal plane). An additional thin region of A_1 phase (not shown) develops in high fields close to T_C.

the Fermi liquid particle states. The states with $p < p_F$ are seen to have their group velocity in the opposite direction to their momentum and are called quasiholes. They represent empty states below the Fermi energy. In fact there are different ways of describing the quasiparticles. Either the Fermi-liquid particle states can be used with energies above and below the Fermi energy, or (as is more conventional and as is used here) particle and hole excitation states are used with only positive energies ξ measured from the Fermi energy. The minimum energy required to break a Cooper pair (the binding energy) is 2Δ as two excitations (particle and hole) must be produced.

In the case of spin triplet pairing, the gap parameter is somewhat more complex. Three independent elements corresponding to the three spin projections (S_z=0, ±1) are now required. These can be described in terms of a vector $\mathbf{d}(\mathbf{k})$ in spin space. The components of the vector $\mathbf{d}(\mathbf{k})$ are chosen so that $\mathbf{d}(\mathbf{k})$ points along the direction of zero spin projection. The quasiparticle energy eigenvalues are then found to be

$$E(\mathbf{k})_{1,2} = \sqrt{\xi^2 + |\Delta(\mathbf{k})_{1,2}|^2} \tag{2.3.4}$$

The energy gap is in general a function of spin, and is related to $\mathbf{d}(\mathbf{k})$. The A-phase and the B-phase in low magnetic fields are well described by unitary states which have $|\Delta(\mathbf{k})| = |\mathbf{d}(\mathbf{k})|$.

In order to describe fully the Cooper pairs of superfluid ^{3}He, the three possible orbital angular momentum projections of the L=1 pairs must be accounted for also. Hence a total of nine elements are required to describe the three spin and three orbital angular momentum projections. The order parameter describing the Cooper pairs is thus a tensor $\underline{\mathbf{d}}$ with nine components $d_{\mu j}$. The vector $\mathbf{d}(\mathbf{k})$ (also known as the order parameter vector) is given in terms of the proper order parameter $\underline{\mathbf{d}}$ and the unit vector $\hat{\mathbf{k}}$ as

$$d_\mu(\mathbf{k}) = \sum_{j=1}^{3} d_{\mu j}\hat{k}_j \qquad \text{or} \qquad \mathbf{d}(\mathbf{k}) = \underline{\mathbf{d}}\hat{\mathbf{k}}\,. \tag{2.3.5}$$

The large number of independent elements in the spin triplet order parameter can of course accommodate a large number of different states. Those which correspond to the known superfluid phases of ^{3}He are those which have the lowest free energy and one might expect these to be the most uniform states in $\mathbf{k}$ space. The actual phase diagram for superfluid ^{3}He is shown in figure 2.3.2.

2.4 Superfluid ^{3}He-B: The BW State

The most uniform state in $\mathbf{k}$ space is the spherically symmetric BW state first discussed by Balian and Werthamer(10) in 1963. It consists of an equal number of pairs having the three spin and orbital angular momentum projections, giving a total angular momentum $\mathbf{J}$=$\mathbf{L}$+$\mathbf{S}$=0. Its order parameter is given by

$$d_{\mu j} = e^{i\phi}\Delta\delta_{\mu j} \tag{2.4.1}$$

giving an order parameter vector

$$\mathbf{d}(\mathbf{k}) = e^{i\phi}\Delta\,\hat{\mathbf{k}} \tag{2.4.2}$$

It has an isotropic energy gap Δ given by the BCS expression Δ=$1.76k_BT_C$ in the weak-coupling limit, as shown in figure 2.4.1. So as far as its thermal

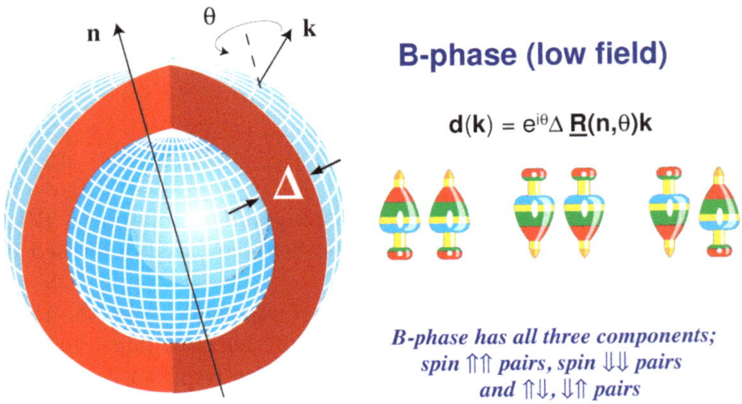

Figure 2.4.1. The superfluid B-phase in low magnetic fields corresponding to the BW state. All possible Cooper pair orientations exist in equal proportions giving an isotropic energy gap.

and mechanical properties are concerned, it is very similar to conventional superconductor, however it's magnetic properties are far more complex.

In the absence of spin-orbit interactions, the original BW state is in fact degenerate with a set of states which have their spin and position axes rotated with respect to each other giving a vector order parameter

$$\mathbf{d}(\mathbf{k}) = e^{i\phi'}\Delta\mathbf{R}(\mathbf{n},\theta)\hat{\mathbf{k}} \tag{2.4.3}$$

where the matrix $\mathbf{R}(\mathbf{n},\theta)$ describes a rotation about an axis $\mathbf{n}$ by an angle θ. The quasiparticle dispersion relation is unaffected by the rotation. However, the Cooper pairs are no longer in a state of definite total angular momentum. There is now a superposition of J=0,1,2 states.

The B-phase in low magnetic fields is described successfully by a BW state with a rotation angle (or Leggett angle),

$$\theta = \cos^{-1}(-1/4) = 104^\circ \tag{2.4.4}$$

as calculated by minimising the dipolar free energy(11). In the weak-coupling limit, the BW state in low magnetic fields is the only stable state for a P-wave superfluid. Experimentally, the B-phase is indeed the stable state in low fields, as shown in figure 2.3.2, except at high pressure and temperature.

Finally we note here that flow properties of the low field B-phase are very simple, and defined by the phase of the order parameter directly analogous to ^{4}He. Thus vortex lines in the B-phase are also particularly simple, resembling those of superfluid ^{4}He, formed by a 2π change of the order parameter around a central core region with a size of order the coherence length.

2.5 ^{3}He-B in a Magnetic Field

The magnetic properties of the B-phase are rather complex and will not be discussed in any detail here. However, we mention a few important points.

When a magnetic field is applied to a normal degenerate Fermi gas, the Fermi momenta for the up and down spin particles are split to allow the Fermi energies to equalise after the added magnetisation energy (there will be more spin up particles than spin down particles). In a spin singlet (S=0) superfluid, up and down spin quasiparticles with equal and opposite momenta form Cooper pairs just below the Fermi-surface. When a field is applied, the binding energy of these pairs acts to resist this splitting of the Fermi momenta and thus leads to a reduction in magnetisation energy. The quasiparticle excitation energies become split by the Zeeman energy, giving rise to a magnetisation that is entirely due to the excitations and vanishes in the T=0 limit.

In the B-phase, the Cooper pairs form in spin-triplet (S=1) states. One third of these have S_z=0 where the z-axis is along the direction of the net magnetisation. These pairs form from a symmetric superposition of spin up and spin down quasiparticles and therefore require the Fermi momenta of the two spin sub-systems to be identical. These pairs will lead to a reduction in magnetisation in much the same way as the pairs of a spin singlet superfluid. Thus one might naively think that at T=0 the static magnetisation is reduced to two thirds of its normal state value. This would be true in the absence of Fermi liquid corrections (in practice the susceptibility is reduced to about one third of its normal state value at T=0).

The suppression of the S_z=0 produces a suppression of the corresponding energy gap (along the predominant momentum axis of the quasiparticles which form these pairs). This leads to an anisotropy of the energy gap (and hence superfluid properties). The anisiotropy is usually very small except in high magnetic fields so, unlike the magnetic properties, the thermal and magnetic properties are only weakly effected. There is an induced orbital angular momentum along the anisotropy axis $\hat{\ell}$. The direction $\hat{\ell}$ and $\mathbf{n}$ are influenced by often competing external boundary conditions such as cell walls, magnetic fields and normal-superfluid counterflow. So generally these vectors vary through-out the experimental cell leading to magnetic and orbital textures.

Spin orbit coupling which arises from the orbital nuclear dipole moments of the Cooper pairs lead to very rich NMR behavior. The coupling may provide an extra torque on the spin which leads to relatively large frequency shifts in NMR experiments. The torque is sensitive to the relative orientations of $\hat{\ell}$ and $\mathbf{n}$. NMR therefore provides a powerful tool for studying textures in the superfluid phases.

2.6 Superfluid ^{3}He-A: The ABM State

At higher pressures and temperatures, the A-phase becomes the stable phase of superfluid ^{3}He as shown by the phase diagram in figure 2.3.2. The magnetic susceptibility of this state is found to be very close to the normal state value. As this suggests, the A-phase is thought to consist of $S_z=\pm1$ pairs only. It is widely believed to correspond to the ABM state first discussed in 1961 by Anderson and Morel(12). Its stability in the zero magnetic field P-T plane of the phase diagram is explained by corrections to the weak-coupling theory.

Since the A-phase has no $S_z=0$ pairs, its free energy is reduced with respect to that of the B-phase by a strong magnetic field, resulting in a 1st order phase transition (in which the $S_z=0$ pairs of the B-phase are suppressed completely) at a pressure-dependent critical field of order 0.5 T.

The ABM state is described by the order parameter

$$\mathbf{d}(\mathbf{k}) = \Delta_0 \hat{\mathbf{d}}(\hat{\mathbf{k}} \cdot \hat{\mathbf{m}} + i\hat{\mathbf{k}} \cdot \hat{\mathbf{n}}) \tag{2.6.1}$$

So this state has a single preferred direction $\hat{\mathbf{d}}$ in spin space perpendicular to $\mathbf{S}$. The unit vectors $\hat{\mathbf{m}}$ and $\hat{\mathbf{n}}$ are perpendicular and give preferred directions in orbital space. They are related to the so-called orbital angular momentum vector unit vector $\mathbf{l}$ by

$$\mathbf{l} = \hat{\mathbf{m}} \times \hat{\mathbf{n}} \tag{2.6.2}$$

The order parameter is seen to be intrinsically complex and a rotation of $\hat{\mathbf{m}}$ and $\hat{\mathbf{n}}$ about $\mathbf{l}$ describes a change of phase. For this reason, superfluid flow in the A-phase can be produced by orbital textures leading to a large variety of different vortex lines, including 'soft core' vortices which are doubly quantised with large cores defined by macroscopic textures.

The energy gap is independent of spin and found to be:

$$|\Delta(\mathbf{k})| = \Delta_0 |\hat{\mathbf{k}} \times \mathbf{l}| \tag{2.6.3}$$

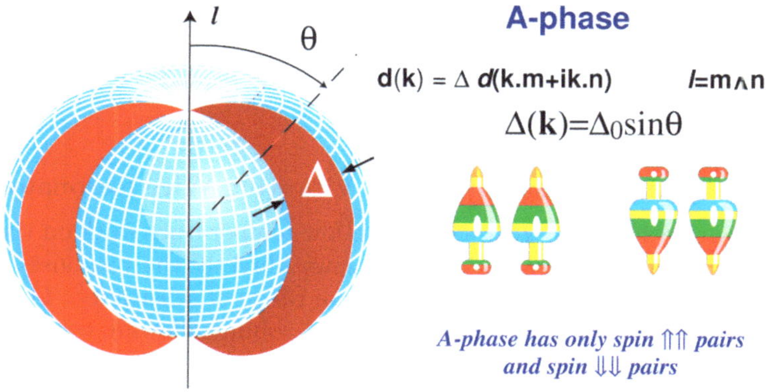

Figure 2.6.1. The superfluid A-phase. Only equal spin pairing occurs leading to a highly anisotropic energy gap.

So the ABM state has polar nodes in its energy gap along the **l** direction as shown in figure 2.6.1. The weak-coupling value for the maximum gap Δ_0 is approximately,

$$\Delta_0 = 2.02 k_B T_C \,. \tag{2.6.4}$$

The low temperature mechanical properties of the A-phase should be very different from those of the B-phase. The node in the energy gap gives rise to a T^3 dependence of the number density of excitations at low temperatures. This contrasts with the number density in the B-phase which falls exponentially due to the isotropic energy gap. Furthermore, at low temperatures the vast majority of excitations in the A-phase will travel along the **l** axis giving rise to strong textural properties.

2.7 Superfluid ^{3}He-A$_1$

The spin up and spin down sub-systems of the A-phase may often be regarded as acting independently. A strong magnetic field increases the number density of the up spin pairs at the expense of the down spin pairs. The form of the energy gap is believed to remain the same for the up and down spins. However, there is a small splitting in the magnitude Δ_0 of the gaps for the two spin sub-systems, corresponding to a small splitting in the transition temperatures. This means that between the two transitions there must be second phase which only has spin up pairs. This is called the A$_1$-phase . With the present range of magnetic fields available to us, the A_1-phase can only exist very close to T_C, the splitting being only a few percent in these fields.

3 Vibrating Wire Techniques

3.1 The Design and Construction of a Vibrating Wire Resonator

Vibrating wire resonators are appealing to low temperature work on the helium liquids for several reasons. They measure directly the response of the fluid to mechanical motion over a very large range of sensitivity. They give no significant heat leak into the experimental volume allowing very low temperatures to be achieved. In pure ^{3}He very fine superconducting wires with high Q factors have been used to measure temperatures down to $T_C/9$ where the damping on the wire resonator is less than that which can be achieved in a vacuum at 1 K. They are able to resolve temperature changes of order 1 nK. They have been used to measure power dissipations in the B-phase with a resolution of order 10 fW, and can be used to measure vortex lines and turbulence.

Typically, two types of wire are used to make the resonators; pure tantalum and niobium-titanium. Both are superconducting under the experimental conditions used so far (magnetic fields no greater than 0.6 T). The niobium-titanium resonators comprise a single filament separated from copper-clad filamentary wire and have diameters ranging from 4.5 μm to 13 μm. The tantalum wire is bare with a diameter of 125 μm.

A NbTi vibrating wire is shown in Figure 3.1.1. It is constructed as follows. First, two pin-holes are are made in a piece of stycast-impregnated paper used to make the inner cells. The copper-clad NbTi filamentary wire is stripped of its insulating material and threaded through the holes, forming a loop on one side. A cylindrical metal bar with a diameter equal to the required leg spacing is passed through the loop and the paper base. The loop is then tightened around the bar to give a well-defined shape of a semicircle with a diameter referred to as the leg spacing, on 'legs' with a length equal to half the leg spacing. Stycast is then run into the pin-holes to hold the wire firm and to form a superleak-tight seal. The bar is then removed and the copper-cladding is stripped off the top of the loop by nitric acid leaving stiff copper clad legs with a length of about one third the leg spacing at the base of the loop. At this point there are sixty or so free filaments of diameter 4.5 or 13 μm looped across the legs. These are picked off, one by one, under a microscope with fine tweezers and an endless amount of patience. Finally, with some luck, a single filament remains to form the vibrating wire. Two blobs of stycast are put over the broken ends of the discarded filaments at the top of each leg to form a seal and to stop any broken filament ends from

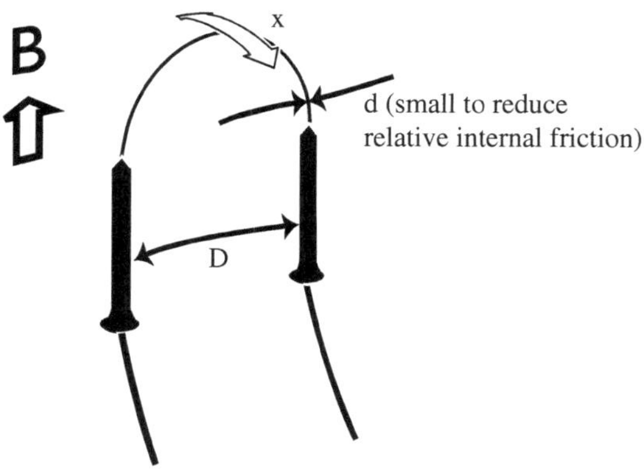

Figure 3.1.1. A schematic diagram of a vibrating wire resonator.

interfering with the motion of the active filament. The resonator is then ready for glueing into the inner cell. The construction of a Ta vibrating wire follows the same lines, although it is much easier of course since there are no filaments.

3.2 Vibrating Wire Theory

The equation of motion for a vibrating wire is modelled as that of a long, uniform cylinder with axis in the y-direction, of mass m per unit length, and free to vibrate rigidly perpendicular to its length in the x-direction. The restoring force on the cylinder is $m\omega_0{}^2$ per unit length per unit displacement x. In a magnetic field B applied in the z direction, the cylinder is forced into oscillation by the Lorentz force exerted on a current $I = I_0 e^{i\omega t}$ passed along it. If the cylinder moves in a fluid, the equation of motion per unit length of the cylinder can be written

$$m\ddot{x} + m\lambda\dot{x} + m\omega_0^2 x = BI_0 e^{i\omega t} \tag{3.2.1}$$

where λ is a complex variable describing the damping force exerted by the fluid on the cylinder. If the damping force is linear with velocity, λ is independent of velocity and may be written as a complex constant

$$\lambda = \lambda_2 + i\lambda_1. \tag{3.2.2}$$

The term λ_1 represents the non-dissipative part of the force (out-of-phase with the velocity) associated with the inertia of the fluid backflow. This effectively gives the moving cylinder a greater mass. The term λ_2 represents the dissipative component of the force (in-phase with the velocity).

To convert the above equation of motion to one which describes a real vibrating loop of wire, the right hand side of equation 1 is multiplied by a constant C of order unity which describes how the spatial maximum displacement of the loop (the displacement of the top of the loop) relates to the applied force. This constant will depend on the shape of the wire and on the way in which it bends. This procedure is valid provided that the equation of motion remains sufficiently linear. The equation of motion for a vibrating wire is thus written as

$$\ddot{x} + \lambda_2 \dot{x} + i\lambda_1 \dot{x} + \omega_0^2 x = C\frac{BI_0}{m}e^{i\omega t} \tag{3.2.3}$$

where x now represents the spatial maximum displacement of the wire loop.

The steady state solution of Equation (3.2.3) takes the form

$$\dot{x} = \dot{x}_0 e^{i\omega t} \tag{3.2.4}$$

Substituting this in Equation (3.2.3) gives:

$$\begin{aligned} \dot{x}_0 &= C\frac{I_0 B}{m}\left(\frac{i\omega}{\omega_0^2 - \omega^2 - \omega\lambda_1 + i\omega\lambda_2}\right) \\ &= C\frac{I_0 B}{m}\left(\frac{\omega^2\lambda_2 + i\omega(\omega_0^2 - \omega^2 - \omega\lambda_1)}{(\omega_0^2 - \omega^2 - \omega\lambda_1)^2 + \omega^2\lambda_2^2}\right) \end{aligned} \tag{3.2.5}$$

The component of the velocity which is in-phase with the driving force $Re\{\dot{x}_0\}$ is seen to be described by the familiar Lorentzian as shown in figure 3.2.1,

$$Re\{\dot{x}_0\} = C\frac{I_0 B}{m}\left(\frac{\omega^2\lambda_2}{(\omega_0^2 - \omega^2 - \omega\lambda_1)^2 + \omega^2\lambda_2^2}\right). \tag{3.2.6}$$

The out-of-phase component (the quadrature) is given by

$$Im\{\dot{x}_0\} = C\frac{I_0 B}{m}\left(\frac{\omega(\omega_0^2 - \omega^2 - \omega\lambda_1)}{(\omega_0^2 - \omega^2 - \omega\lambda_1)^2 + \omega^2\lambda_2^2}\right). \tag{3.2.7}$$

The maximum value of the in-phase velocity is thus

$$Re\{\dot{x}_0\}_{max} = C\frac{I_0 B}{m\lambda_2} \tag{3.2.8}$$

and occurs at the resonant frequency ω_0' given by

$$\omega_0^2 - {\omega_0'}^2 - \omega_0'\lambda_1 = 0$$

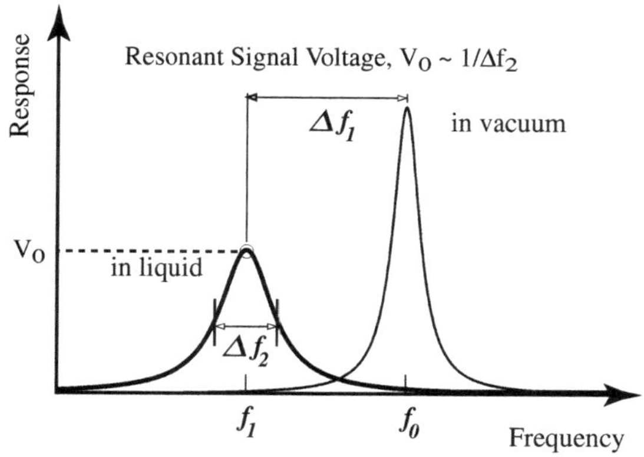

Figure 3.2.1. The VWR response (voltage or velocity) in-phase with the driving current (force), showing the resonant lineshape in vacuum (with some vacuum or 'intrinsic' damping) and in a fluid. The resonant frequency is reduced in the fluid and the damping is higher.

$$\Rightarrow \omega_0 - \omega_0' \simeq \lambda_1/2 \tag{3.2.9}$$

where the last step is justified provided that $(\lambda_1/\omega_0) \ll 1$, which is always valid for the wires which we use. The wire density is large compared with that of the ^{3}He so the inertia of the backflow has only a small effect. The resonant frequency of the wire is therefore shifted by

$$\Delta\omega_1 = \omega_0' - \omega_0 = -\frac{\lambda_1}{2} \quad \text{or} \quad \Delta f_1 = -\frac{\lambda_1}{4\pi}\,. \tag{3.2.10}$$

At this frequency the quadrature term is zero.

The in-phase velocity has half the maximum value and the quadrature its maximum or minimum value at a frequency $\omega_{1/2}$ given by

$$\left(\omega_0^2 - \omega_{1/2}^2 - \omega_{1/2}\lambda_1\right)^2 = \omega_{1/2}^2\lambda_2^2\,. \tag{3.2.11}$$

With the additional assumption that $\lambda_2 \ll \omega_0'$, which is true for the majority of our experiments,

$$\omega_0 - \omega_{1/2} \simeq (\lambda_1 \pm \lambda_2)/2\,. \tag{3.2.12}$$

Hence the (in-phase half height) frequency width of the resonance is given by

$$\Delta\omega_2 = \omega_{1/2}^+ - \omega_{1/2}^- = \lambda_2 \quad \text{or} \quad \Delta f_2 = \frac{\lambda_2}{2\pi}\,. \tag{3.2.13}$$

To convert the spatial maximum of the wire velocity to the quantity which is actually measured, the voltage generated across the wire loop, another factor dependent on the wire geometry must be introduced. A voltage is induced as the wire moves through the magnetic field, as given by Faraday's law

$$V = -\frac{d(\mathbf{B} \cdot \mathbf{A})}{dt} \tag{3.2.14}$$

where $\mathbf{A}$ is the vector area bounded by the wire loop and its support. For a goal-post shaped wire which vibrates rigidly (i.e. hinged), the rate of change of area parallel to the field is simply the leg spacing D, multiplied by the spatial maximum of the velocity $\dot{x}$ (the velocity of the cross bar). Thus the voltage amplitude generated is

$$V_0 = BD\dot{x}_0 \qquad \text{(goal post).} \tag{3.2.15}$$

For a rigid semi-circular loop the area bounded by the wire and its support is $\pi D^2/8$ and the rate of change of the angle which it makes to the field is $2\dot{x}/D$, therefore

$$V_0 = \frac{\pi}{4} BD\dot{x}_0 \qquad \text{(rigid semi circle).} \tag{3.2.16}$$

The voltage generated by the actual wires is thus written generally as

$$\dot{x}_0 = K\frac{V_0}{BD}\,. \tag{3.2.17}$$

Since no two vibrating wires can be made identical, the constant K will be different for each wire. However, we expect K to be close to that of a rigid semi circle, since as discussed in the last section, the wires are approximate semi-circles on small 'stilts' such that their height is about 20 per cent greater than their width. All the data presented in this thesis therefore uses the rigid semi-circle value, $K = \pi/4$. We estimate the uncertainty on this in be about 10 per cent.

An important quantity to note is the in-phase voltage amplitude multiplied by the frequency half width of the resonant lineshape, as this gives a measure of the magnetic field perpendicular to the motion of the wire loop. Using Equations (3.2.8), (3.2.13) and (3.2.17), this is found to be

$$V_{0max}\Delta f_2 = \frac{C}{K}\frac{I_0 B^2 D}{2\pi m} \tag{3.2.18}$$

where V_{0max} is the amplitude of the voltage at the resonant frequency. The value of C/K is empirically found to be close to unity, lying typically between 0.85 and 1.05 for the NbTi vibrating wires.

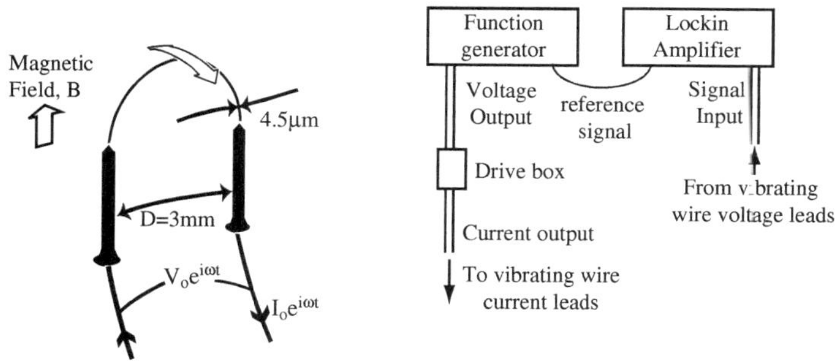

Figure 3.3.1. Left: a schematic diagram of a typical vibrating wire resonator used for low temperature measurements. Right: the measurement circuit and electronics

3.3 Vibrating Wire Measurements and Circuits

Figure 3.3.1 shows the usual set-up for making vibrating wire measurements. Also shown is a schematic drawing of the leads used in the cryostat. Owing to the paranoia over superleaks, the vibrating wire leads are always brought out of the cell as bare single core (superconducting) wires. In the case of Ta vibrating wires, the leads are brought straight out of the cell and spot welded to copper-clad single core NbTi to allow solder joints to be made further up the cryostat. The NbTi vibrating wires, made from either copper or cupro-nickel clad filamentary wire, are soldered to the single core NbTi wire inside the outer cell and then brought out of the cell as bare NbTi.

Unfortunately, the Ta wires occasionally break while cooling owing to differential contraction where they pass through the plastic outer cell. It is practically impossible to repair such a break and some wires have been lost for this reason. The obvious solution is to make the spot weld joints to the NbTi inside the outer cell, however such joints are rather bulky for the outer cell and also have a risk of breaking under the stresses of thermal cycling.

The leads from the cell are twisted in pairs, thermally anchored to the mixing chamber and then taken up to the 20 mK ring. Each pair of leads is soldered to two more twisted pairs (0.05 mm cupro-nickel clad filamentary)

on the 20 mK ring, forming current and voltage leads. Voltage lead connections are sometimes made in the outer cell to reduce the 'background' voltage from the lead inductance. These twisted pairs pass through thermal anchors on the still and 1 K flanges, before reaching vacuum seals made from black stycast on the 4 K flange. Twin coaxial cables connect the leads from the top of the cryostat to the measuring equipment.

The vibrating wires are driven by a frequency synthesiser (or function generator) acting as an AC voltage source with a large frequency range. This is coupled to the vibrating wire current leads by a 'drive box' (see figure 3.3.1) consisting of a high quality 6.3:1 ratio transformer and a set of series resistances . The resistances are used to set the range of drive current accessible from the voltage output range of the synthesiser.

The voltage generated across the vibrating wire is measured by a two-phase lockin amplifier, either directly or via a pre-amplifier, usually a transformer or sometimes a SQUID situated below the 4 K flange. The synthesiser provides a frequency and phase reference for the lockin amplifier. Both the amplitude of the signal at the reference frequency and its phase (or alternatively, the in-phase and out-of-phase components) relative to the reference signal are measured. These quantities may also be measured for the signal component at the twice the reference frequency.

The lockin and the synthesiser are controlled by computer which is programmed to take the required measurements which are output to a data file.

There are three common measuring techniques. These may be referred to as 'frequency sweep', 'drive sweep' and 'time sweep'. These are described below.

Frequency Sweep For a frequency sweep, the synthesiser is swept through the resonant frequency of the vibrating wire resonator at a fixed output voltage (fixed drive). For each frequency the voltage generated across the wire (both the in-phase and out-of-phase components) is read from the lockin and stored by the desktop computer. The data are then fitted (allowing for linear resistive and inductive 'background' components) to the Lorentzian lineshape of a linear oscillator as shown in figure 3.2.1. Details of the fitting procedure can be found in reference (13). The fit is then displayed along with the deviation from the ideal lineshape. The fitted values for the resonant frequency, the frequency width and the height of the resonance are obtained.

Drive Sweep In the drive sweep method, the driving current (force) to the vibrating wire is swept from zero up to the desired value at (usually) the resonant frequency of the wire. Both in-phase and out-of-phase components of the resulting voltage (velocity) generated across the wire are measured for each drive level and stored by the computer. The program can adjust the frequency to maintain resonance by keeping the out-of-phase component (which is zero at resonance) below some small (specified) fraction of the in-phase component. If the magnetic field B is known, the velocity-force relationship for the wire can be inferred using

$$\text{peak wire velocity, } v = K\frac{V_0}{BD} \tag{3.3.1}$$

$$\text{peak applied force, } F = BDI_0 \tag{3.3.2}$$

where D is the vibrating wire leg spacing, I_0 is the amplitude of the applied current and the constant K is taken to be $K = 4/\pi$.

Time Sweep For a time sweep, the vibrating wire is driven with a constant driving force, usually at the resonant frequency while the voltage generated is periodically read from the lockin. The frequency can either be fixed, or the resonant frequency can be tracked by maintaining zero out-of-phase signal. Since the voltage across the vibrating wire is inversely proportional to the damping force for a linear oscillator, this provides a comparatively fast measurement of changes in the damping. The absolute damping (that is the effective Δf_2 of the resonance) can, of course, only be obtained by a knowledge of the factor which relates the height to the frequency width. This factor is obtained from a frequency sweep. The damping can also be inferred when the wire is driven in the non-linear regime, at higher velocities (to increase the signal strength), provided that the non-linearity is measured and well characterized.

4 Thermal Damping of Vibrating Wire Resonators in the B-phase of superfluid ^{3}He at low temperatures

4.1 Thermal Damping Measurements

The thermal damping of a vibrating wire resonator as a function of temperature in ^{3}He in low magnetic fields is shown in figure 4.1.1. At high temperatures, in the normal state, ^{3}He is a thick viscous fluid. Here, vibrating wires act as viscometers which can be used to infer the fluid viscosity(14) (the theory of which was worked out by G.G. Stokes more than a century ago(15)). Below the transition temperature, the damping can again be calculated from hydrodynamic equations, but incorporating the two fluid model(16). The damping drops rapidly as only the normal fluid component

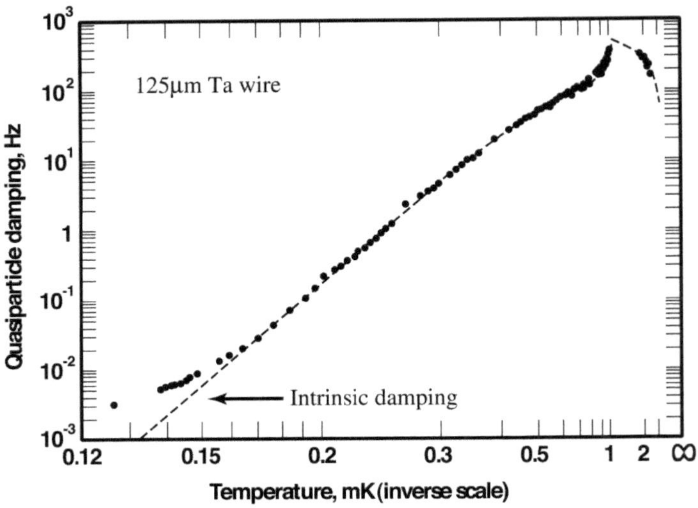

Figure 4.1.1. The temperature dependence of the damping (frequency width) of a vibrating wire resonator in superfluid ^{3}He-B.

contributes. However, at temperatures below $\sim 0.4\,T_C$ the quasiparticle mean free path becomes larger than the size of the experimental chamber. In this low temperature limit, the damping on a vibrating wire can no longer be described by hydrodynamic equations and a ballistic approach to model the damping must be adopted in which quasiparticle-quasiparticle scattering may be neglected.

As one would expect, the temperature dependence of the damping is dominated at low temperatures by the Boltzmann factor $\exp(-\Delta/k_BT)$. The number density of thermally excited quasiparticles falls exponentially due to the isotropic energy gap Δ in the B-phase. The exponential dependence on $1/T$ was observed (13; 17; 18) in measurements taken some years ago using platinum NMR thermometry and a vibrating wire resonator. These measurements, shown in figure 4.1.1, form the basis of our vibrating wire thermometry. However, as discussed below, the magnitude of the damping was found to be much larger than expected and was found to decreases at higher wire velocities. These effects were later shown to be a consequence of Andreev scattering of quasiparticles by the superfluid flow around the wire. Andreev Scattering leads to some counter-intuitive behavior which we discuss below.

4.2 Simple Kinetic Theory

It was originally expected that simple kinetic theory would suffice to describe the damping of an object (in this case a wire resonator) moving through the gas of quasiparticle excitations. The damping on a vibrating wire in ^{3}He/^{4}He solutions is described well by such arguments in the ballistic regime (19). However, the simple model predicts a damping in ^{3}He-B which is some three orders of magnitude too small. The simple kinetic theory model is as follows.

Consider a paddle-shaped object of area A moving through the superfluid with velocity $\mathbf{v}$ parallel to its surface normal. For simplicity suppose the quasiparticles are confined to move forwards or backwards along this direction only. The simple model treats the quasiparticles as a classical gas of atoms having an average speed v_g and a momentum p_F. If there are n per unit volume, then the number which hit the paddle on the forward facing side is $An(v_g + v)/2$ per unit time and the number hitting the rear is $An(v_g - v)/2$ per unit time ($v_g \gg v$). Each quasiparticle exchanges a momentum $2p_F$ with the paddle, and so the opposing force exerted per unit area is

$$F = p_F n(v_g + v) - p_F n(v_g - v) = 2np_F v \qquad (4.2.1)$$

The damping coefficient G is defined as the force per unit velocity F/v and is therefore independent of the paddle velocity for this model (the frequency width of a vibrating wire is proportional to G and will be referred to simply as the damping in the following). The temperature dependence of the damping should therefore be dominated by the Boltzmann factor $\exp(-\Delta/k_B T)$ which is consistent with the measurements. However, the magnitude of the damping force given by Equation (4.2.1) is three orders of magnitude too small.

4.3 Beyond the Two Fluid Model

When the vibrating wires were driven to higher velocities, the thermal damping force was found to become non-linear with velocity (20). The thermal damping fell with increasing velocity. This lead to thoughts that the superfluid velocity field around the wire may be playing an important role in the damping mechanism.

To understand better the damping process, we need to consider the quasiparticle excitation dispersion curve. The dispersion curve in the sta-

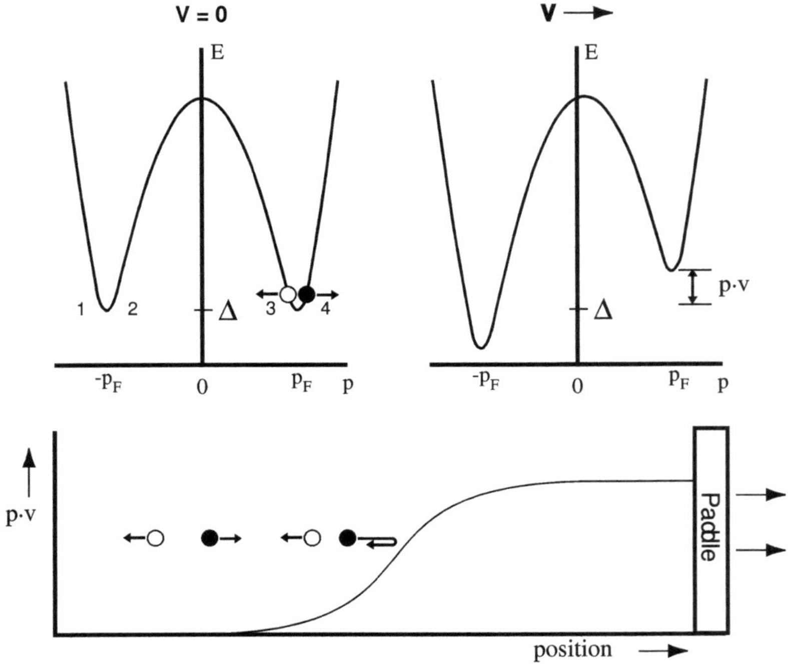

Figure 4.3.1. Top: dispersion curves for excitations incident on the rear of a paddle in the laboratory frame (Left: far away where there is no flow. Right: at paddle surface where the flow equals the paddle velocity). Bottom: the corresponding effective potential: low energy incident particles on branch 4 are Andreev reflected whereas incident holes on branch 2 may freely hit the paddle, pulling it to the left, leading to a large damping.

tionary reference frame of the superfluid condensate (zero flow) is shown in figure 4.3.1, top left. The hole states have their group velocities and momenta in opposite directions so at first sight it appears that the damping force due to the normal scattering of holes will almost completely cancel that due to the particles; the particles push whereas the holes pull.

The quasiparticle dispersion curve as viewed in a frame of reference in which the superfluid is moving with velocity v is shown in figure 4.3.1, top right. The quasiparticle states have undergone a Galilean transformation; the group velocity $\mathbf{v}_g$ of each state is transformed to $(\mathbf{v}+\mathbf{v}_g)$ and the energy of each state is shifted by $\mathbf{p}_F \cdot \mathbf{v}$ (all the states which are thermally occupied have momenta close to the Fermi momentum p_F).

For simplicity, consider again a paddle moving through the superfluid and consider only the quasiparticle states moving in the one dimension along which the paddle is moving. At a large distance from the paddle the superfluid is stationary with respect to the walls of the container, whereas close to the paddle surface the superfluid is pushed along with the paddle velocity $\mathbf{v}$. The dispersion curves for quasiparticle states both close to the paddle surface and at a large distance are shown figure 4.3.1, top right and left respectively. The excitations are assumed to have a thermal equilibrium distribution at a large distance from the paddle (where the superfluid is stationary). Since we are dealing with the ballistic temperature regime, we assume that there is no re-thermalisation of quasiparticle energies in the neighbourhood of the paddle surface.

A normal scattering process is one in which an excitation is scattered from one side of the Fermi surface to the other without changing its character i.e. particles do not become holes. An excitation which is scattered normally at the paddle surface will exchange an energy $-2\mathbf{p}_F \cdot \mathbf{v}$ with the paddle in the lab frame, so that it is elastic (no energy transfer) as viewed from the paddle's frame of reference. Therefore, referring to figure 4.3.1, any excitation which has reached the surface of the paddle has an un-bound state to scatter in to via a normal scattering process. It is assumed in the following that excitations will always scatter via a normal process on collision with the paddle surface.

As shown in figure 4.3.1, the excitations in branches labelled 1 and 2 in the stationary superfluid can reach the paddle surface, whereas those in branches labelled 3 and 4 which have an excitation energy less than $\Delta + \mathbf{p}_F \cdot \mathbf{v}$ have no available states close to the paddle (the excitations must, of course, propagate with a constant energy). As these excitations approach the paddle, their energy as measured from the superfluid frame of reference decreases and so does their group velocity. There will be a point where the excitations do not have enough energy to continue moving forward and their group velocity approaches zero. They then change character, holes become particles and particles become holes, their group velocity is reversed, whereas their momentum changes only by an amount of order $(\Delta/E_F)p_F$. This is known as Andreev scattering. Since $\Delta/E_F{\sim}10^{-3}$, the force exerted from Andreev scattering is neglected in the following. The Andreev reflection process is illustrated in the low part of figure 4.3.1. The flow velocity may be considered as providing a potential $\mathbf{p}_F \cdot \mathbf{v}$ for the quasipartcle dynamics.

Therefore, at large velocity or low temperature, only excitations which occupy branches 1 and 2 in figure 4.3.1 may reach the paddle surface where they transfer a momentum $2p_F$. In other words, only particles on branch 1 (which push) may hit the paddle from the front, whereas only holes on branch 2 (which pull) may hit the paddle from the rear. This gives a velocity-independent drag force with a much larger magnitude than one would obtain from classical arguments. In fact it is the maximum force which could possibly be exerted by the gas of excitations. The force corresponds to that which a classical gas would exert if the paddle were pushing against the full pressure of the gas while having a vacuum behind.

At lower velocities or higher temperatures some of the excitations in branches 3 and 4 will have enough thermal energy to reach the paddle. These will partially cancel the force due to the excitations in branches 1 and 2. To be a little more quantitative, all the quasiparticles in branches 1 and 2 are able to reach the paddle surface and we assume that all of these are normally scattered. Therefore the force exerted from these on unit area of the paddle is

$$F_{1,2} = p_F \int_{\Delta}^{\infty} (v_g(E) + v)G(E)f(E)dE + p_F \int_{\Delta}^{\infty} (v_g(E) - v)G(E)f(E)dE \tag{4.3.1}$$

where $G(E)$ is the density of particle or hole states with an energy E measured from the Fermi energy and $f(E)$ is the mean occupation number of a state with that energy. The first integral on the right hand side of the equation is the contribution from the particles in branch 1. The second is that from the holes in branch 2 which hit the paddle from the opposite direction. Adding the terms gives

$$F_{1,2} = 2p_F \int_{\Delta}^{\infty} v_g(E)G(E)f(E)dE \tag{4.3.2}$$

The integral on the right hand side of this equation is equal to the 'quasiparticle flux' defined as,

$$\langle nv_g \rangle \equiv \int_{\Delta}^{\infty} v_g(E)G(E)f(E)dE \tag{4.3.3}$$

where n is the total number density of holes or particles (in one dimension, $\langle nv_g \rangle/2$ is the total number of holes or particles passing through unit area, in a particular direction, per unit time). So Equation (4.3.1) becomes

$$F_{1,2} = 2p_F \langle nv_g \rangle \tag{4.3.4}$$

Now consider the force due to the normal scattering of excitations in branches labelled 3 and 4 in figure 4.3.1. This is given by an expression similar to that of Equation (4.3.1) except it is of opposite sign (it opposes the damping force $F_{1,2}$) and also the lower bound of the integration is now $\Delta + p_F v$, since those states with energy less than this can not reach the paddle surface. So

$$F_{3,4} = -2p_F \int_{\Delta + p_F v}^{\infty} v_g(E) G(E) f(E) dE. \tag{4.3.5}$$

Now, $G(E)v_g(E)$ is equal to the density of states in momentum space $G(p)$ and the quasiparticle momentum is always close to p_F, so

$$G(E)v_g(E) \simeq G(p_F). \tag{4.3.6}$$

The density of states in momentum space $G(p_F)$ is a constant equal to that of the normal fluid. Substituting this into Equation (4.3.5) gives

$$F_{3,4} = -2p_F G(p_F) \int_{\Delta + p_F v}^{\infty} f(E) dE \tag{4.3.7}$$

In this ballistic regime, the Fermi distribution function $f(E)$ approximates to the Maxwell Boltzmann function $\exp(-E/k_B T)$ so that

$$\begin{aligned} F_{3,4} &= -2p_F G(p_F) \int_{\Delta + p_F v}^{\infty} \exp(-E/k_B T) dE \\ &= -2p_F G(p_F) k_B T \exp(-\Delta/k_B T) \exp(-p_F v/k_B T) \end{aligned} \tag{4.3.8}$$

Also using Equation (4.3.6), the 'quasiparticle flux' becomes

$$\begin{aligned} \langle n v_g \rangle &= G(p_F) \int_{\Delta}^{\infty} \exp(-E/k_B T) dE \\ &= G(p_F) k_B T \exp(-\Delta/k_B T) . \end{aligned} \tag{4.3.9}$$

So we can rewrite Equation (4.3.8) as

$$F_{3,4} = -2p_F \langle n v_g \rangle \exp(-p_F v/k_B T) . \tag{4.3.10}$$

Adding the two contributions (Equations (4.3.4) and (4.3.10)) gives the total damping force per unit area on the paddle as

$$\begin{aligned} F &= F_{1,2} + F_{3,4} \\ &= 2p_F \langle n v_g \rangle \big[1 - \exp(-p_F v/k_B T)\big] . \end{aligned} \tag{4.3.11}$$

Therefore, at low velocity the force per unit area becomes linear in $(p_F v/k_B T)$:

$$F = 2p_F \langle n v_g \rangle \left(\frac{p_F v}{k_B T} \right) \qquad (p_F v \ll k_B T) , \tag{4.3.12}$$

whereas at high velocity it tends to a constant, dominated by $F_{1,2}$:

$$F = 2p_F\langle nv_g\rangle \qquad (p_F v \gg k_B T)\,. \tag{4.3.13}$$

To compare this simple model with a real vibrating wire we introduce two fitting parameters. An overall numerical pre-factor γ (expected to be of order unity) is included together with a numerical pre-factor λ to $p_F v/k_B T$ in the exponential term. The values of λ and γ will be determined by many things such as the geometry of the wire, the type of surface scattering (diffuse or specular) and of course the superfluid velocity profile around the wire. So we hope to describe the force per unit length on a vibrating wire by the equation

$$F = 2d\gamma p_F\langle nv_g\rangle\left[1 - \exp(-\lambda p_F v/k_B T)\right], \tag{4.3.14}$$

where d is the wire diameter. A discussion of the arguments presented above may also be found in references (21) and (22).

Full three dimensional calculations have been made of the damping force on a smooth infinite cylinder assuming both specular (23; 24) and diffuse (27) scattering of thermal quasiparticles. The details of which will not be given here as the physics involved is essentially the same as outlined above, and the models gives results which are well described by Equation (4.3.14).

4.4 Non-Linear Thermal Quasiparticle Damping Measurements

The non-linearity of the velocity-force relationship for a vibrating wire in the B-phase was studied by a series of 'drive sweeps' at various temperatures. The damping force on the wire as a function of its velocity is shown in figure 4.4.1 for various temperatures. The temperature of the helium in the experimental cell is varied by another vibrating wire used as a heater.

To extract the thermal contribution to the damping force on the vibrating wire, it is assumed that the total damping force may be expressed as a sum:

$$F = F_I + F_C + F_T \tag{4.4.1}$$

where the various terms have the following meanings

F_I: The intrinsic damping force of the wire plus effects due to the circuit.
F_C: The damping force due to pair-breaking.
F_T: The thermal damping force due to collisions with existing quasiparticles.

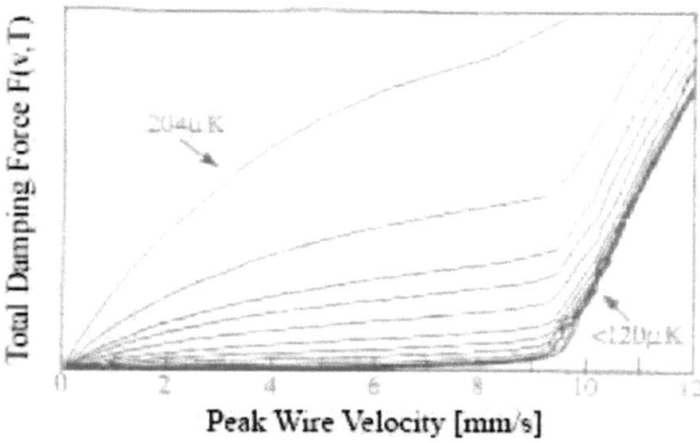

Figure 4.4.1. The total damping force measured on a vibrating wire moving through the B-phase for various temperatures.

The F_I term is the 'vacuum damping' (which may be assumed to be independent of temperature in our range), whereas the pair-breaking term, F_C, rises rapidly above the critical velocity v_C and quickly becomes the dominating term. The measurements taken at the base temperature are dominated by F_I below v_C and by F_C above v_C. Therefore, assuming that F_C is independent of temperature, we can use the data taken at the base temperature to represent $F_I(v)+F_C(v)$ and then subtract this from the data taken at higher temperatures to leave only the thermal damping force, $F_T(v,T)$.

Figure 4.4.2 shows the normalised thermal damping force, F_{norm}, plotted against $p_F v/k_B T$. The normalisation has been made by dividing each data set for $F_T(v,T)$ by their initial gradients with respect to $p_F v/k_B T$. The one-dimensional model developed in the previous section (Equation (4.3.14)) predicts an initial gradient

$$\left.\frac{dF}{d\left(\frac{p_F v}{k_B T}\right)}\right|_{v=0} = 2\gamma d p_F \langle n v_g \rangle \lambda \,. \tag{4.4.2}$$

Therefore the normalised force becomes

$$F_{norm} = \frac{1-\exp(\lambda p_F v/k_B T)}{\lambda} \,. \tag{4.4.3}$$

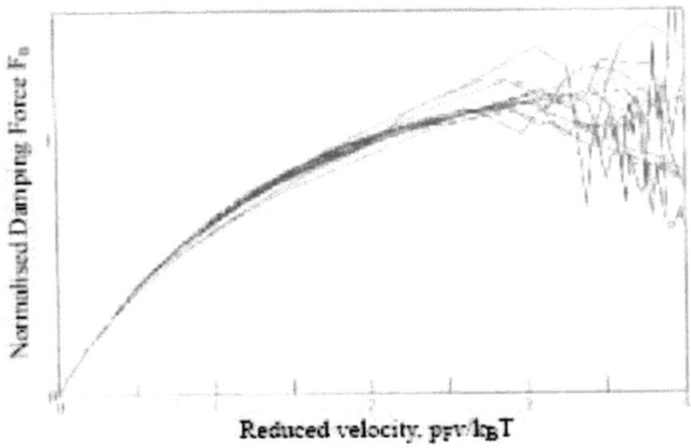

Figure 4.4.2. The normalised thermal damping force for various temperatures as a function of $p_F v/k_B T$. Note the data collapse onto a single line. Dashed line gives the fit to the one dimensional model.

This function is shown by the dashed line in figure 4.4.2 with $\lambda = 0.85$. The data is rather poor at velocities above v_C since the total damping force is dominated by the pair-breaking force which has been subtracted. However, at smaller velocities the simple theory is seen to give a reasonably good fit to the data. The magnitude of the force is in good agreement with the simple model, giving a value for γ of order 1. A more accurate measurement of γ can be obtained using black-body radiator techniques (25; 26). The data are also in equally good agreement with more sophisticated three-dimensional models(23; 24) without any fitting parameters provided that the quasiparticles are assumed to scatter diffusely at the wire surface (27).

5 The Pair-Breaking Critical Velocity of a Resonator in ^{3}He-B at low temperatures

A microscopic object moving through the B-phase, which does not disturb the local gap parameter or create backflow, is predicted to create quasiparticle excitations at a critical velocity given by the Landau velocity,

$$v_L = \Delta/p_F \tag{5.0.4}$$

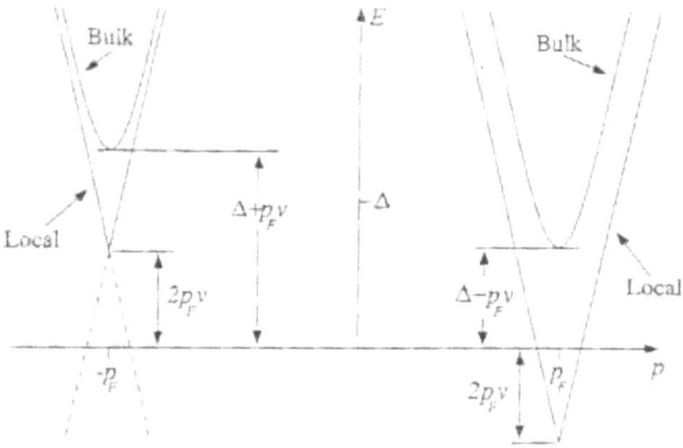

Figure 5.0.3. Dispersion curves in the rest frame of a cylindrical wire moving through the B-phase. The bulk superfluid is approaching the wire at a velocity v (from the left, corresponding to the wire moving to the right in the lab frame), while the local superfluid at the surface has a maximum velocity $2v$. The dotted lines of the local superfluid dispersion curve corresponds to filled states below the Fermi-surface, these are shown only to clarify the range of energies available for excitations created by pair-breaking.

This is easily realised by considering the dispersion curve of the excitations in the stationary reference frame of the object, as shown by the solid 'bulk' curves in figure 5.0.3. In this frame of reference, momentum is transferred to the helium at constant energy, assuming that only elastic processes occur. At the Landau velocity, quasiparticle states with the Fermi momentum opposite to the direction of flow (parallel to the velocity of the object moving in the lab. frame) have zero energy in the object frame (corresponding to the RHS of the bulk curve in figure 5.0.3 dipping down to zero energy). When a Cooper pair is broken, a momentum $2p_F$ is supplied corresponding to an energy $2p_F v_L$=2Δ in the lab. frame. A Cooper pair is broken, producing a quasi-particle and a quasi-hole excitation (at zero energy in the object frame) which move in opposite directions and have almost equal momenta. Experimental confirmation of this has been given by Ahonen *et. al.*(28), although a sharp critical velocity was not observed.

A macroscopic object presents a more complex problem. The energy gap is suppressed within a few coherence lengths from a solid surface and also the superfluid backflow around the object must be taken into account. Figure 5.0.4 shows a set of force-velocity curves, taken from a 13 μm NbTi vibrating wire at various pressures (Landau velocities) in a low magnetic field (32 mT) and at $T \simeq 0.1T_C$. The velocity axis is scaled by the Landau velocity. The critical velocity is marked by the rapid rise in the gradient of the curves and occurs at a pressure-independent value of approximately $v_L/3$. The uncertainty in the absolute maximum wire velocity is about 10 per cent.

Zhang *et. al.*(29) have studied the distortion of the B-phase energy gap close to surfaces of varying roughness using quasiclassical theory. They find that the order parameter is reduced within a few coherence lengths from the wall as a result of quasiparticle scattering at the surface. The energy gap becomes anisotropic, with the perpendicular component falling to zero at the surface. The gap in the plane of the wall is more strongly dependent on the surface roughness; for a specular surface it increases slightly, while for a rough surface it is also suppressed although at a shorter distance from the surface. The effect of flow was also examined and they found that the parallel gap is further suppressed mainly in the direction of flow. The surface of a vibrating wire is found to be very rough on the sub-micron scale, we therefore assume that the B-phase energy gap is suppressed to zero at the surface of the wire in all directions.

A value of $v_L/3$ for the critical velocity can be derived if we make the additional assumption that the backflow around the wire is that of pure potential flow around a perfect cylinder (at least for velocities up to the critical velocity). Following the arguments given by Lambert(30), the observed pair-breaking critical velocity is not the minimum velocity required to break a Cooper pair. Rather, it is the minimum velocity at which the quasiparticles from a broken pair can escape into the bulk superfluid. The dispersion curves are drawn in figure 5.0.3 for excitations both at the wire surface and in the bulk as viewed in the stationary reference frame of the wire. Since the energy gap is zero at the surface of the wire, excitations are created from zero velocity. However, they will be bound to the wire surface unless they have enough energy to move out into the bulk. Pure potential flow past a cylinder has a maximum backflow velocity $2v$ in the rest frame of the cylinder at its top and bottom surfaces as shown in figure 5.0.3. Therefore, assuming this to be a good approximation for a vibrating wire, filled states are raised in energy by $2\mathbf{p}_F \cdot \mathbf{v}$ at the top and bottom surface of the wire as shown in figure 5.0.3. Surface roughness scatters some of the quasiparticles in these states across the Fermi surface into the available empty states producing quasiparticle and quasihole excitations with equal and opposite energies in the wire frame of reference (we assume that the

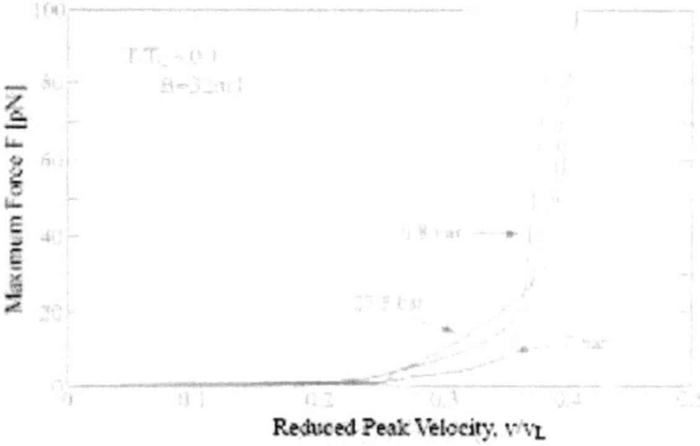

Figure 5.0.4. The response curves for a vibrating wire at the lowest temperatures for various pressures. The velocity axis is scaled by the Landau velocity v_L. The pair-breaking critical velocity is clearly seen at around $v_L/3$.

scattering is elastic). Alternatively, one can consider that Cooper pairs at the origin of the dispersion curve are broken producing a particle and a hole with equal and opposite energy in the wire reference frame. These excitations therefore have a maximum energy of $2p_F v$. Excitations with an energy greater than $\Delta - \mathbf{p}_F \cdot \mathbf{v}$ are able to move out into the bulk. This suggests a critical velocity given by,

$$\begin{aligned} 2p_F v_C &= \Delta - p_F v_C \\ v_C &= \Delta / 3p_F \end{aligned} \tag{5.0.5}$$

which agrees well with the measured value.

The assumption that the superfluid backflow is given by that of a smooth cylinder might seem a little bizarre considering the surface roughness of the vibrating wire. However, superflow can only change on the scale of a coherence length and so only roughness on a scale larger than this will have any effect. The coherence length decreases from about 0.06 μm at 0 bar to around 0.014 μm at 27 bar, the highest pressure shown in figure 5.0.4. The vibrating wires used have a surface roughness on a similar scale and so the surface might appear somewhat smoother as far as the superflow is concerned, at least at the lower pressures. As shown in the figure 5.0.4, some extra damping is observed below v_C which becomes more pronounced at the higher pressures. This could be a consequence of the smaller coherence length at the higher pressures. This would allow the surface roughness to have a greater influence on the superfluid backflow and would increase the local superfluid velocity in some places, thus allowing excitations to escape at lower velocities.

6 Detection of Vortex Lines in ^{3}He-B at very low temperatures

6.1 Andreev Scattering of Quasiparticles by a vortex line

Andreev scattering of quasiparticles around a vortex lead provides a very powerful tool for studying vortices and turbulence in superfluid ^{3}He-B in the low temperature regime. Since the quasiparticle dispersion curve is tied to the reference frame of the superfluid, the curve becomes tilted (via a Galilean transformation) for a moving superfluid, $E(\mathbf{p}) \rightarrow E(\mathbf{p}) + \mathbf{p}_F \cdot \mathbf{v}$. The dispersion curves are thus tilted by the circulating superflow around a vortex line as shown in figure 6.1.1.

Referring to the figure 6.1.1, first consider excitations incident on the top side of the vortex from the left (top of the figure). As the particles (with momentum to the right) approach the vortex line from the top, the superfluid flow increases to the left, which thus tilts down the dispersion curve to the right. The excitations move with constant energy in the laboratory reference frame. Hence the quasiparticle is able to pass through the flow with out difficulty (it's group velocity actually increases as it passes through the flow since it has a higher energy in the reference frame of the superfluid). So the full flux of quasiparticles incident on the top side of the vortex line are transmitted (here we neglect the core region of the vortex). However, the quasiholes incident from the left are on the opposite side of the dispersion curve which is lifted up on the top side of the vortex. In this case, the flow presents an effective energy barrier and the excitation will not have enough energy to pass if it passes too close to the core (where the flow is higher). As the quasihole approaches, it's energy in the superfluid reference frame decreases towards the energy minimum (the energy gap), and its group velocity decreases. If it has insufficient energy, then its energy in the superfluid frame will become equal to the energy gap at some point, here it is at the bottom of the dispersion curve and its group velocity is zero. The excitation is effectively pushed around the bottom of the dispersion curve by the flow gradient and so it emerges as a quasiparticle with a group ve-

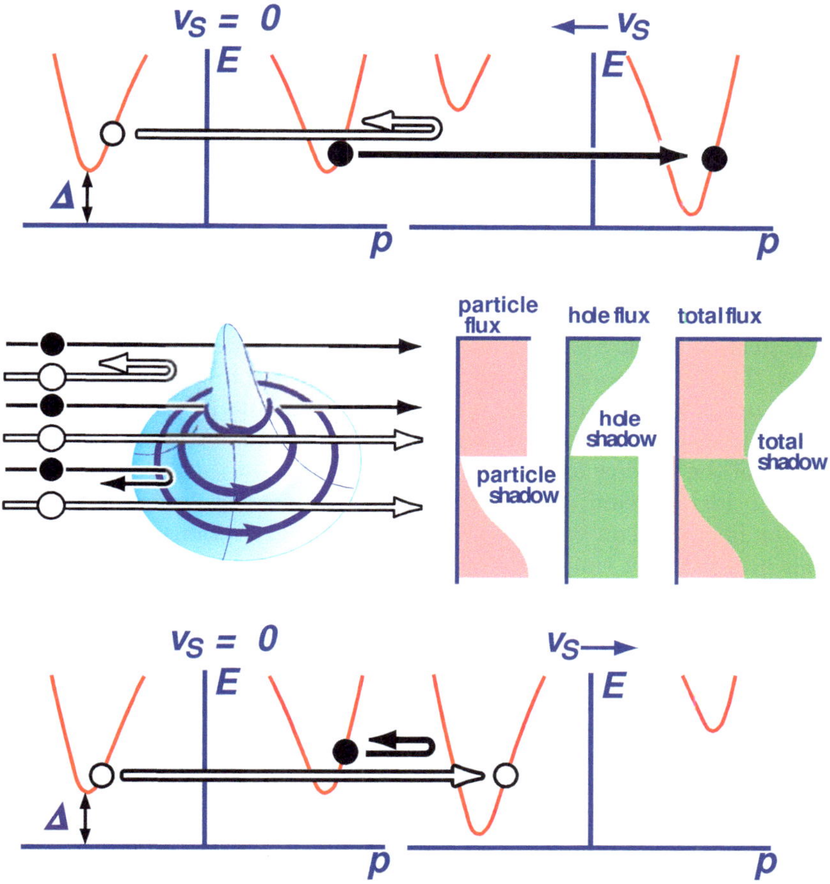

Figure 6.1.1. The excitation dispersion curves in the vicinity of a vortex line. Top and bottom left: un-tilted dispersion curves for excitations far from the vortex. Top/bottom left: the tilted curve for excitations passing the top/bottom side of of the vortex. Center: the transmitted flux of excitations incident from the left side of the vortex line, showing an excitation shadow due to the Andreev reflection of excitations.

locity in the opposite direction. The momentum transfer is extremely small (negligible) but the excitation trajectory is reversed. This, almost perfect retro-reflection, is called Andreev reflection (or Andreev scattering). The overall effect is that quasiholes passing close to the top side of the vortex are reflected, thus casting a quasihole shadow behind the vortex.

Referring to the low half of figure 6.1.1, on the bottom side of the vortex, the flow is in the opposite direction, so in this case all quasiholes are trasmitted but quasiparticles which pass too close to the vortex are Andreev reflected, producing a particle shadow.

Overall, the vortex casts a symmetric shadow (holes at the top and particles at the bottom). If we can measure the excitation flux (which we effectively do with vibrating wire techniques), then we can detect the vortex lines. The is the essence of the technique. The important point to note (as we show below) is that the width of this shadow is very large, making quasiparticles very sensitive vortex line 'detectors'. We can calculate the width of the shadow as follows:

Consider a beam of thermal excitations incident from the left of the vortex line. We can write the flux of incident excitations (number passing unit area per unit time) as:

$$\langle n v_g \rangle_i = \int_{\Delta}^{\infty} v_g(E) G(E) f(E) dE \tag{6.1.1}$$

$$= G(p_F) \int_{\Delta}^{\infty} \exp(-E/k_B T) dE$$

$$= G(p_F) k_B T \exp(-\Delta/k_B T) \, . \tag{6.1.2}$$

Consider a quasihole trajectory incident on the top side of the vortex with a distance of closest approach r (the impact parameter). The maximum tilt of the dispersion curve corresponding to the flow $\mathbf{v}(r)$ at this point is simply $\mathbf{p}_F \cdot \mathbf{v}(r) = p_F \hbar/(2m_3 r)$. Incident quasiholes will only be transmitted if they have an energy greater than $\Delta + \mathbf{p}_F \cdot \mathbf{v}(r)$ so the flux transmitted is:

$$\langle n v_g \rangle_t = \int_{\Delta + p_F \hbar/2m_3 r}^{\infty} v_g(E) G(E) f(E) dE \tag{6.1.3}$$

$$= G(p_F) \int_{\Delta + p_F \hbar/2m_3 r}^{\infty} \exp(-E/k_B T) dE$$

$$= G(p_F) k_B T \exp[-(\Delta + p_F \hbar/2m_3 r)/k_B T] \, . \tag{6.1.4}$$

The fraction of the quasihole flux transmitted on the top side of the vortex is thus

$$\begin{aligned} f_t &= \langle n v_g \rangle_t / \langle n v_g \rangle_i \\ &= \exp(-p_F \hbar / 2 m_3 r k_B T). \end{aligned} \tag{6.1.5}$$

Similarly, this is also the fraction of quasiparticles transmitted on the bottom side of the vortex. (Note that all quasiparticles are transmitted on the top side of the vortex and all quasiholes are transmitted on the bottom side of the vortex). The half width of the combined shadow $d = 2r$ is then obtained by setting $f_t = 1/2$ giving:

$$d = \frac{p_F \hbar}{m_3 k_B T \ln(2)} \tag{6.1.6}$$

Or in terms of the circulation quantum $\kappa = h/2m_3$,

$$d = \frac{p_F \kappa}{\pi k_B T \ln(2)} \tag{6.1.7}$$

This equates to around 20 μm at T =100 μK, which is of order a thousand times larger than the vortex core size. So vortices have an enormous cross section for Andreev scattering quasiparticle exciations which makes them relatively simple to observe using quasiparticle detection techniques (i.e. with vibrating wire resonators).

6.2 Early Vibrating wire measurements

Turbulence in superfluid ^{3}He-B(31) was first discovered accidentally at very low temperatures using the arrangement shown in figure 6.2.1. There were 2 vibrating wires spaced 1mm apart which were intended for use as a heater and thermometer for an NMR experiment located further up the cell. For the current purposes, we will refer to these as the detector and source wire. The wires are located deep with the refrigerant sinter stack.

We would normally expect that when heat is injected into the cell, by driving the heater (source) wire above its pair-breaking critical velocity, the thermometer will show an increase in damping corresponding to the rise in quasiparticle density/ temperature. This was indeed found to be the case at the lowest temperatures where the background quasiparticle density (before driving the source wire) was immeasurably small. However, as shown in figure 6.2.2 (top) in the presence of a significant background of thermal excitations (at slightly higher temperatures), the detector wire damping actually decreased despite the production of near-by quasiparticles. At first this was a very surprising result as it was very well known that the wire damping arises from quasiparticles. The only feasible explanation for such

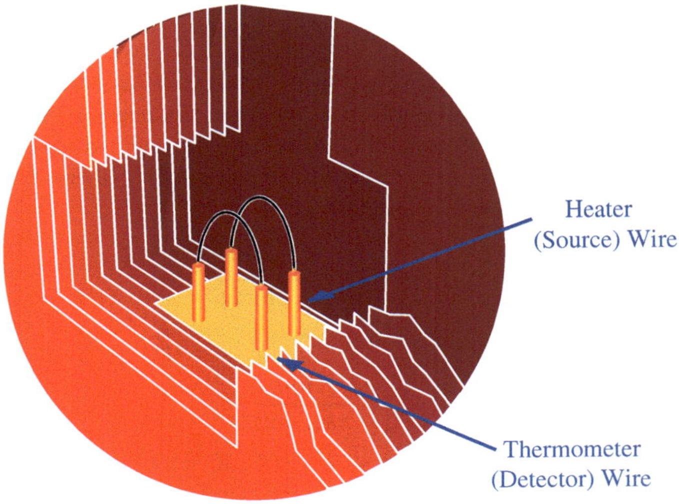

Figure 6.2.1. The experimental arrangement used for the first observation of turbulence in superfluid ^{3}He in low temperatures.

a behavior is that the source wire also produces something in addition to quasiparticles which is able to screen the detector wire from the background thermal quasiparticle excitations. The only objects which could do this in the B-phase are vortices. The damping was also observed to be quite 'noisy' in the presence of the generated vorticity, implying that the source wire was generating an erratic vortex tangle, i.e. quantum turbulence.

The fractional change in the thermal damping of the detector wire is equal to the fraction of incident thermal quasiparticles which are Andreev reflected by the turbulent flow field. This may therefore be considered as a measure of the surrounding line density. The fractional change in the detector wire thermal damping is shown in the lower half of figure 6.2.2 as a function of the source wire velocity for various temperatures.

The data clearly show that turbulence from a vibrating wire is only produced above it's pair-breaking critical velocity (equal to one third of the Landau critical velocity, $\sim$ 10mm/s at low pressures). This is an interesting result, especially when compared with correspond critical velocities in ^{4}He. Typical critical velocities for macroscopic objects to produce turbulence in ^{4}He are a few tens of mm/s, roughly a thousand times smaller than the Landau critical velocity in ^{4}He. The very different behaviors can be explained

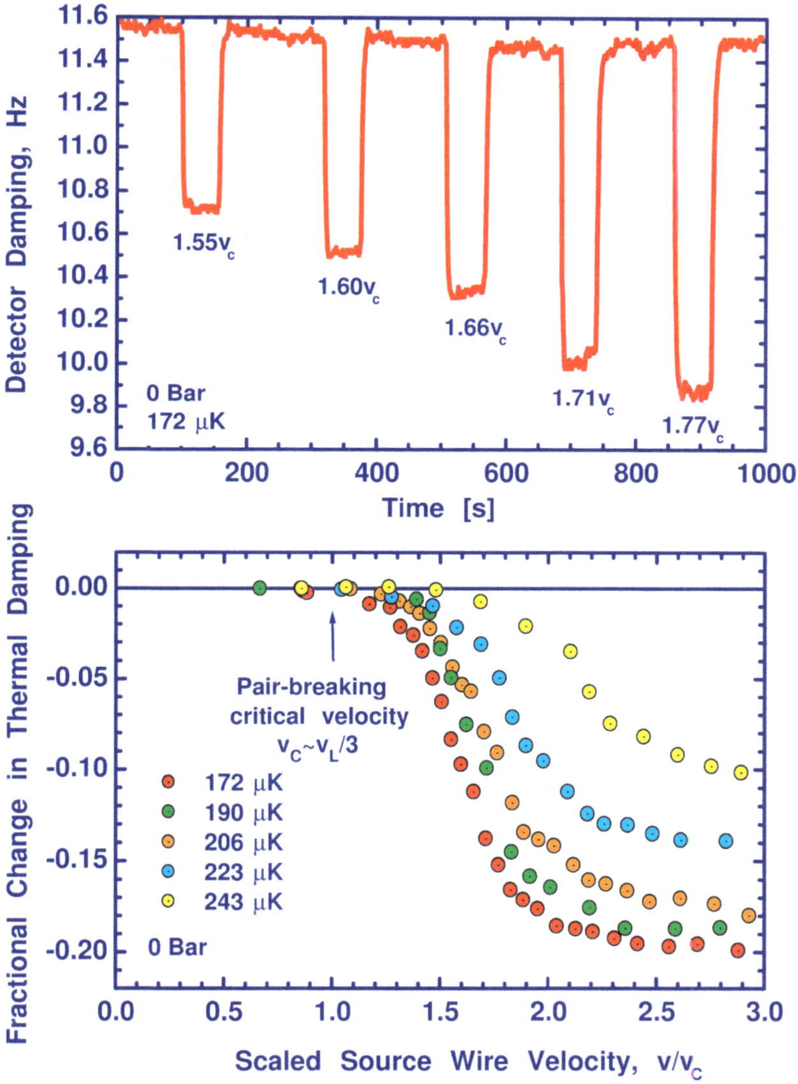

Figure 6.2.2. The first observations of turbulence in superfluid ^{3}He-B. Top: the damping on the detector wire as the source wire is driven to various velocities (given in terms of the pair-breaking critical velocity). Bottom: the fractional reduction in the thermal damping of the detector wire (due to turbulence) as a function of the source wire velocity.

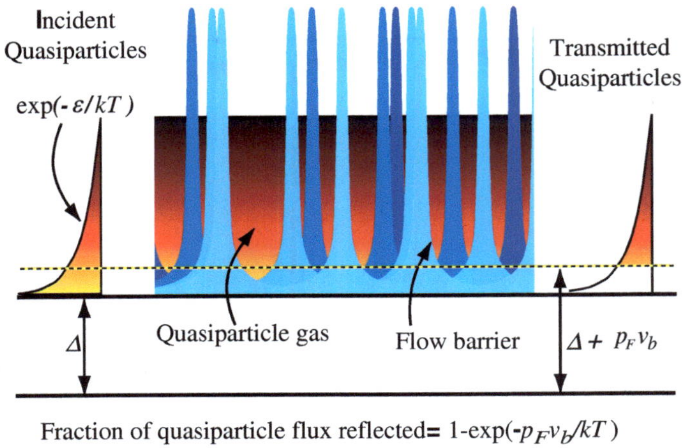

Figure 6.2.3. The effective flow barrier experience by thermal quasiparticle excitations incident on quantum turbulence.

by the size of the coherence length and surface roughness. In superfluid ^{4}He the coherence length is on the atomic scale (0.1 nm) so the superfluid is sensitive to atomic roughness. Any macroscopic object will in practice be very rough on this scale, and may therefore hold vortices pinned to surface roughness. In relatively small flows, these trapped remnant vorticies may grow, and self re-connect producing turbulence. So in this case, vortices do not have to be nucleated and the Landau critical velocity is irrelevant. In the case of superfluid ^{3}He however, the coherence length is much larger (∼60 nm at low pressures), smaller scale roughness is 'smoothed out' on the scale of the pair-breaking superfluid. Vibrating wires look quite smooth on this larger scale and so are unlikely to trap vortices. So to generate turbulence in this case, new vortices have to be nucleated and this requires a velocity close to that of the pair-breaking critical velocity.

Also apparent for the data in figure 6.2.2, is that the fractional screening of quasiparticles is smaller at higher temperatures. This is partly due to the fact that the average energy of quasiparticles is higher at higher temperature so they are more likely propagate over the energy barriers presented by the flow field.

To be more quantitative, the turbulent flow field can be considered as some random terrain of peaks and troughs as depicted in figure 6.2.3. The

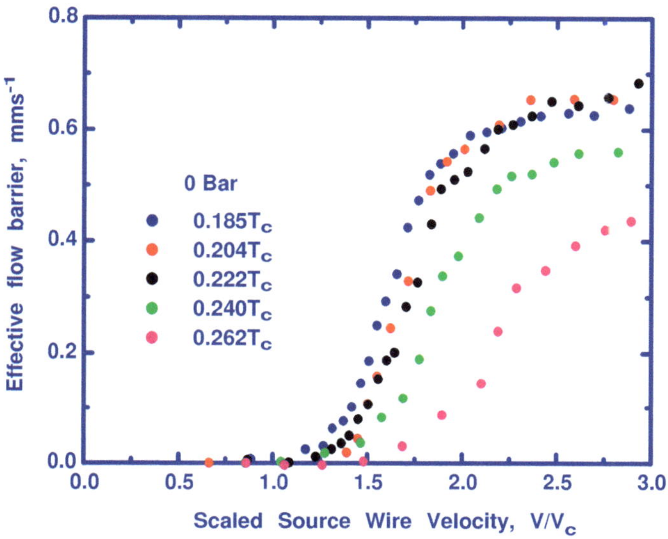

Figure 6.2.4. The effective flow barrier measurements for turbulence produce by a vibrating wire resonator.

peaks correspond to the large flow around the vortex cores. Quasiparticles are incident with a thermal energy distribution shown at the left. Low energy excitations are Andreev reflected, so only higher energy excitations are transmitted. The effect of the flow can then be modelled as an effective flow barrier $p_F v_b$ corresponding to the angular averaged maximum flow v_b along quasiparticle trajectories incident on the detector. The fractional screening (reduction in thermal damping) can then be written as

$$f = 1 - \exp(-p_F v_b / k_B T) \qquad (6.2.1)$$

The inferred flow barrier as a function of source wire velocity is shown in figure 6.2.4.

As shown in figure 6.2.4 the maximum flow barriers observed are of order 1 mm/s or less. To put this in perspective, this is the flow that would be observed at a distance of order 10 to 100 μm from a vortex core, so this sets a lower limit to the vortex line spacing in the turbulent tangle. It is not possible to obtain a precise number for the line density in these measurements as the fractional screening also depends on the size of the turbulent tangle, which is not known from these measurements.

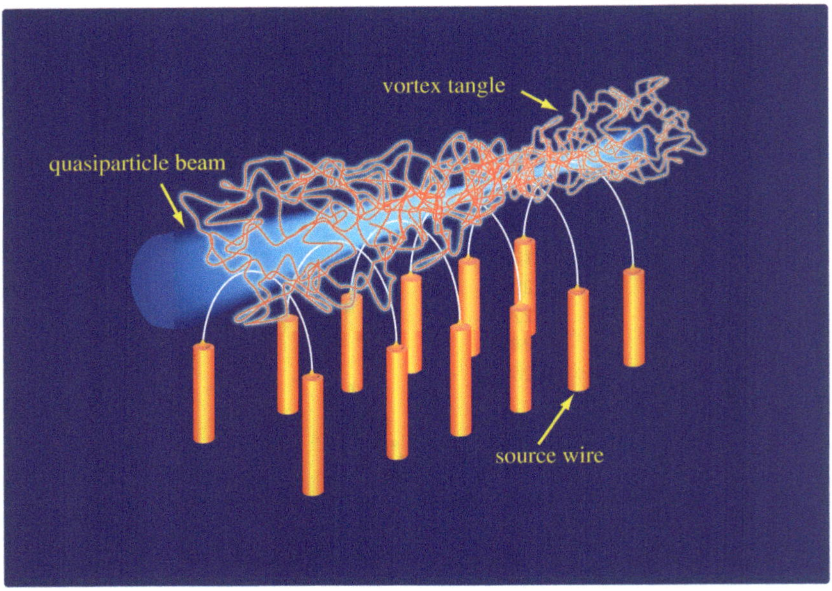

Figure 6.3.1. Schematic diagram showing the experimental arrangement for measuring the spatial extent of turbulence produced by a vibrating wire resonator.

Note also that the effective flow barrier is measured to become independent of temperature below about $0.22\,T_c$. This indicates that the low temperature regime has been reached in which the dissipation is temperature independent, no longer governed by thermal excitations (normal fluid component) and so some new (quantum) dissipation mechanism must now operate.

6.3 Spatial Distribution of Turbulence produced from a Vibrating Wire Resonator in ^{3}He-B

After the first observations of turbulence, discussed above, purpose-built experiments were run to further investigate its properties. One of the first experiments set out to measure the spatial extent of the turbulence produced by a vibrating wire resonator(55). The experimental arrangement used for this is shown in figure 6.3.1.

Six vibrating wire resonators are positioned in a row, with roughly 1 mm separation. One wire is chosen as the source wire, while the damping on

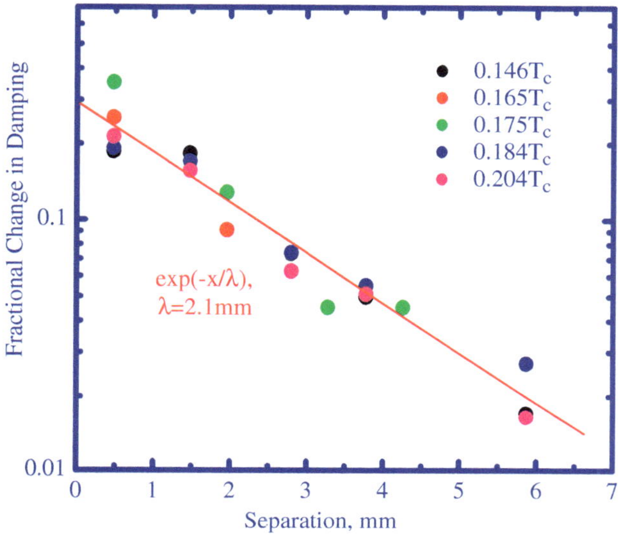

Figure 6.3.2. Measurements of the spatial extent of turbulence produced by a vibrating wire resonator. Data show the fractional reduction in damping (fraction of Andreev screening by vortices) as a function of distance from the source wire at various temperatures.

the other (detector) wires are monitored continuously. Vortices are detected via the associated Andreev scattering of excitations which causes a proportionate decrease in the thermal damping of the detector wires as discussed above. The fractional reduction in damping gives a measure of the surrounding vortex line density. The figure also illustrates that the vibrating wire produces both turbulence and a beam of ballistic quasiparticles when driven above its critical velocity. The quasiparticle beam is known, from other experiments, to be quite focussed along the axis of motion.

The fractional reduction in damping as a function of distance from the source wire (corresponding to different detector wires) is shown in figure 6.3.2. The solid line in the figure indicated that the spatial decay of the vortex tangle is consistent with an exponential decay with a decay length of around 2 mm. Note that this is enormous compared to the diameter of the vibrating wires ($\sim 5\,\mu$m). Note also that the decay is temperature independent at these low temperatures.

This experiment also shows that the turbulence produced is focussed along the axis of motion of the source wire (as is the quasiparticle beam).

This is illustrated in figure 6.3.1. Not shown in the figure is an additional vibrating wire resonator located just to the side of the wires shown. This 'side' wire shows very little vortex signal. One possible explanation for this is that the excitation beam produced by the source wire is pushing the turbulence outwards along the beam axis. There are however alternative explanations.

Given that we now know the spatial extent of the turbulence, we should now be able to infer the density of vortex lines in the turbulence observed. Before doing this, we describe below another experiment which was designed to observe the Andreev reflection directly. Here a beam of quasiparticles is fired at the vortex tangle while the reflected component of the beam is measured. This is done using a quasiparticle black-body radiator described below.

6.4 Quasiparticle Beam Experiments using Black-body Radiator

In order to produce and observe a ballistic beam of quasiparticle excitations in the B-phase, two criteria must be met. First, the temperature must be low enough for the beam to propagate without significant loss due to scattering with existing equilibrium quasiparticles i.e. the quasiparticle mean free path must be sufficiently large. Secondly, the intensity of the beam must be comparable, at least, to that of the background quasiparticles otherwise the beam cannot be resolved.

The quasiparticle mean free path is thought to become comparable to a typical length of the experimental region ($\simeq$1 cm) at $T \sim T_C/4$ and rises to several hundred meters for $T < T_C/10$. Most of the experiments which we have performed with ballistic quasiparticle (BQP) beams were operated at $T < T_C/6$, so that any loss due to scattering with existing quasiparticles ought to be negligible. The second criterion for observing a BQP beam is met by the design of the Lancaster style nulear cooling stages, giving excellent thermal contact to the near-by refrigerant.

A quasiparticle black-body radiator is shown in figure 6.4.1 (along side an external vibrating wire resonator which will be used to generate turbulence (described later). The radiator is simply a box with a small hole (the radiator orifice) on one side. The box contains two vibrating wire resonators, one used as a heater, the other as a thermometer. Quasiparticles, injected into the box by the heater wire ballistically bounce from wall to wall, thermalising before eventually leaving the box through the orifice, forming a quasiparticle beam. The time constant of the box is governed by the hole size and is typically less than 1 s. So a steady state is quickly reached inside the box in which the total power Q_T entering the the box

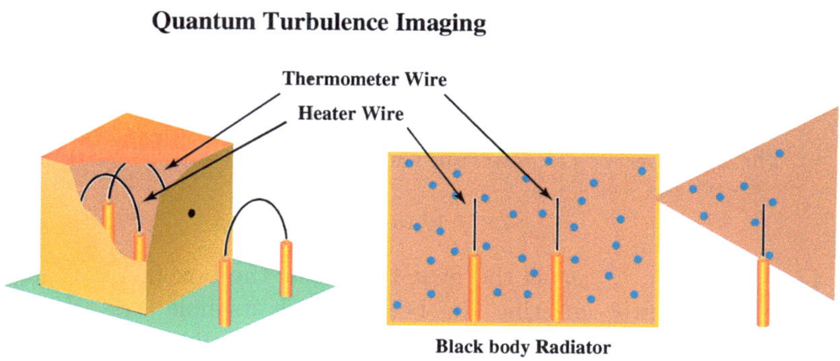

Figure 6.4.1. The quasiparticle black body radiator used for ballistic quasiparticle beam experiments.

must be balanced by the power emitted from the hole (the beam power).

Assuming that thermal equilibrium is always achieved inside the box, the power Q_T emitted from the hole is given by

$$Q_T = 2\frac{\langle nv_g \rangle}{4}\tilde{E}_T A_h \tag{6.4.1}$$

where the factor of 2 accounts for both quasiparticle and quasihole contributions, A_h is the area of the hole and $\tilde{E}_T$ is the average thermal quasiparticle energy. This average $\tilde{E}_T$ must be taken over the quasiparticle flux and can be calculated as follows:

$$\begin{aligned} \tilde{E}_T &\equiv \frac{\langle nv_g E \rangle}{\langle nv_g \rangle} \\ &\equiv \frac{\int_\Delta^\infty G(E)f(E)v_g(E)EdE}{\int_\Delta^\infty G(E)f(E)v_g(E)dE} \\ &= \Delta + k_B T \end{aligned} \tag{6.4.2}$$

where the last step makes use of the identity, $G(E)v_g(E) = G(p)$ which may be replaced by $G(p_F)$ in the equation since all the excitations have, approximately, the Fermi momentum.

Substituting for the quasiparticle flux (derived earlier) in Equation (4.3.9) and rearranging, then gives the following relationship between the frequency width inside a box (neglecting the intrinsic damping) and the total power entering the box(25):

$$\Delta f_2 T \tilde{E}_T = \gamma' \frac{2d}{\pi m A_h} \frac{p_F^2}{k_B} Q_T. \tag{6.4.3}$$

The total power Q_T entering the radiator has three contributions:

$$Q_T = Q_{wall} + Q_{ex} + Q_{ap} \tag{6.4.4}$$

The first term, Q_{wall}, is the heat leak in to the box from its paper walls. Q_{ex} is the power entering the box through its hole. It arises from the quasiparticles in the rest of the experimental volume and is presumably determined by the outside temperature. Q_{ap} is the power applied inside the box from the heater wire.

The power entering the through the holes, Q_{ex}, can be estimated from the frequency width of a thermometer wire outside of the radiator. It is usually found to be very small and will therefore be set to zero in the following discussion. The heat leak into the boxes, Q_{wall}, dominates the frequency widths of the thermometer wire inside the radiator when there is no applied power. The intrinsic widths of the wires can therefore also be assumed to be negligible in what follows, although in fact even if the intrinsic widths happened to be moderately large, there would be no significant error in the interpretation of beam experiments discussed below.

Hence, setting also $Q_{ap} = 0$, Equation (6.4.3) reduces to

$$\Delta f_2^0 T^0 \tilde{E}_T^0 = \gamma' \frac{2d}{\pi m A_h} \frac{p_F^2}{k_B} Q_{wall} \tag{6.4.5}$$

The superscripts '0' refer to the measured values of Δf_2 and T, when there is no applied heat. The value of $\tilde{E}_T$ is inferred from Equation (6.4.2) using the BCS expression for the energy gap, $\Delta = 1.76 k_B T_C$. The effect of the applied heat is then observed by measuring the change in the quantity on the left hand side of the equation. Thus the model predicts:

$$\begin{aligned} \delta \Delta f_2 T \tilde{E}_T &\equiv \Delta f_2 T \tilde{E}_T - \Delta f_2^0 T^0 \tilde{E}_T^0 \\ &= \gamma' \frac{2d}{\pi m A_h} \frac{p_F^2}{k_B} Q_{ap} \end{aligned} \tag{6.4.6}$$

The quantity on the left hand side of this equation is obtained directly from the measurements of the thermometer wire and is often refer to as the 'width parameter', W. So Equation (6.4.6) therefore implies that the width parameter of the radiator is simply proportional to the applied heat. Furthermore the constant of proportionality, the 'radiator calibration constant'

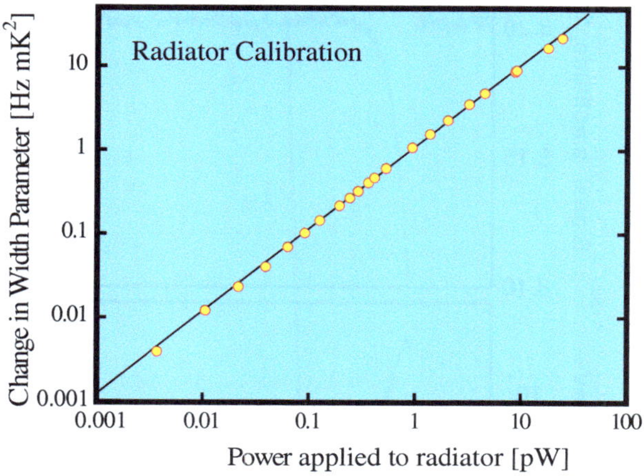

Figure 6.4.2. Calibration of a blackbody radiator. Note the linearity over several orders of magnitude in the applied power, and the high sensitive to low power inputs

can be easily measured by applying heat with the heater wire. An example of such measurements are shown in figure 6.4.2.

As shown in figure 6.4.2 the radiator response (width parameter) is linear in the applied power over several decades. For the width parameter, T and $\tilde{E}_T$ are expressed in units of mK, using the temperature scale of Greywall(9). The values used for p_F are those given by Wheatley(32). Note that the logarithmic scale extends over 5 orders of magnitude and therefore provides an good test of the model. The value for γ' is found to be around 0.3 (33).

Once calibrated, the black-body radiator is able to detect incoming power from any source (provided that the power is dissipated in the form of quasiparticle excitations). This has led to a very wide variety of applications, including measurements of the pair-breaking beam emiitted from a vibrating wire (25), direct observations of Andreev reflection (34; 35), measurements of thermodynamic properties (36) and phase nucleation (37). It has also been used for particle and vortex detection as described below.

6.5 Particle detection using a Black-body radiator.

A small energy deposition E into a black-body radiator produces a (near) instantaneous temperature rise given by:

$$\Delta T = E/C(T) \tag{6.5.1}$$

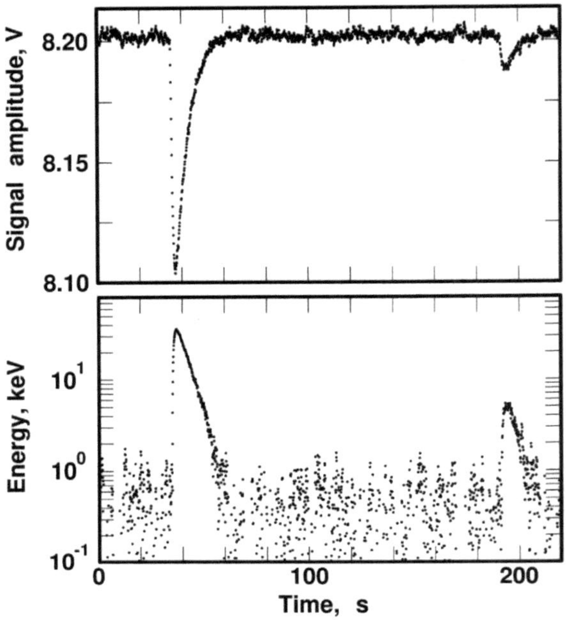

Figure 6.5.1. Particle detection using a black-body radiator. The top plot shows the raw measurement: events are due to background radioactivity. The lower plot shows the same data converted to energy deposition.

where $C(T)$ is the heat capacity of the superfluid ^{3}He-B contained within the radiator, due to thermal quasiparticle excitations. The heat capacity is roughly proportional to the quasiparticle density and falls exponentially at low temperatures. Consequently a small energy deposition can produce a relatively large temperature rise. This combined with the very high sensitivity of a vibrating wire thermometer makes for an very sensitive energy detector (bolometer).

Figure 6.5.1 shows an example of how the radiator can be used for particle detection. The top plot in figure 6.5.1 shows the raw measurement of the resonant signal height of the thermometer wire inside a radiator as a function of time. In this case, the signal has been amplified using a SQUID to reduced signal noise. Occasional events are observed due to background activity arising from cosmic rays and remnant radioactivity in the construction materials used for the refrigerator and surrounding laboratory. The events correspond to a high energy particle (e.g. a cosmic ray muon) colliding with one or more of the ^{3}He atoms and depositing energy. The energy is quickly converted to quasiparticle excitations resulting in a transient temperature rise and a corresponding reduction in the thermometer signal. The lower plot shows the same data converted to energy deposition. The conversion may be done by calibration using heater wire pulses, or using Equation (6.5.1). Alternatively, the energy may be found by integrating the extra power ΔQ leaving the radiator,

$$E = \int \Delta Q dt \tag{6.5.2}$$

where the extra power is obtained from the increase in width parameter using the radiator calibration as given by Equation (6.4.6).

Note the very good energy resolution, in this case of order 1 keV. This is comparable to other state-of-the-art cryogenic particle detectors and it has been proposed that this may be useful as a potential dark matter detector (38). Indeed, this application is currently being pursued and developed by the ULT group at the CRTBT, CNRS, Grenoble.

6.6 Measurements of the Kibble-Zurek Mechanism for Defect creation via a rapid phase transition.

Superfluid ^{3}He has many analogies with cosmology. In particular, various predicted defects in space-time have direct analogies with known defects in the superfluid ^{3}He condensate (with various types of monopoles, vortices and domain walls). Quantum vortices in the superfluid are analogous to cosmic strings (line defects in space-time). The accepted theory for the formation mechanism of defects in cosmology is known as the Kibble mechanism. This has has been modified slightly in more recent years and applied to more general systems, including condensed matter systems, by Zurek and is thus commonly known as the Kibble-Zurek Mechanism. The mechanism basically describes the formation of defects during rapid symmetry-breaking phase transitions. In the universe, such transitions occurred in the very first moments of the big bang, as the four fundamental forces became separated and distinct. In a superfluid the corresponding transition is from the normal fluid phase.

In the case of superfluid ^{3}He we can produce a very rapid phase transition by irradiating the fluid with low energy neutrons (from a source placed outside of the cryostat). ^{3}He has an extremely large cross-section for neutron capture, forming a tritium nucleus and a proton:

$$^3\text{He}+\text{n} \rightarrow ^3\text{H} + \text{p} + 764\text{keV}$$

the proton and triton fly apart, back to back, carrying the combined kinetic energy of 764 keV. This energy is quickly transferred to the surrounding atoms forming a very localised intense hot-spot of normal fluid which then subsequently cools very quickly as the quasiparticles disperse into the surrounding cold superfluid. This produces very rapid phase transition as the hot-spot cools through the superfluid transition.

The low temperature, ordered phase of a system generally has associated broken symmetries (e.g. a solid crystal breaks translational and orientational symmetries). This leads to a degeneracy in the symmetry variable or 'order parameter'(e.g. the crystal is just as likely to form in one orientation as another). The ordering in a simple superfluid breaks gauge symmetry as the phase of the wavefunction must be chosen, but the choice is arbitrary. In ^{3}He, more complex symmetries are also broken associated with the spin and orbital momentum orientations of the Cooper pairs (the broken symmetries are actually remarkably similar to those of space-time). The type of symmetry which is broken determines the type of defects that are possible in the ordered system. For instance, a discrete symmetry (eg up/down for spins in a ferromagnet) leads to domain walls as the order parameter changes from one value (e.g. up) to another (e.g. down). If the order parameter has a phase then line defects (vortices/strings) are possible in which the phase changes by 2π around the core. Note, that in all cases, the core of the defect has an undefined order parameter so the ordering is at least partially destroyed. The core thus resembles the higher temperature phase (vortex cores can be considered to contain normal fluid, whereas cosmic strings have enormously heavy cores corresponding to the enormous energy of the high temperature phase).

The Kibble-Zurek mechanism is depicted in figure 6.6.1 for the case of our superfluid ^{3}He experiment (although the extension to other systems is self-evident). In caption 'A' of the figure, we show the initial energy deposition following the neutron capture. This produces a localised hot spot of normal fluid (B), analogous to early moment of the big bang when all forces were unified. As the system rapids cools (universe expands) the rapid phase transition(s) occurs. During a rapid phase transition, different regions or 'proto-domains' form in the system (C). Each domain must choose the order parameter independently (D), since there simply isn't enough time for the ordering information to be transmitted from one proto-domain to the next, i.e. the regions are casually disconnected. As the system further cools, the gradients in the order parameter are smoothed out. However, in the case of an order parameter described by a phase factor, regions may form in which the phase evolves by 2π around some central core (E). This corresponds to the formation of a line defect (vortex/string). Many such

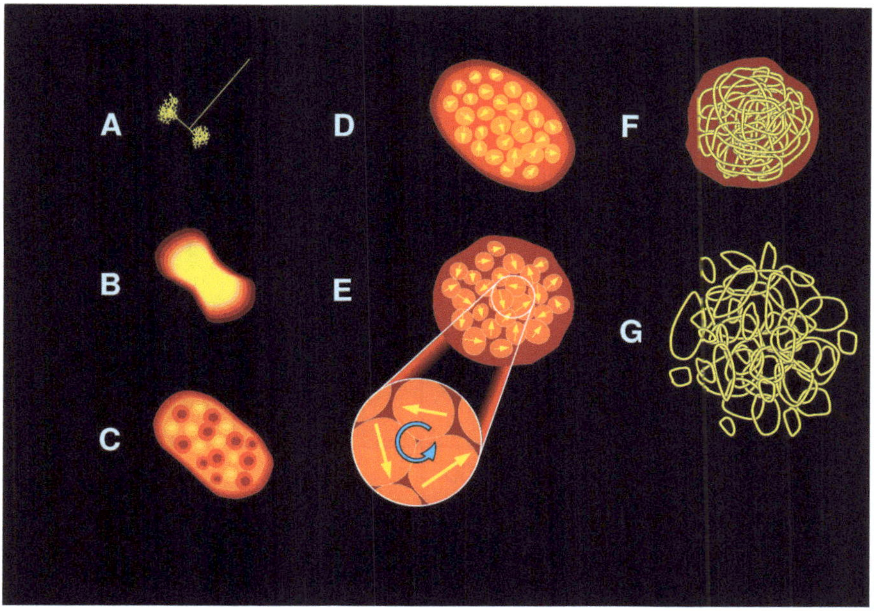

Figure 6.6.1. Schematic of the Kibble-Zurek mechanism, applied in this case to vortex creation via a rapid neutron-induced phase transition in ^{3}He-B. See text.

defects may form in random locations producing a tangle (F), which then subsequently evolves/decays on much longer time scales (G). The density of the defects generated depends simply on the size of the proto-domains which depends on the rate of the transition (the faster the transition, the smaller the regions which are casually disconnected, hence the higher the density of vortex lines).

Measurements of the neutron events observed in a black-body radiator are shown in figure 6.6.2. Further details can be found in ref. (39). The arrows indicate the events observed, most of which arise from background radiation and gamma rays which are also emitted from the source. The neutron absorption events produce the largest signals and are easily resolved.

The energy of each of the observed events can be measured and then plotted as a histogram of number of events versus energy. This results in a clear peak associated with the neutron absorption events with the background events mostly at much lower energies. The neutron peaks are shown in figure 6.6.3 for three different fluid pressures (the pressure changes

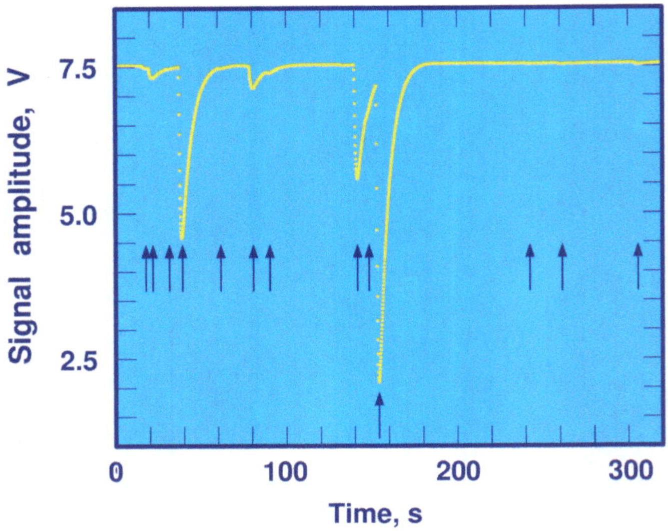

Figure 6.6.2. Measurements of events in a black-body radiator in the presence of a neutron source.

the superfluid properties which should effect the number of defects predicted by the Kibble-Zurek mechanism).

As shown in figure 6.6.3, the energy of the neutron peak is significantly lower than the 764 keV energy which is released in the nuclear reactions. Part of the discrepancy can be attributed to the emmission of uv scintillation (some of the energy is converted to uv). The scintillation losses have been measured by other experiments. The remaining energy deficit is pressure dependent and is in excellent quantitative agreement with the amount of vorticity predicted to occur in the Kibble-Zurek mechanism(39) (the vorticies produced will decay on much longer timescales so their energy is not observed in these measurements). It is very seldom that a cosmological theory can be tested so rigorously in the laboratory, thus these experiments received a great deal of interest and has led to much further research and the formation of a European network on further developing analogies between various areas of cosmology and condensed matter systems.

6.7 Direct Measurements of Andreev Reflection from Vortices

Precise measurements of the Andreev reflection of quasiparticles from a vortex tangle are relatively simple using the black-body radiator techniques.

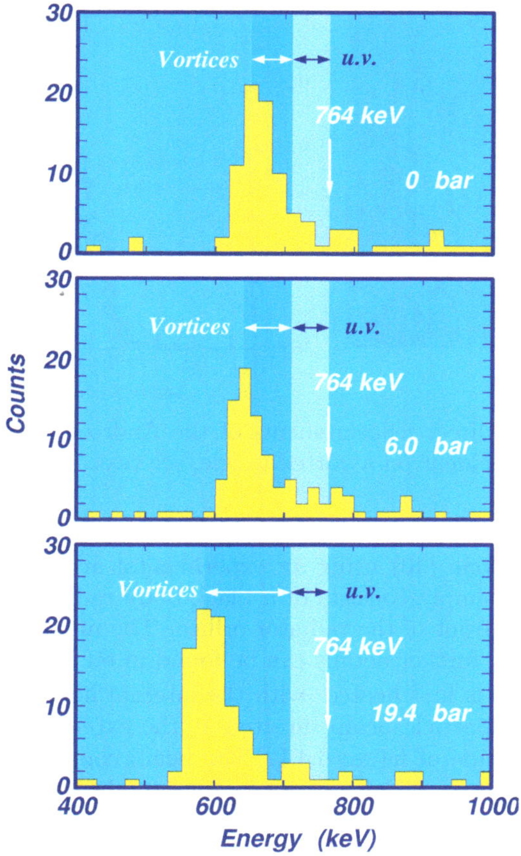

Figure 6.6.3. The measured neutron absorption peak for three different fluid pressures, clearly revealing a 'missing energy', as expected for vortex production via the Kibble-Zurek mechanism.

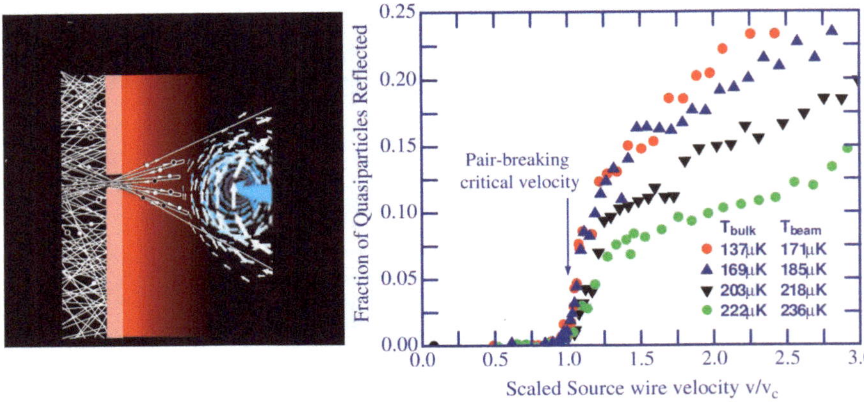

Figure 6.7.1. Direct measurements of the Andreev reflection of a quasiparticle beam incident on a vortex tangle, see text.

This was first done (40) using an experimental arrangement is shown in figure 6.4.1. It simply consists of a black-body radiator, with a vibrating wire outside, in front of the radiator orifice. The measurements are breifly described below, further details can be found in ref. (40).

The radiator is first heated, with the internal heater vibrating wire, to produce a quasiparticle beam directed at the external wire, as shown on the right hand side of figure 6.4.1. The beam contains a thermal energy distribution of quasiparticles corresponding to the temperature of the radiator. The external wire is then driven to some, relatively high, velocity to produce a vortex tangle (quantum turbulence). The flow field around the vortex Andreev reflects low energy quasiparticles in the beam as shown schematically on the left hand side of figure 6.7.1. The beauty of Andreev reflection is that the quasiparticles are near perfectly retro reflected, so regardless of the angle of incidence, an reflected excitation will accurately retrace its path back through the orifice and into the radiator. So the effect of the tangle, is to effective restrict the excitations when can escape the radiator. However, the beam flux (power) is fixed by the heat applied to the radiator, therefore the Andreev reflection produces an increase in the quasipartcle density (width parameter) in the radiator.

To be a more quantitative, in the absence of the tangle (when the external wire is not driven, $v = 0$), the measured width parameter in the box is

$$W(0) = CQ \tag{6.7.1}$$

where C is the box calibration constant described above, and Q is the applied power which may be taken as the beam power since the heat leak into the radiator may be neglected. In the presence of vortices, some fraction f of the beam power incident of the tangle Q_i is reflected back into the box (shown schematically in figure 6.7.1, left), resulting in a higher box temperature and width parameter $W(v)$. The beam power which is transmitted through the tangle must balance the power supplied so:

$$Q = Q_i - fQ_i = Q_i(1 - f) = W(0)/C \tag{6.7.2}$$

where the last equality is obtained form Equation (6.7.1). However, the power emerging from the radiator Q_i is simply determined by the radiator width parameter, $W(v) = CQ_i$. Substituting this in Equation (6.7.2) and then re-arranging gives the fraction of the beam which is Andreev reflected:

$$f = 1 - W(0)/W(v) \tag{6.7.3}$$

Note that if the fraction reflected was unity then the radiator would effectively be closed, and the width parameter would rise to infinity.

Equation (6.7.3) now allows the fraction of the beam Andreev Reflected by the turbulence to be obtained from the measured width parameters. The result is shown in figure 6.7.1. The data is very similar to that obtained by measuring the fractional reduction in damping of a wire inside a vortex tangle. As found previously, clearly the critical velocity for the production of the tangle (turbulence) is very close to the pair-breaking critical velocity. Also the fractional Andreev reflection becomes temperature independent at low temperatures indicating that the tangle density becomes limited by some non-thermal (quantum) mechanism.

6.8 Measurement of the Vortex Line Density

Given the fraction of excitations Andreev reflected from a vortex tangle, and the spatial extent of the tangle, we should be able to infer the vortex line density. This is done, using a simple model(40) described below.

Consider a thin slab of unit cross section and thickness Δx containing a homogeneous vortex tangle of line density (length of line per unit volume) L, as depicted in figure 6.8.1. The total length of line in the slab is $L\Delta x$. Taking the vortex as a tube of radius r and projecting its area onto the face of the slab gives an area of order $2rL\Delta x$, which gives the probability of a quasiparticle normally incident on the slab of coming within a distance r of the vortex core. As argued above, excitations of energy $E = \Delta + k_B T$ (and momentum in the appropriate direction) will be retro reflected if they approach to within a distance $p_F\hbar/2m_3k_BT$ of a vortex

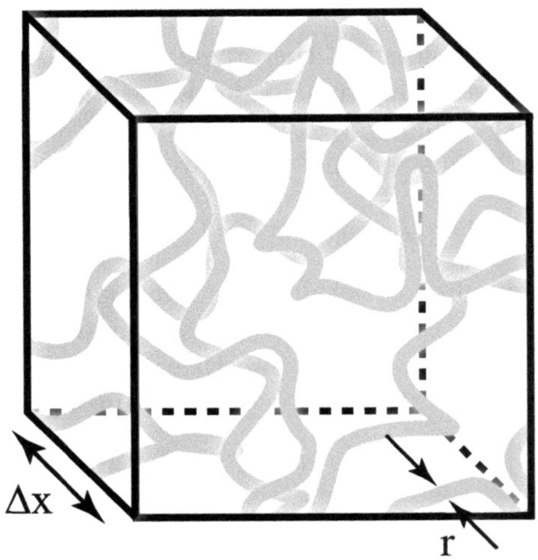

Figure 6.8.1. Simple model for estimating the vortex line density from the Andreev reflection measurments, see text.

core (assuming that the vortices are well separated so that their flow fields do not overlap significantly). Thus the fraction of excitations reflected is of order $p_F \hbar L \Delta x / 2 m_3 k_B T$. Integrating, the fraction of excitations Andreev reflected by a tangle of thickness D is

$$f \sim 1 - \exp(-D/\lambda) \tag{6.8.1}$$

i.e. the flux of quasiparticles transmitted through homogeneous turbulence decays exponentially with distance with a decay length given by

$$\lambda \sim \frac{2 m_3 k_B T}{p_F \hbar L} \tag{6.8.2}$$

Using these equations and setting $D = 2\,\mathrm{mm}$ as measured above, we can convert the fractional Andreev reflection data of figure 6.7.1 into a vortex line density. The results are shown in figure 6.8.2.

The line densities are estimated to be correct to within a factor of 2. A more accurate estimation would require a more precise knowledge of the spatial and angular distribution of the tangle (the values obtained should be

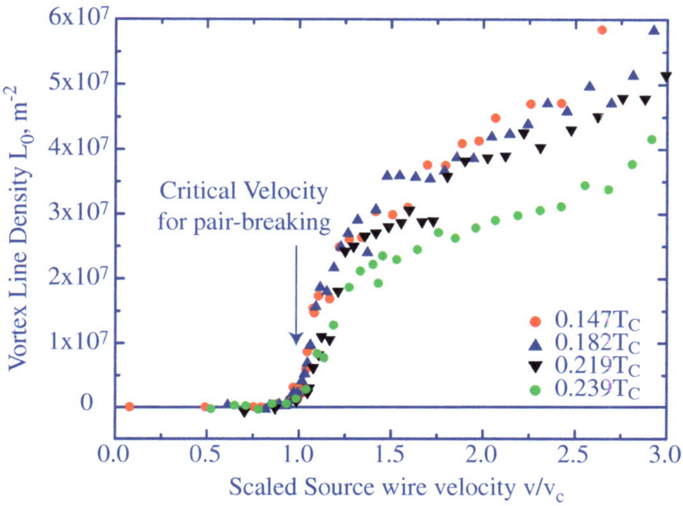

Figure 6.8.2. The inferred vortex line density of turbulence produced by a vibrating wire as a function of the wire velocity. See text.

considered as spatial/angular averages). As inferred previously, the tangle density becomes temperature independent below around $0.2T_C$ suggesting a temperature independent decay mechanism operates. This is not too surprising as the density of thermal excitations is so small that they should have very little effect on the tangle dynamics. Also note that the maximum observed line densities are of order $10^7\,\mathrm{m}^{-2}$ to $10^8\,\mathrm{m}^{-2}$ corresponding to a minimum vortex line spacing $l = L^{-2} \sim 200\,\mu\mathrm{m}$.

6.9 Vibrating Grid experiments

The initial vibrating wire experiments were quite successful observing the basic properties of turbulence in the B-phase in the low temperature regime. In order to produce more homogenous turbulence with a greater intensity (line density) a vibrating grid resonator was used. Grids are commonly used in studies of classical turbulence, usually the turbulence is generated by pass a steady flow through the grid and observing the turbulence down stream. This is not practical for a low temperature superfluid experiment. However, a similar arrangement has been realised in superfluid ^{4}He at relatively high temperatures by towing a grid through the superfluid(41; 42). At the temperatures required for superfluid ^{3}He experiments, a towed grid set-up would be very difficult to achieve without producing excessive heating. However, we can easily vibrate a grid through the superfluid at the lowest achievable temperatures.

The experimental arrangement for the first vibrating grid experiments(56) in ^{3}He-B at low temperatures is shown in figure 6.9.1. The grid is made from a 5.1×2.8 mm mesh of fine copper wires. The wires have an approximately 11 μm square cross-section and are spaced 50 μm apart leaving 40 μm square holes. A 125 μm diameter Ta wire is bent into a 5 mm square frame and glued through Stycast-impregnated paper which forms the inner cell wall of a Lancaster style nuclear cooling stage. The mesh is glued to the Ta wire over thin strips of cigarette paper for electrical insulation.

Facing the grid are two vibrating wire resonators made from 2.5 mm diameter loops of 4.5 μm NbTi wire. The 'near' and 'far' wires are positioned 1 mm and 2 mm from the grid respectively. An additional wire resonator (not shown in the figure) is used as a background thermometer. This wire, not shown in the figure, is located about 4 mm to the side of the grid and enclosed in a mesh cage to ensure that its response is not influenced by any stray turbulence.

The grid is operated similarly to a wire resonator. It is situated in a vertical applied magnetic field and driven by the Lorentz force generated by passing an ac current through the Ta wire. As the grid moves, the Ta wire develops a Faraday voltage proportional to its velocity. The grid resonates at a frequency of ~1300 Hz, determined by the stiffness of the Ta wire and the mass of the grid.

Measurements of the resonant peak velocity v of the grid as a function of the peak driving force F are shown in figure 6.9.2 (bottom). The measurements were taken as the cell was slowly warming under its external heat leak. Also, significant warming occurred at the highest velocities due to the power required to drive the grid. The temperatures indicated in the figure cover the range of cell temperatures as measured by the thermometer wire over the given set of data. At low velocities the grid response is quite linear and is due to a combination of the intrinsic (vacuum) damping of the resonator and the thermal quasiparticle damping. At higher velocities, the grid response becomes non-linear due to turbulence, discussed further below. We can subtract the thermal contribution to the damping force as follows.

The frequency width of the grid resonance (at low velocities) was measured as a function of the thermometer resonant width as the cell warmed though-out the experiments. The frequency width is proportional to the damping force per unit velocity, $(dF/dv)_{v=0}$. After subtracting the intrinsic damping (0.22 Hz for the grid and 0.05 Hz for the thermometer), we

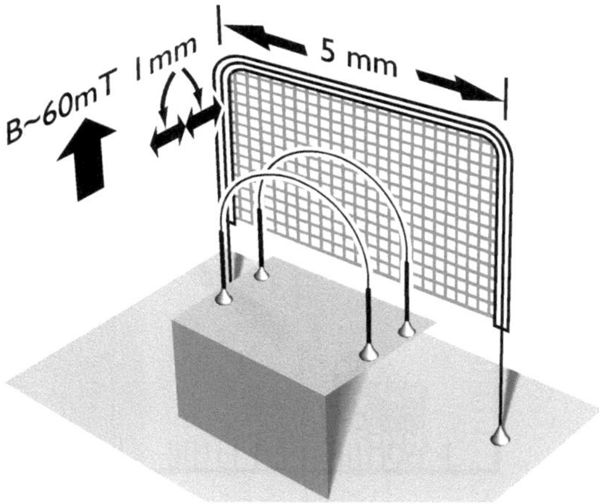

Figure 6.9.1. The arrangement of the grid and associated vorticity detector wires.

obtain the thermal damping at low velocities for the two wires which is plotted in the in figure 6.9.2 (top).

The thermal damping of the grid and the thermometer vibrating wire are seen to be proportional to each other over the range of the measurements. We can now use the measured damping on the thermometer wire to infer directly the thermal damping force on the grid at low velocities. The thermal damping force at higher velocities becomes non-linear. However the non-linearity for the vibrating wires is known to be proportional to $1 - \exp(-\lambda p_F v_0/kT)$ where the constant λ is found to be 0.85(56). Assuming that the thermal damping on the grid has this same form then the thermal contribution to the damping force can be found with no free parameters. The thermal damping force for the grid is thus inferred and can then be removed from the data of figure 6.9.2 to leave the inferred zero temperature response curve shown in figure 6.9.3. The corrected data collapse to a single curve, verifying the assumptions made.

The zero temperature response of the grid shows linear behavior below about 0.5 mm/s. This corresponds to the intrinsic damping of the resonator. At higher velocities the response tends to a power law behavior, approximately proportional to the square-root of the force. This is the expected variation for turbulent dominated damping(43). The corresponding behavior for conventional vibrating wire resonators, also shown in the figure, is very different. These show a pair-breaking critical velocity of around 8 mm/s

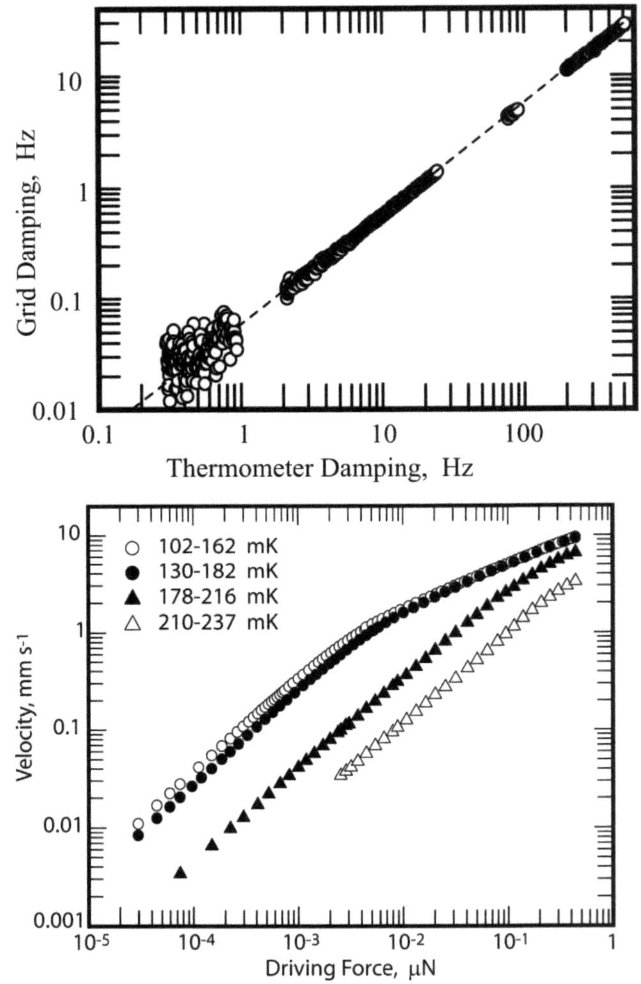

Figure 6.9.2. Top: The inferred thermal damping of the grid resonator as a function of the thermal damping of a conventional wire resonator. A linear relationship is found as expected for thermal quasiparticle damping. Bottom: The grid response curve for various temperatures.

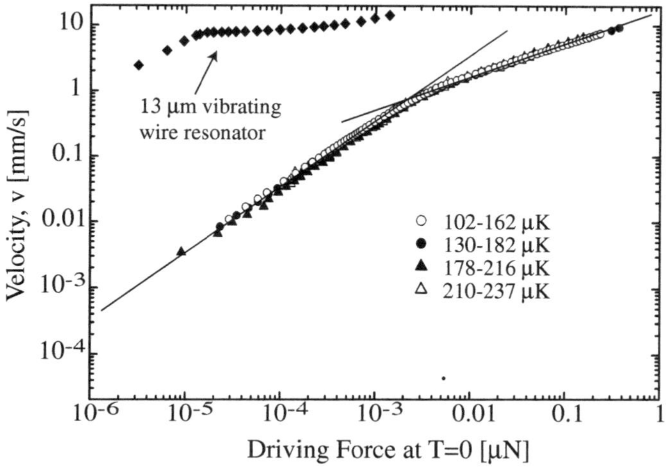

Figure 6.9.3. The inferred $T = 0$ grid response and the response of a conventional wire resonator. Lines show linear and $v \alpha \sqrt{F}$ behavior

at 0 Bar above which the response is seen to be very flat. This indicates that the pair-breaking contribution to the damping on the grid is much smaller and may well be negligible. This makes the grid very advantageous for studying turbulence in ^{3}He-B since the pair-breaking beam not only causes massive heating, but it may also have a significant effect on the vortices generated by conventional wires as discussed above.

6.10 Detection of Grid Turbulence

Vortices generated by the grid are detected by the two facing vibrating wire resonators via the screening effect on the thermal damping as discussed above. Briefly, the two resonators and the thermometer resonator are driven on resonance at relatively low velocity. The resulting induced voltages across the wires are continuously monitored, allowing us to deduce the quasiparticle damping (frequency width of the resonance) $\Delta f_2(T)$ for all three wires. The grid is then driven to some velocity v generating vortex lines. This vorticity Andreev-reflects some fraction f of quasiparticles approaching a vibrating wire, giving rise to a reduced damping

$$\Delta f_2(v, T) = (1 - f)\Delta f_2^0(T) \tag{6.10.1}$$

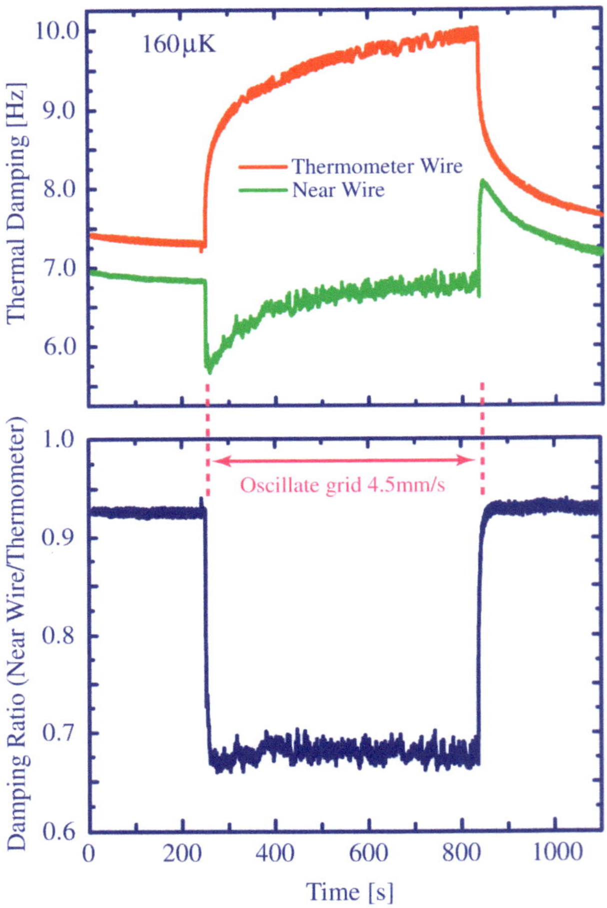

Figure 6.10.1. The thermal damping on the thermometer wire (top) and the nearest detector wire (bottom) as a function of time. The is driven to some velocity producing vorticity during the interval shown.

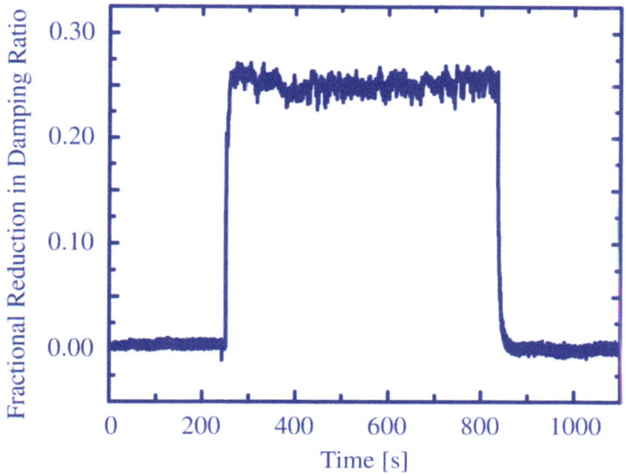

Figure 6.10.2. The fractional screening of quasiparticles from the nearest detector wire, due to vorticity produced by the grid, inferred from the data of figure 6.10.1.

In practice, significant power is required to drive the grid, resulting in an overall warming of the cell, as measured by the thermometer wire damping shown in the upper plot of figure 6.10.1. The detector wires show this same overall warming effect, but in addition show a reduction in damping due to vortices produced by the grid, as also shown in the upper plot in figure 6.10.1. The damping in the absence of turbulence $\Delta f_2^0(T)$ is simply proportional to the thermometer wire damping (ideally, they would be equal, but in practice a scale factor, typically less than 10%, is needed to account for the fact that, even with no turbulence, the quasiparticle damping on each of the three wires is slightly different owing to differing wire geometries). So the factor $1 - f$ in Equation (6.10.1) is simply given by the scaled ratio of the detector and thermometer wire dampings. The ratio of the detector to thermometer wire damping, corresponding to the data in the upper plot of figure 6.10.1, is shown in the lower plot of figure 6.10.1. The scaling is applied to make this ratio equal to unity before the grid is driven. The corresponding inferred fractional screening f due to the vortices emitted by the grid is shown in figure 6.10.2. The steady state average values of the fractional screening f are found to increase roughly as v^2. The 'far' wire, 2 mm from the grid, has roughly a factor of two less screening than the 'near' wire, 1 mm from the grid, over the entire velocity range. If the variation with distance followed an exponential decay, as found previously

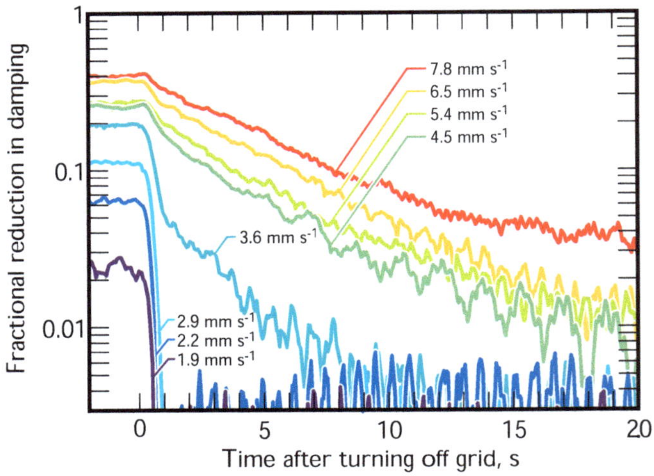

Figure 6.11.1. The fractional screening of quasiparticles from the nearest detector wire as a function of time after turning the grid off from various initial grid velocities

for turbulence generated by vibrating wires(55) (see above), then this would correspond to a spatial decay length of $D \sim 1.5\,\mathrm{mm}$. This allows the approximate vortex line density to be inferred from these measurements using the arguments given above (Equations (6.8.1) and (6.8.2)).

6.11 Vortex ring production

The transient behavior of the inferred line density(46) after the drive to the grid is turned off is shown in figure 6.11.1 for the wire nearest to the grid. Data are shown for various initial grid velocities. Evidently there is a very clear transition in the behavior at grid velocities of around 3.5 mm/s (depending a little on pressure). At lower velocities the recovery is very much faster. The initial decay time constants are shown in figure 6.11.2 as a function of the initial grid velocity.

As shown in figure 6.11.2, there is a very sharp transition in the initial decay time constant of the observed vorticity. Note that there are two points for the 3.6 mm/s data corresponding to the short fast decay followed by the longer slow tail of the data shown in figure 6.11.1. At low grid velocities the decay time is actually limited by the response time of the measuring equipment and that of the grid resonator, so here the observed decay time of $\sim 0.1\,\mathrm{s}$ should be considered as an upper limit to the decay time of the

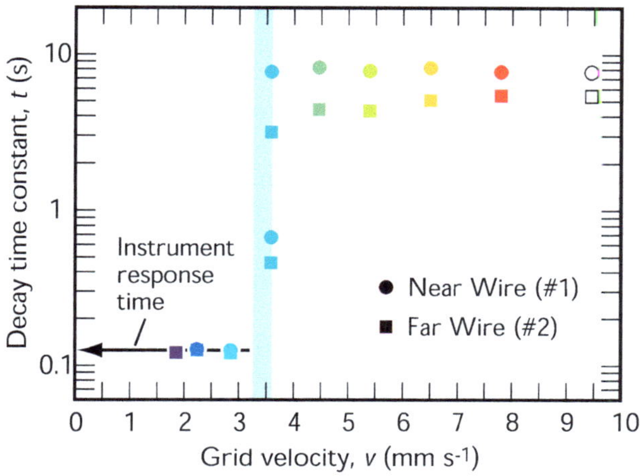

Figure 6.11.2. The initial decay time constant as a function of initial grid velocity for the data shown in figure 22.

observed vorticity. At higher initial grid velocities the decay is much slower, on the scale of 10 s.

The slow decay is due to the decay of a vortex tangle (quantum turbulence) as will be discussed further below. But what causes the very rapid decay at lower grid velocities ? In this case the vorticity must disperse from the grid region with a velocity in excess of 10 mm/s. This can only happen if the vorticity is being produced in the form of individual vortex rings. (We note that vortex rings have also been found to be emitted from vibrating wire resonators under certain conditions(44).) An isolated vortex ring of radius R travels at its self-induced velocity(45)

$$v(R) \simeq \frac{\kappa}{4\pi R}(\ln(8R/\xi) - 0.5) \tag{6.11.1}$$

where ξ is the core size, approximately equal to the coherence length. The ring velocity exceeds 10 mm/s only for ring diameters less than 5 μm. We therefore must conclude that at low grid velocities the emitted vorticity is in the form of vortex rings with diameters no larger than $\sim 5\,\mu$m.

Hanninen *et. al.*(47) have made computer simulations of ring production by a vortex line pinned at its ends and exposed to a perpendicular ac superflow. They find that, provided the ac flow exceeds some critical velocity amplitude, large amplitude oscillations of the vortex lines are excited leading eventually to reconnections and the emission of vortex rings.

A range of ring diameters is produced but typically of order the initial line length. The critical velocity becomes very small when the flow frequency matches the frequency of the first Kelvin wave resonance of the vortex line, Kelvin waves being helical excitations propagating along a vortex line(45). At the operating grid frequency (~1250 Hz), the first Kelvin wave resonance will be excited on vortices with initial line lengths of around 5 μm, consistent with our upper limit for the vortex ring diameters.

Since the grid is very rough on the μm scale, there will be a wide range of trapped vortex line lengths with an upper limit given approximately by the mesh size. On oscillating the grid, Kelvin waves are preferentially excited on the vortex lines which are initially around 5μm long. This leads to the production of vortex rings with diameters centered on 5μm. These propagate ballistically from the grid producing the very rapid response observed.

6.12 Thermal decay of vortex rings

We can confirm the interpretation of the data discussed above by reference to the temperature dependence of the vortex ring signals. Vortex rings gradually shrink owing to mutual friction and eventually dissipate. Thus at high enough temperatures the rings should not be able to reach the nearby VWRs. Since the cross-section for Andreev scattering by a straight vortex line is inversely proportional to the temperature T (Equation (6.1.6) above), the length of vortex line surrounding the two detector wires should be proportional to the quantity fT, which we denote the 'vorticity signal' (the argument is only approximately valid for vortex rings with diameters of a few μm). The steady state value of the vortex signal while the grid is driven at 2.5 mm/s is plotted in figure 6.12.1 as a function of the thermal quasiparticle damping (and corresponding temperature).

The range of a vortex ring with initial radius R_0 can be estimated as(45)

$$x \simeq R_0/q \tag{6.12.1}$$

at low temperatures. The mutual friction parameter q (which is roughly proportional to the quasiparticle damping at low temperatures) has been measured down to temperatures of around $0.35T_C$ by Bevan *et al*(57). Using their fitting formula for q extrapolated to the lower temperatures, we find that a 5 μm diameter vortex ring should have a range of 1 mm and 2 mm at temperatures of $0.178T_C$ and $0.194T_C$, respectively. As indicated by the arrows in the figure, these temperatures are consistent with the temperatures at which the vortex signal for the two wires begins to fall with increasing temperature. This effectively measures the range of the vortex rings. Note that, at the lowest temperatures the range of the rings

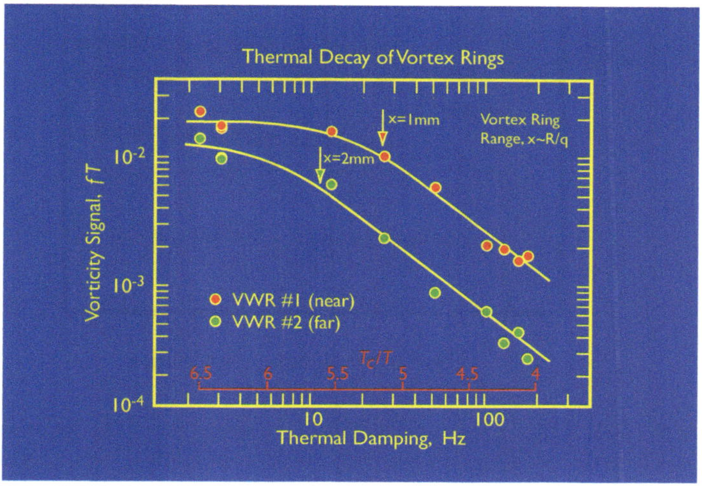

Figure 6.12.1. The vortex signal as a function of temperature for the two vibrating wire detectors. The arrows indicate the at temperature at which the range of 5 μm diameter rings is expected to correspond to the distance to the detector wires

should exceed the size of the cell and the vorticity signal is temperature independent. Presumably in this case the rings are effectively absorbed (and dissipate) on collision with the cell wall.

6.13 The transition to turbulence

So at low grid velocities, we know that small vortex rings are emitted which travel ballistically (independently) away from the grid. As the grid velocity increases, the density of the emitted rings also increases. Eventually the emitted ring density will become so high that the rings can no longer remain independent. When two rings collide, a larger ring may form which moves at a lower velocity and is therefore more likely to collide with further rings. So a cascade process is likely in which collisions and reconnections suddenly form a vortex tangle which disperses on much longer timescales. This will explain the sudden transition observed in the decay data of figures 6.11.1 and 6.11.2. This picture is confirmed in recent simulations by Tsubota's group in collaboration with the Lancaster group(48).

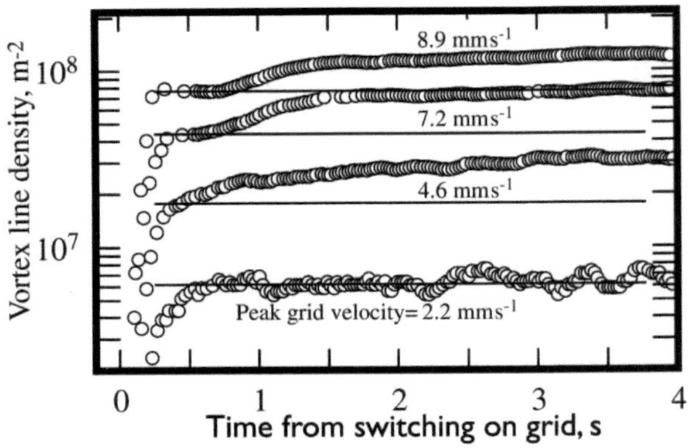

Figure 6.14.1. The vortex line density as inferred from the nearest detector time as a function of time after switching on the grid to various velocities.

6.14 Evolution of Quantum turbulence

The vortex line density can be inferred from the measured fractional screening f and the estimated vortex tangle size D as discussed above. In figure 6.14.1 we show the inferred Line density surrounding the wire nearest to the grid as a function of time after turning on the grid to various velocities as indicated. The initial fast rise time is limited by that of the measurement equipment and the response time of the grid itself (which is velocity dependent). We can interpret this initial fast rise as being due to ballistic vortex rings, discussed above. The slower evolution which follows at the higher grid velocities is then due to the formation of the vortex tangle. The equilibrium line density is ultimately given by the balance between the ring injection and the dissipation (see below). We can estimate the timescale for the development of the turbulent tangle by noting that the largest timescale in the problem corresponds to the largest vortex rings. These have sizes approaching that of the turbulent region D, giving a timescale $\tau \sim D/v(D)$. This is of order a few seconds, consistent with the measurements shown.

6.15 Decay of ^{3}He Quantum Turbulence at low temperatures

We return now to the behavior observed at higher grid velocities(49). Here the emitted rings collide and form a vortex tangle, quantum turbulence, which decays relatively slowly. The inferred vortex line density as a function of time after switching off the grid is shown in figure 6.15.1. Note the log-log scale. Also shown in the figure is data obtained from a towed grid in superfluid ^{4}He at relatively high temperatures using second sound techniques(41; 42).

In the case of superfluid ^{4}He at high temperatures, the superfluid and normal fluid components are thought to be effectively clamped together by mutual friction, so both flow fields are locked together. In this case we have combine superfluid and normal fluid turbulence. This led the authors of this work to consider the application of the ideas of classical turbulence to their measurements(41; 42). Classically, turbulence decays via a Richardson cascade in which energy flows from the largest eddies to smaller and smaller eddies, eventually being dissipated by some effective viscosity. Energy is only dissipated at the smallest length scales. Simple dimensional arguments lead to the famous Kolmogorov spectrum for the energy content of the turbulent energy as a function of the length scale, or waveneumber k,

$$E(k) = C\xi^{2/3}k^{-5/3} \tag{6.15.1}$$

where C is the dimensionless Kolmogorov constant, close to unity, and ξ is the energy dissipation rate, which for a classical fluid is given by $\nu'\omega^2$. The authors argued that in the case of a superfluid, the effective kinematic viscosity ν' can be taken as the normal fluid viscosity divided by the total fluid density$\nu' = \eta/\rho$, and the vorticity ω can be simply related to the average vortex line density L by $\omega = \kappa L$. It was further assumed that the maximum eddie size (large scale cut-off of the Kolmogorov spectrum) should be given by the size of the turbulent region, of the size of the experimental cell d. The details will not be given here, but the model leads to the prediction that the Line density at late times should be given by

$$L = (d/2\pi\kappa)\sqrt{(27C^3/\nu')}\,t^{-3/2} \tag{6.15.2}$$

This equation agrees well with the measured decay, as shown in figure 6.15.1, taking an effective kinematic viscosity which is around half of the measured value. This can be considered as very good agreement considering the simplicity of the model and the assumptions made.

If we take a similar approach and naively use this classical expression for the late-time line density, substituting the appropriate numbers for our experiment then we obtain the lower in figure 6.15.1 labeled '^{3}He classical'. The line lies lower than that of ^{4}He partly since dimension d is smaller (d =1.5 mm in our case against d=10 mm for the ^{4}He experiments) but mainly because the normal fluid viscosity(50) is orders of magnitude larger for ^{3}He. Clearly our measurements, even though they are similar to those

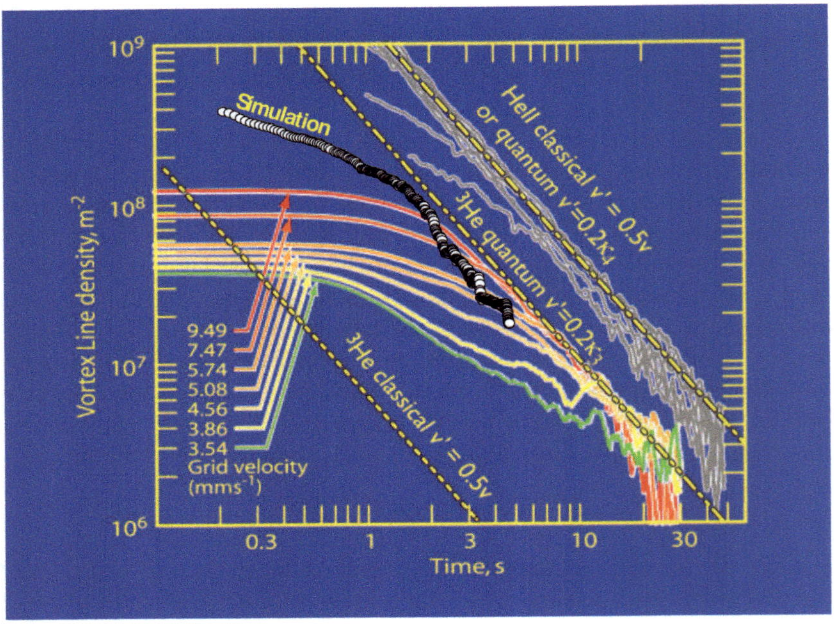

Figure 6.15.1. Colored lines show the vortex line density inferred from the nearest detector wire as a function of time after stopping the grid from various velocities. Grey lines show towed grid measurements in superfluid ^{4}He at relatively high temperatures. Straight lines are discussed in the text. Circles correspond to a computer simulation - see text.

in superfluid ^{4}He, cannot be explained by the classical decay mechanism. This is as expected since the ^{3}He measurements were made in the low temperature limit, where the normal fluid is virtually absent, so there is no significant mutual friction and viscosity is completely redundant. In fact, the entire concept of a normal fluid component in the spirit of the two fluid model breaks down in this temperature regime since the quasiparticle excitations behave independently of each other (so e.g. a normal fluid velocity is meaning less).

The Kolmogorov energy cascade in classical turbulence is a consequence of dissipation being negligible on large length scales. As suggested by Vinen(52), it seems reasonable to expect that superfluid turbulence as generated by grids may display a similar cascade process owing to the similar absence of large length scale dissipation mechanisms. This expectation supported by numerical simulations(51) which show evidence of a Kolmogorov-like cascade in pure superfluid turbulence in the absence of any normal-fluid component. At the higher temperatures, mutual friction will now couple the superfluid turbulence to the highly viscous normal ^{3}He. In this case turbulence is actually found to be completely suppressed(53) owing to the very high viscosity. At intermediate temperatures a different energy spectrum for the turbulence is predicted since mutual friction acts over all length scales(54).

So the power law decay behavior can be explained by a Richardson type cascade with energy dissipated at the smallest length scales. But in the absence of viscosity, what determines the decay rate? As argued by Vinen(52) on purely dimensional grounds, any dissipation process associated with pure superfluid turbulence is likely to give a dissipation of order $\rho\kappa(\kappa L)^2$. The effective kinematic viscosity in the decay equation should therefore be replaced by a term $\zeta\kappa$ where ζ is a dimensionless constant, presumably of order unity. The line density at late times of the turbulent decay should thus be described by

$$L = (d/2\pi\kappa)\sqrt{(27C^3/\zeta\kappa)}\, t^{-3/2} \tag{6.15.3}$$

In He-II the kinematic viscosity and circulation quantum are numerically similar, $\nu \approx 0.1\kappa$, so the ^{4}He data are consistent with a dissipation of this form with $\zeta \approx 0.2$. If we apply this form for our data, with $d = 1.5\,\text{mm}$ and $\zeta = 0.2$, then we obtain the expected late-time behavior shown by the line labelled '^{3}He quantum' in the figure. (This is equivalent to scaling the late-time He-II data by d and κ). The agreement is very remarkable, considering that not only is the superfluidity in the two systems based on completely different mechanisms, but the temperature regimes and normal fluid viscosities are at opposite extremes.

Finally we comment on the circles in figure 6.15.1. These correspond to the results of a computer simulation of the vibrating grid experiments(48). This also shows very close agreement with the experimental data and the model discussed above. This is remarkable because the simulations were i) based on the local induction approximation in which intervortex coupling is neglected and ii) the simulation is done for zero temperature and has no physical dissipation mechanism. In practice the simulation does have a loss mechanism associated with its finite space resolution - energy transferred to smaller length scales is effectively lost resulting in the observed decay. The full consequences of the observed agreement between this simulation and the experiments are currently on-going, but it suggests that the decay of quantum turbulence at low temperatures simply relies on the loss of energy at small scales and that the details of the mechanism are not too important.

6.16 Summary

The recent experiments on quantum turbulence in superfluid ^{3}He-B have already revealed a wealth of information on the properties of quantum turbulence in the low temperature limit, where the normal fluid fraction is virtually absent. This is a particularly interesting regime since viscosity is completely absent and we have a very simple system of quantised vortex lines moving in their self-induced flow field. Nevertheless, the behavior shows remarkable similarity with turbulence in superfluid ^{4}He at much higher temperatures where the normal fluid component should play an important role. The basic decay properties also appear to agree well with a simple model based on the classical Richardson cascade. It seems that the basic ideas of turbulence may be far more wide-reaching and applicable to a wider range of (exotic) systems than previously anticipated. Furthermore, we now have access to a regime in which turbulence is conceptually much simpler and so may lead to a clearer understanding of turbulence in general.

Bibliography

[1] D.I. Bradley, T.W. Bradshaw, A.M. Guénault, V. Keith, B.G. Locke-Scobie, I.E. Miller, G.R. Pickett, and W.P. Pratt, Jr., *Cryogenics* **6**, 296 (1982).

[2] D.I. Bradley, A.M. Guénault, V. Keith, C.J. Kennedy, I.E. Miller, S.G. Mussett, G.R. Pickett, and W.P. Pratt, Jr., *J. Low Temp. Phys.* **57**, 359 (1984).

[3] D.J. Cousins, S.N. Fisher, A.M. Guénault, R.P. Haley, I.E. Miller, G.R. Pickett, G.N. Plenderleith, P. Skyba, P.Y.A. Thibault, and M.B. Ward, *J. Low Temp. Phys.* **114**, 547 (1999).

[4] D.D. Osheroff, R.C. Richardson, and D.M. Lee, *Phys. Rev. Lett.* **29**, 920 (1972).

[5] J. Bardeen, L.N. Cooper, and J.R. Schrieffer, *Phys. Rev.* **108**, 1175 (1957).

[6] L.D. Landau, *Sov. Phys. JETP* **3**, 920 (1957).

[7] A.J. Leggett, *Reviews of Modern Physics* **47**, 331 (1975).

[8] D. Vollhardt, and P. Wölfle, *The Superfluid Phases of Helium 3* (Taylor & Francis, 1990).

[9] D.S. Greywall, *Phys. Rev. B* **33**, 7520 (1985).

[10] R. Balian, N.R. Werthamer, *Phys. Rev.* **131**, 1553 (1963).

[11] A.J. Leggett, *Phys. Rev. Lett.* **31**, 352 (1973).

[12] P.W. Anderson and P. Morel, *Phys. Rev.* **123** 1911 (1961).

[13] A.M. Guénault, V. Keith, C.J. Kennedy, S.G. Mussett, and G.R. Pickett, *J. Low Temp. Phys.* **62**, 511 (1986).
[14] D.C. Carless, H.E. Hall, and J.R. Hook, *J. Low Temp. Phys.* **50**, 583 (1982).
[15] G.G. Stokes, *Mathematical and Physical Papers* (Cambridge University Press, London, 1901), vol. 3, p.38.
[16] D.C. Carless, H.E. Hall, and J.R. Hook, *J. Low Temp. Phys.* **50**, 605 (1982).
[17] A.M. Guénault, V. Keith, C.J. Kennedy, I.E. Miller, and G.R. Pickett, *Nature* **302**, 695 (1983).
[18] A.M. Guénault, V. Keith, C.J. Kennedy, and G.R. Pickett, *Quantum Fluids and Solids* (AIP Conference Proceedings) **103**, 273 (1983).
[19] A.M. Guénault, V. Keith, C.J. Kennedy, and G.R. Pickett, *Phys. Rev. Lett.* **50**, 522 (1983).
[20] J.P. Carney, A.M. Guénault, G.R. Pickett, and G.F. Spencer, *Phys. Rev. Lett.* **62**, 3042 (1989).
[21] S.N. Fisher, A.M. Guénault, C.J. Kennedy, and G.R. Pickett, *Phys. Rev. Lett.* **63**, 2566 (1989).
[22] S.N. Fisher, A.M. Guénault, C.J. Kennedy, and G.R. Pickett, *Physica B* **165-166**, 651 (1990).
[23] S.N. Fisher, G.R. Pickett, and R.J. Watts-Tobin, *J. Low Temp. Phys.* **83**, 225 (1991).
[24] R.J. Watts-Tobin and S.N. Fisher, *Physica B* **165-166**, 683 (1990).
[25] S.N. Fisher, A.M. Guénault, C.J. Kennedy, and G.R. Pickett, *Phys. Rev. Lett.* **69**, 1073 (1992).
[26] C. Bäuerle, Y.M. Bunkov, S.N. Fisher, and H. Godfrin, *Phys. Rev. B.* **57**, 14381 (1997)
[27] M.P. Enrico, S.N. Fisher, and R.J. Watts-Tobin, *J. Low Temp. Phys.* **98**, 81 (1995).
[28] A.I. Ahohen, J. Kokko, M.A. Paalanen, R.C. Richardson, W. Schoepe, and Y. Takano, *J. Low Temp. Phys.* **30**, 205 (1978).
[29] W. Zhang, J. Kurkijarvi, and E.V. Thuneburg, *Phys. Rev. B* **36**, 1987 (1987).
[30] C.J. Lambert, *Physica B* **165-166**, 653 (1990).
[31] S.N. Fisher, A.J. Hale, A.M.Guénault, and G.R.Pickett, *Phys. Rev. Lett.* **86**, 244 (2001).
[32] J.C. Wheatley, *Reviews of Modern Physics* **47**, 415 (1975).
[33] C. Bäuerle, Y.M. Bunkov, S.N. Fisher, and H. Godfrin, *Phys. Rev. B.* **57**, 14381 (1997)

[34] M.P. Enrico, S.N. Fisher, A.M. Guénault, G.R. Pickett, and K. Torizuka, *Phys. Rev. Lett.* **70**, 1846 (1993).

[35] D.J. Cousins, M.P. Enrico, S.N. Fisher, S.L. Phillipson, G.R. Pickett, N.S. Shaw, and P.J.Y. Thibault, *Phys. Rev. Lett.* **77**, 5245 (1996).

[36] M. Bartkowiak, S.W.J. Daley, S.N. Fisher, A.M. Guénault, G.N. Plenderleith, R.P. Haley, G.R. Pickett, and P. Skyba, *Phys. Rev. Lett.* **83**, 3462 (1999).

[37] M. Bartkowiak, S.N. Fisher, A M. Guénault, R.P. Haley, G.R. Pickett, G.N. Plenderleith, and P. Skyba, *Phys. Rev. Lett.* **85**, 4321 (2000).

[38] D.I. Bradley, Yu.M. Bunkov, D.J. Cousins, M.P. Enrico, S.N. Fisher, M.R. Follows, A.M. Guénault, W.M. Hayes, G.R. Pickett, and T. Sloan, *Phys. Rev. Lett.* **75**, 1887 (1995).

[39] C. Bäuerle, Y.M. Bunkov, S.N. Fisher, H. Godfrin, and G.R. Pickett, *Nature* **382**, 332 (1996).

[40] D.I. Bradley, S.N. Fisher, A.M. Guénault, M.R. Lowe, G.R. Pickett, A. Rahm, and R.C.V. Whitehead, *Phys. Rev. Lett.* **93**, 235302 (2004)

[41] S.R. Stalp, L. Skrbek, and R.J. Donnelly, *Phys. Rev. Lett.* **82**, 4831 (1999).

[42] L. Skrbek, J.J. Niemela, and R.J. Donnelly, *Phys. Rev. Lett.* **85**, 2973 (2000).

[43] L.D. Landau and E.M. Lifshitz, *Fluid Mechanics* (Pergamon Press London, 1966).

[44] D.I. Bradley, *Phys. Rev. Lett.* **84** 1252 (2000).

[45] R.J. Donnelly, *Quantized Vortices in Helium II* (Cambridge University Press, Cambridge, 1991).

[46] D.I. Bradley, D.O. Clubb, S.N. Fisher, A.M. Guénault, R.P. Haley, C.J. Matthews, G. R. Pickett, V. Tseplin, and K. Zaki *Phys. Rev. Lett.* **95**, 305302 (2005).

[47] R. Hänninen, A. Mitani, and M. Tsubota, *AIP Conference Proceedings* **850**, 217 (2006).

[48] A. Mitani *et al.*, to be published.

[49] D.I. Bradley, D.O. Clubb, S.N. Fisher, A.M. Guénault, R.P. Haley, C.J. Matthews, G.R. Pickett, V. Tsepelin, and K. Zaki *Phys. Rev. Lett.* **96**, 035301 (2006).

[50] D.C. Carless, H.E. Hall, and J.R. Hook, *J. Low Temp. Phys.* **50**, 605 (1983).

[51] T. Araki, M. Tsubota, and S.K. Nemirovskii, *Phys. Rev. Lett.* **89**, 145301 (2002).

[52] W.F. Vinen and J.J. Niemela, *J. Low Temp. Phys.* **128**, 167 (2002).

[53] A.P. Finne, T. Araki, R. Blaauwgeers, V.B. Eltsov, N.B. Kopnin, M. Krusius, L. Skrbek, M. Tsubota, and G.E. Volovik, *Nature* **424**, 1022 (2003).

[54] G.E. Volovik, *J. Low Temp. Phys.* **136**, 309 (2004).

[55] D.I. Bradley, S.N. Fisher, A.M. Guénault, M.R. Lowe, G.R. Pickett, and A. Rahm, *Physica B* **329**, 104 (2003).

[56] D.I. Bradley, D.O. Clubb, S.N. Fisher, A.M. Guénault, C.J. Matthews, and G.R. Pickett, *J. Low Temp. Phys.* **134**, 381 (2004)

[57] T.D.C.Bevan *et al.*, *J. Low Temp. Phys.* **109**, 423 (1997).

Classical Turbulence in Cryogenic Helium

Joseph J. Niemela*
The Abdus Salam International Centre for Theoretical Physics
Strada Costiera 11, 34014 Trieste, Italy

Abstract Fluid turbulence is important in numerous applications, and is a model system for nonlinear systems with many degrees of freedom. The equations of motion are not easily amenable to analysis and for this reason progress in the field requires significant effort in experiments and in numerical simulations. In recent years, some progress has been made utilizing helium at low temperatures, enabling high-Reynolds number turbulence to be generated in laboratory experiments. With high Reynolds numbers, however, come new challenges for measurement.

1 Introduction

This chapter will address the various aspects of classical hydrodynamic turbulence and its realization in laboratory flows using low temperature helium, either in liquid or gaseous form, as the working fluid. While the emphasis here will be on turbulent flows of a classical nature, other chapters will address quantum aspects of these flows (see in particular the chapter by L. Skrbek). We refer the reader to a review article by Niemela and Sreenivasan (2007), from which these lecture notes are largely based, for a more complete account.

1.1 Background

Turbulence is an important consideration in a wide range of contexts from large scale flows in nature to engineering applications, from geophysical and astrophysical flows to dispersal of pollutants in the atmosphere or oceans, to the movement of ships and aircraft (Tritton (1988)). Turbulence in natural contexts is often associated with angular motion, magnetic/electric fields and/or particulate matter (Davidson (2004)); in fact,

*The author is indebted to K.R. Sreenivasan for his significant contributions to the main themes of this essay and to the directors of the CISM Advanced School on "Vortices and Turbulence at Very Low Temperature" for the oppotunity to present there the material on which it is based.

it is a necessary concern to those studying interstellar dust and energy transport (Spiegel (1971)) and weather prediction and planetary magnetic fields(Glatzmaier, *et al* (1999)). The turbulence problem is also an often useful model for strongly nonlinear systems, having many coupled degrees of freedom (Falkovich and Sreenivasan (2006)), for which other areas of study as distant as fracture dynamics(Marder (2000))—and perhaps even the "chaos" of fluctuating stock markets (Mandelbrot (1999))—might benefit from a deeper understanding of it (beyond the more trivial level of benefitting from the methods used to extract statistical measures of the turbulence). The subject is a particularly useful paradigm for a wide variety of non-equilibrium systems largely because the equations of motion are known exactly and can be simulated with precision. However, the complexity of the underlying equations has precluded much analytical progress, and the demands of computing power are such that routine simulations of large turbulent flows has not yet been possible (NRC report (1999); Yakhot and Sreenivasan (2005)). Hence, experimental input takes on a more important role. This has resulted in a search for optimal test fluids, and the development of new instrumentation for measurement. The use of low temperature helium as a test fluid for the study of classical turbulent flows is the subject of this chapter. As we shall show, the use of low temperature helium can yield benefits, but also new challenges that must be met for its complete implementation in the study of classical turbulence.

This chapter is organized as follows. In the rest of this section, we consider the fluid dynamical equations and show how the relevant non-dimensional control parameters emerge from them. We will then discuss ways in which the use of low temperature helium can help address classical problems in turbulence. Following this we will discuss some representative problems in classical turbulence and show specifically how low temperature fluids and techniques have been applied to address them; what are the successes and what are the limitations.

1.2 Fluid equations and non-dimensional parameters

In writing the equations of motion we assume a coordinate system fixed in space, referred to as an "Eulerian" description. Even though the relevant length scales in turbulence can be microscopic we will assume that that a hydrodynamic approximation remains valid. Let us consider a fluid element of density ρ moving at a velocity $\mathbf{u}$; then mass conservation can be written as

$$\frac{\partial \rho}{\partial t} + \nabla \cdot (\rho \mathbf{u}) = 0, \tag{1.2.1}$$

which, for the usual case of constant density, reduces to

$$\nabla \cdot \mathbf{u} = 0. \tag{1.2.2}$$

For Newtonian fluids in which the stress and strain tensors are linearly related, the momentum equation can be expressed as

$$\frac{D\mathbf{u}}{Dt} = -\frac{1}{\rho}\nabla p + \nu \nabla^2 \mathbf{u} + \mathbf{F}_{ext}, \tag{1.2.3}$$

in terms of the pressure p and kinematic viscosity $\nu = \mu/\rho$, where μ is the dynamic or shear viscosity. For notational convenience we have used the so-called substantive (or convective) derivative $\frac{D}{Dt} = \frac{\partial}{\partial t} + \nabla \cdot \mathbf{u}$. The term $\mathbf{F}_{ext}$ represents an external body force.

The body force could arise from a number of sources, such as gravity, rotation and magnetic field. We shall be concerned mainly here with buoyancy in thermally driven flows, for which there is only one component, in the direction of gravity, given by $F_{ext} = g\alpha\Delta T$, where g is the acceleration of gravity, α is the isobaric coefficient of thermal expansion and ΔT is the temperature difference across a layer of fluid in the direction of gravity. If we assume further that the system satisfies the Oberbeck-Boussinesq, or simply Boussinesq, approximation (Oberbeck (1879)), we can neglect the effect of the buoyancy-induced motion on pressure or the incompressibility condition. The equation for energy conservation in this same approximation is

$$\frac{\partial T}{\partial t} = \kappa \nabla^2 T - \mathbf{u} \cdot \nabla T. \tag{1.2.4}$$

Here κ is the thermal diffusivity of the fluid at temperature T.

Equations (1.2.1)-(1.2.4) describe the motion of the fluid and determine its velocity, pressure and temperature. Naturally, these partial differential equations are incomplete unless we also specify the appropriate initial and boundary conditions (e.g., rigid or "free" surfaces, conducting or insulating surfaces, rough or smooth surfaces, etc.). Then, the the solutions of these equations should correspond to all flows, including turbulence. We can remark at this point that it is not clear that small differences in boundary conditions produce only small effects in the resulting flows. We shall have more to say on this aspect later in the chapter.

It is useful to non-dimensionalize the equations by scaling the variables in terms of a characteristic velocity U, length L, time L/U, and pressure ρU^2. Then Equation (1.2.3) for the case of simple pressure-driven flow becomes

$$\frac{\partial \mathbf{u}}{\partial t} + \mathbf{u} \cdot \nabla \mathbf{u} = -\nabla p + \frac{1}{Re}\nabla^2 \mathbf{u}, \tag{1.2.5}$$

where single variable now appears, namely the Reynolds number $Re = UL/\nu$ in place of the multiple variables of the dimensional equations. We may say that the Reynolds number is a manifestation of dynamical similarity in the sense that if we consider two geometrically similar flows, then they are also *dynamically* identical if the corresponding Re is the same, regardless of the specific values of velocity, length or fluid viscosity. This follows directly from the fact that the equation of motion (1.2.5) has only a single adjustable parameter in it, namely Re. Dynamical similarity is the principle which allows aerodynamic testing in which a scale model may provide direct information on the actual full-scale prototype. A physical interpretation of the Reynolds number is as a measure of the ratio of the inertial forces (the term $\mathbf{u} \cdot \nabla \mathbf{u}$ in the substantive derivative, see Equation (1.2.3)) and viscous forces (the term $\nu \nabla^2 \mathbf{u}$).

In the same way, adding the buoyancy term for describing thermal flows leads to another dynamical similarity parameter, namely the Rayleigh number

$$Ra = g\alpha \Delta T L^3 / \nu\kappa. \tag{1.2.6}$$

Here g is the acceleration due to gravity, α and κ are, respectively, the isobaric thermal expansion coefficientand thermal diffusivity of the fluid, and ΔT and L are the characteristic temperature difference and the length scale of the flow. In this case the physical interpretation of the Rayleigh number is as a measure of the ratio of the rate of potential energy release due to buoyancy to the rate of its dissipation due to thermal and viscous diffusion (Busse (1978)). A third similarity parameter is the Prandtl number

$$Pr = \nu / \kappa \tag{1.2.7}$$

interpreted as the ratio of time scales due to thermal diffusion ($\tau_\theta = L^2/\kappa$) and momentum diffusion ($\tau_v = L^2/\nu$).

For now, we will keep he focus on isothermal flows driven by gradients in pressure, if only for simplicity. In such systems turbulence develops if the ability of the fluid to dissipate fluctuations is not sufficient to counteract the large inertial forces; a situation which occurs when the Reynolds number is large. The subsequent form of turbulence can be described as being composed of fluctuations on many different length scales (or "eddies") between the largest scale comparable to the physical size of the system or apparatus in which the flow exists and the smallest scale for which viscosity acts strongly to dissipate the kinetic energy of the turbulence. Eddies of a given characteristic length scale are associated with their own characteristic velocity and therefore their own characteristic Reynolds number. The large eddies are more energetic and their Reynolds number is correspondingly large; by the principle of dynamical similarity, this can be considered equivalent to having a small viscosity. Thus, viscous effects are unimportant for large eddies and so their energy is not lost by viscosity; instead, the loss of energy occurs to other scales by the nonlinear interaction with them. It is conventionally thought that the most effective interaction occurs with eddies of comparable sizes (local interaction). This gives rise to the notion of a cascade of energy from one scale of eddies to those of the neighboring scales which has important consequences for the standard model of turbulence. There is an average direction to the energy cascade because the energy is eventually dissipated in small scales. The Reynolds number for the small scales gets progressively smaller and so the viscous effects become important below a nominal scale; the smallest dynamically important scale is the one for which the Reynolds number is of the order unity. This is called the dissipation, or the Kolmogorov, scale, η.

As long as the energy is supplied steadily at a large scale ℓ, a statistically steady state is obtained; in this state, the most important dynamical phenomenon that occurs in the range of scales between the "large" and "small" scales—the so-called *inertial* range—is the scale-to-scale energy transfer without viscous dissipation. The kinetic energy is dissipated at the small scales and the rate of this dissipation per unit mass, ϵ, is clearly the same as the rate of energy input. Elevating Re (reducing viscosity), results in the transfer of energy continuing to smaller scales before being converted to heat. That is, for a given viscosity, the Reynolds number of the eddies for which dissipation occurs is of order unity. As we shall see this poses challenges for experiments at high Re, using fluids having small viscosity, where such small scales need to be well resolved. A similar difficulty will arise for numerical experiments, except that here it is the number of total scales needing to be resolved which poses the challenge.

The energy dissipation rate ϵ is proportional to the input power, u^3/ℓ, where u is a characteristic measure of the fluctuating velocity (such as its root-mean-square value) at the scale ℓ at which the energy is injected. Since the smallest scale η is responsive entirely to the viscosity and the energy dissipation rate, dimensional analysis shows that

$$\eta = (\nu^3/\epsilon)^{1/4} = \ell Re_\ell^{-3/4}, \qquad (1.2.8)$$

where $Re_\ell = u\ell/\nu$ is the characteristic Reynolds number of the large scale.

It may seem more familiar if we consider the energy spectrum $E(k)$, where k is the fluctuation wavenumber and $E(k)dk$ is the kinetic energy contained in the wavenumber shell between k and $k+dk$. From dimensional analysis and considering what was discussed above (Kolmogorov (1941)),the energy spectrum can be written as

$$E(k) = C_k \epsilon^{2/3} k^{-5/3}, \tag{1.2.9}$$

where the proportionality constant C_k is of order unity and has been determined from various experiments (Monin and Yaglom (1975); Sreenivasan (1995)) to have a average value of about 1.5.

1.3 Universal behavior

From Kolmogorov's hypothesis of local isotropy (Kolmogorov (1941)) we can assume that the turbulent flow, at least for wavenumbers within the interial range discussed above, is nearly homogeneous and isotropic. Therefore, if any of the statistical properties of turbulence are to exhibit universality it seems reasonable to look for such behavior there. In particular, as the Reynolds number becomes very large, the scales in the intertial range should be largely decoupled from both the scale at which energy is injected and that at which it is dissipated, assuming of course that the scale-to-scale interaction is *local*, which is reasonable.

From an experimental point of view, then, the impetus is to obtain as high *Re* as possible in order to separate the scales. We need to define what is high enough for *Re*. One definition is simply an empirical one: it is that *Re* for which statistical properties cease to vary as *Re* is increased even further. Can we meet this requirement? Well, with common fluids the answer is "not quite"– there are many flows for which the asymptotic value of *Re* is larger than can be easily obtained or the asympototic state is "just" atttained, with no further range available for discerning scaling relationships. Indeed, in this case we need to demand that experiments allow us to obtain data extending over some range of high *Re*, measured in decades, and further it would be preferable to do this in an apparatus for which the boundary conditions remain the same in some essential way. Since we don't have any precise idea on how sensitive turbulent flows are to small changes in the boundary conditions, it is further preferable to perform experiments in a single apparatus if possible. We will now consider why a particular fluid (or fluids) namely, cryogenic helium, might be suitable for pushing experimental turbulence research forward.

1.4 Helium as a test fluid

Low temperature helium has a kinematic viscosity that is very low compared with other fluids. This is perhaps most widely appreciated for the case of helium II, in which there is a superfluid component. However, even as we approach absolute zero, there is always an effective kinematic viscosity (Vinen and Niemela (2002)) for turbulent flows which precludes the possibility of having a test fluid with zero kinematic viscosity and hence capable of producing infinite *Re*.

Table 1. Kinematic viscosities for some test fluids. The liquid phases of helium are evaluated at the saturated vapor pressure (SVP).

Fluid	$T(K)$	$P(Bar)$	$\nu(cm^2/s)$
Air	293	1	0.15
Water	293	1	0.01
Helium I	2.2	SVP	1.8×10^{-4}
Helium II	1.8	SVP	8.9×10^{-5}
Helium gas	5.5	2.8	3.2×10^{-4}

Table 1 shows some typical values of kinematic viscosity belonging to various test fluids. The advantage of helium is clear, although there are many considerations to an experimental program and one must weigh them all, as we will explore below.

For isothermal flows, for which the principal control parameter is *Re*, it is enough to consider the kinematic viscosity. For other types of flows this is not obvious from the outset and we must examine the fluid properties more carefully.

For instance, we can consider thermal turbulence, generated through buoyant forcing under gravity. This is likely one of the most common forms of turbulent fluid flow we know of since it plays a major role in stellar energy transport. For this type of turbulence we have to consider the Rayeigh number *Ra* and Prandtl number *Pr*, both introduced earlier. Our goal of obtaining higher *Re* above translates into a similar need to obtain high *Ra* in laboratory experiments; extreme values of *Pr* affect the turbulent state but it is not a control parameter in the same sense. *Pr* is, however, a similarity parameter and we must be careful when comparing flows to either match it, or to show that it plays no role.

From our previous description of *Ra* it is clear that here too the low kinematic viscosity of helium can lead to its increase in value. There are two other fluid properties that we must consider, however: the thermal diffusivity and the isobaric thermal expansion coefficient. It turns out that the gaseous form of low temperature helium, while not necessarily having the lowest viscosity, possesses the the most favorable *combination* of properties for obtaining large *Ra* as we shall show below.

The thermal diffusivity $\kappa = k/\rho C_P$, where k is the thermal conductivity of the fluid, ρ its density, and C_P the constant pressure specific heat. It becomes vanishingly small near the critical point because C_P diverges there, and this helps to boost the value of the Rayleigh number since it appears in the denominator. The third fluid property important for the Rayleigh number is the isobaric thermal expansion coefficient $\alpha \equiv -\rho^{-1}\partial\rho/\partial T$, where the derivative is evaluated at constant pressure. This property of the fluid directly determines the buoyancy, rather than the dissipation, as is the case for the other two properties we considered. Clearly, the sign of α determines the direction of the temperature gradient that is needed to induce buoyancy; in most cases it is positive, meaning that one has to heat from below to induce convective flow and hence to initiate thermal turbulence. This is not always the case; for instance, the isobaric expansion of water changes sign below 4 degrees Celsius, as does that of liquid helium just above the lambda point. In these cases, of course, one has to heat from *above* to cause convection. For all ideal (non-interacting) gases, we can define $\alpha = 1/T$. With this definition, simply having a gas at low temperature is enough to ensure large α. In addition, if we take the critical gas near the critical point α diverges just like the specific heat; in fact, the two are thermodynamically related. Naturally, since it is associated with buoyancy (thus appearing in the numerator of Ra) one wishes to have α as large as possible for obtaining large Ra.

So as we have seen above, not only can Re be made large using low temperature helium. but also Ra. For the latter, it is the combination $\alpha/\nu\kappa$ that matters and for comparison we show in Table 2 this ratio for helium, as well as for the common test fluids air and water.

When the cryogenic helium gas is approximately ideal, the Prandtl number $Pr = \nu/\kappa$ maintains a constant value near 0.7. This changes only near the critical point, where the gas can no longer be assumed to be non-interacting.

Table 2. Values of the combination of fluid properties $\alpha/\nu\kappa$ for air, water and helium

Fluid	$T(K)$	$P(Bar)$	$\alpha/\nu\kappa$
Air	293	1	0.12
Water	293	1	14
Helium I	2.2	SVP	2.3×10^5
Helium II	1.8	SVP	– – –
Helium gas	5.25	2.36	6×10^9
Helium gas	4.4	2×10^{-4}	6×10^{-3}

How large a value of Ra can we achieve? Taking advantage of critical point divergences, and considering a fluid layer some 10 meters tall (difficult and expensive, but well within the capacity of presently available cryogenic capability) and a reasonable temperature difference of, say, $0.5K$, Rayleigh numbers of the order 10^{21} are possible. Indeed, such large values are characteristic of turbulent solar convection (Sreenivasan and Donnelly (2001)).

Tunability and scaling

It may seem that Table 2 has rather conflicting information: the combination of $\alpha/\nu\kappa$ is not only very large for helium gas but can also be quite small, smaller than for either water or air! This is in fact one of the "bonuses" of using helium gas rather than the liquid (although it too can produce very high Ra); the flexibility or tunablity, by varying slightly the pressure and or temperature of the gas, is such that one can obtain roughly a factor of 10^{12} difference in this ratio and hence Ra. Since we have already indicated that for obtaining useful scaling information, there is a need for a wide range of control parameter in addition to large values, low temperature helium gas brings us closer to an ideal test fluid for turbulence studies. It only then remains to choose L to be as large as possible so that the upper end of the range of Ra is as high as possible, while at the same time keeping ΔT reasonably small so that one does not have to worry about introducing non-Boussinesq effects through the temperature dependence of fluid properties between the "hot" and cold boundaries.

It is useful to expand the fluid parameters and rewrite Ra as

$$Ra = (\rho^2 \alpha C_P) \frac{g \Delta T L^3}{\mu k}. \tag{1.4.1}$$

Then we see that the quantity in parentheses is that which varies substantially and is the real tuning parameter that one adjusts while conducting an experiment. For low densities (or pressures, with roughly $P < 1$ *bar*), we operate more or less in the non-interacting gas limit where both the expansion coefficient and specific heat remain essentially constant. To a good approximation, $Ra \propto \rho^2$ under these conditions. By adding more gas to an experimental sample chamber, increasing the density a little at a time, we can generate a continuous and very wide range of Ra. The gas cannot attain arbitrarily large density; near the critical temperature and pressure its density becomes about half that of the liquid state at the normal boiling point. Subsequent increase of Ra is through the other factors in the parentheses, namely the isobaric expansion coefficient and the heat capacity which begin to diverge. The fine tuning (at fixed density) can be accomplished by varying slightly the temperature difference.

For an experimental program in turbulence, whether of thermal or isothermal flows, we have concentrated on the test fluid itself. Naturally, this is the first consideration, but it is not the only one. While Re and Ra appear in the Navier-Stokes equations and determine the dynamical state of the fluid, the details are also dependent on the boundary conditions as we emphasized above. The most important boundary condition for thermal flows is specifying the temperature and heat flow conditions at solid surfaces; constant temperature is the most common boundary condition for the horizontal boundaries. In practice, this condition is effected by using highly conductive plates with constant heating (or with PID temperature control in case the temperature is directly controlled). Annealed and oxygen-free copper has a thermal conductivity of order 1 $kWm^{-1}\ K^{-1}$ to be compared with the nominal thermal conductivity of helium, which is of order $10^{-2}\ Wm^{-1}\ K^{-1}$ so the low temperature *environment* itself has advantages for studying thermal turbulence by providing nearly ideal boundary conditions.

We should mention that the *effective* thermal conductivity of the fluid at very high Ra can be quite large because of turbulence and, at an Ra of about 10^{17}, be as high as 10,000 times that of the quiescent fluid. This brings it considerably closer to that of the plates and necessitates some form of correction. This is true for all experiments; however, for low temperature experiments in general the correction is negligible (Verzicco (2004); Niemela and Sreenivasan (2006)) except at the very highest partial decade of Ra where it is of the order of a few percent. By comparison, teh needed correction can be of the order of 10-20% for conventional fluids even when the Ra is only moderately high (Ahlers *et al* (2005)).

Finally, we touch on one additional advantage of the use of low temperature helium as a test fluid: the size of the apparatus. It is well known that there exist large machines to do physics, from accelerators to synchrotron facilities to observatories and so forth. These machines are very expensive, if only from consideration of the amount of land and materials needed for their construction, let alone the technical personnel needed to operate them. Even in turbulence, such large facilities exist; e.g., the large wind tunnels used for aerodynamic testing. We will consider these structures further below, but it is sufficient at this point to emphasize that they not easily converted to academic use.

If we were to obtain the same Re as in a large wind tunnel, but were able to do it in a structure that could fit within a standard laboratory rather than requiring several buildings, it could greatly aid the scientific effort. In this light, we can focus as well on the length factor in the Reynolds number, $Re = UL/\nu$, and easily see that for a *given* Re, one can obviously get away with much smaller L by using a fluid (i.e. helium) hwith a smaller kinematic viscosity. Since costs can scale with L, this presents an opportunity for the academic. It may seem possible to circumvent all of these problems by finding ways to increase the velocity U appearing in Re. It must be kept in mind, however, that U cannot be made arbitrarily large without introducing compressibility effects, and thus another similarity parameter, namely the Mach number (i.e., the ratio of flow velocity to the speed of sound in the medium).

For thermal flows the situation is similar. For instance, increasing the temperature difference ΔT has a fundamental limit: It cannot be made larger than would cause a significant variation of fluid properties across the layer. And in liquids, he problem of phase change has to be taken into consideration.

By way of example for isothermal flows, a 100 cm cross-section helium wind tunnel with feasible cooling power (of order 1 kW at the operating temperature) could match the Reynolds number of any wind tunnel in existence (usually tens of meters in its smallest dimension and hundreds of meters in the longer dimension (Sreenivasan and Donnelly (2001)). The former could be envisioned for a campus facility, but not the latter. Finally, we should note that in turbulent pipe flow it has been possible to generate high Re using air compressed to about 200 atmospheres (Zagarola and Smits (1997)). Highly compressed air has both advantages and disadvantages compared to helium.

Since we have mentioned testing facilities as an example of size vs Re considerations, we should also point out that such aerodynamic testing was actually one of the principal motivations for investigating the use of helium as a test fluid. In fact, hoping to capitalize on the vanishing viscosity of the superfluid phase, a small "superfluid wind tunnel" was constructed (Craig and Pellam (1957)) in 1957 at Caltech. The presence of superfluid vortices led to a rather finite Re flow, and the needed technology development presumably added to its eventually being abandoned.

1.5 Between the laboratory and the blackboard

Before pursuing further laboratory efforts, we should discuss there various alternative methods that could be used to extract useful data.

First of all, it may have occurred to the reader that nature itself presents some rather high Re and/or Ra flows and so why not simply place measurement probes in the atmosphere or the ocean in lieu of constructing complicated apparatus and using exotic fluids. In fact, these measurements are made (Kaimal *et al* (1976); Antonia *et al* (1978); Sreenivasan and Dhruva

(1998)), but are of limited utility when a deeper understanding of a subject is being sought or when the questions are more precise.

Another obvious question concerns the role of direct numerical simulations (DNS) in which the appropriate equations are solved on a computer without making any approximation. This has been a rapidly growing area of research tied to the development of more and more powerful computers. In fact, there has been substantial progress made with this technique as new generations of computers become available, but still there are serious limitations for applications to typical turbulent flows of interest at high Reynolds and Rayleigh numbers. Specifically, DNS faces hurdles put forward by the large number of degrees of freedom in turbulence that grows with Re. Indeed, the number of degree of freedom can be seen to grow as $Re^{9/4}$ in three dimensions, by considering that the range of scales needing to be well resolved, ℓ/η, grows like $Re^{3/4}$, which follows directly from Equation (1.2.8) above. In terms of grid points, a uniform grid would then require the same minimum number of points.

Of course, depending on the questions one is interested in, it may not be necessary to use uniform grids, nor even to resolve all scales. For instance, many useful results have been achieved by resorting to large eddy simulations (LES) which compute the large scales only, while modelling the small scales. Other type of models have also been used successfully, but unfortunately they are not universally applicable. It appears that the state of the art in simulations is still many years away from being able to deal with most very large Re flows occurring in nature and engineering (NRC report (1999)).

The final remark is the following: we have emphasized the need to push Re and Ra to high values, mainly with the idea of gaining better understanding of those aspect of turbulence which are universal in nature. However, there are many unresolved questions that do not belong to this line of study and do not concern or require very high Re or Ra. For those questions, we are better served using conventional fluids and technology.

We now consider a couple of examples, one of isothermal turbulence and another of thermal turbulence to illustrate the role of low temperature helium in experimental efforts in classical turbulence research.

2 Applications of cryogenic techniques and fluids

Here some notable attempts to use cryogenic helium for turbulence research will be discussed. Turbulence in helium II will be largely discussed in other lectures of this series and we will concern ourselves with classical flows, even if they happen to utilize helium at temperatures and pressures below the lambda line.

2.1 The "French washing-machine"

One of the most important experiments to date in low temperature classical turbulence involves the so-called von Kármán swirling flow, which develops between two counter-rotating disks immersed in a fluid and separated by a vertical distance H. These flows typically generate rather intense turbulence. We use this example, not only because it is a representative and important cryogenic contribution to turbulence research, but also because it illustrates at the same time some of the methodology of turbulence measurement as well as some of the recent issues connected to it.

One of the conventional ideas in turbulence is that the rate of energy dissipation remains finite even as the viscosity vanishes (Batchelor (1953); Sreenivasan (1984, 1998); Kaneda *et al* (2003)) (recall that vanishing viscosity only expands the inertial range but that the energy dissipation rate remains the same as the rate of energy input). A reliable test would require that we are looking at universal behavior (presuming of course that such behavior really exists) and not details of any particular experiment. This requires, in turn, high Re, as well as a wide range of that parameter that would allow for observation of certain trends in the data. Since any deviations from assumed behavior could be subtle, we would rather not have to untangle spurious effects due to differing apparatus or boundary conditions, and so it would also be desirable to do the study in single experiment (apparatus, protocol, etc). Tabeling and co-workers, realizing that helium gas offered this possibility obtained measurements of turbulent energy dissipation at very high Re using the apparatus shown in Fig. 2.1.1. Briefly we can describe this apparatus: the counter-rotating disks had a radius $R = 20\ cm$ with radial "fins" added to the plates in order to produce stronger stirring of the fluid. The experiment was operated near 5 K with pressures ranging from a few mbar to about 2 bar.

For this flow, the Reynolds number, defined as $Re = \Omega R^2/\nu$, where Ω is the angular velocity of the disks, obtained values between 10^4 and about 10^7. For reference, the upper value is roughly within a decade of characteristic atmospheric flows.

In this type of experiment, the azimuthal velocity (in the direction of the disk rotation) is rather strong near to the disks, especially with the added fins enhancing the entrainment of the fluid. Tabeling and co-workers (Tabeling *et al* (1996)) employed a cryogenic hot-wire to measure the turbulence; this technique requires a large mean flow which was easily obtained from the entrained flow near the disks. In particular, the velocity is obtained from

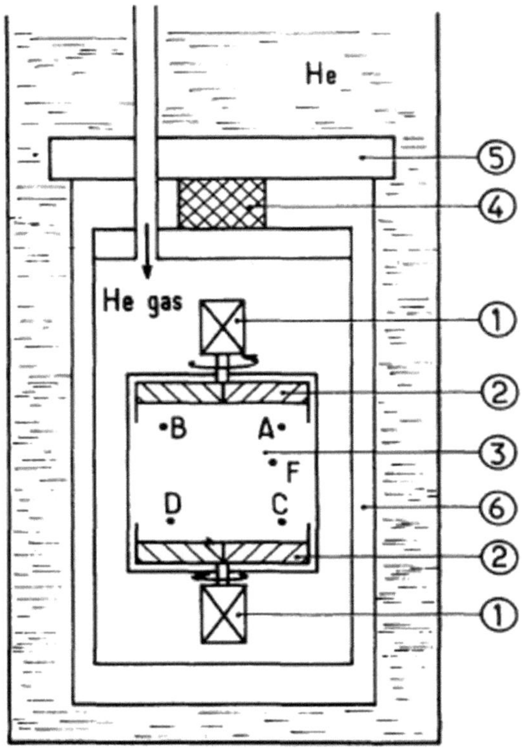

Figure 2.1.1. Swirling flow apparatus. (1) dc motors. (2) Blades fixed on the rotating disks. (3) Working fluid. (4) Thermal link. (5) Copper plate. (6) vacuum. After Tabeling *et al.* (Tabeling *et al* (1996)).

the forced convective cooling of the wire, which is held at some temperature higher than the ambient temperature. The cryogenic hot wire is shown in Fig. 2.1.2. In order to calibrate the wire one must then compare the average flow rates measured simultaneously by some other means. Clearly, hot-wires are designed primarily for isothermal flow conditions. When there are temperature fluctuations present in the flow, this has to be built in the the calibration, a task that is difficult but doable. In practice one might, for instance, alternately measure the background temperature signal from a nearby thermometer and apply the necessary corrections to the velocity signal, or perhaps to even operate a single wire in both the "hot" and "cold" (i.e., as a resistance thermometer) mode.

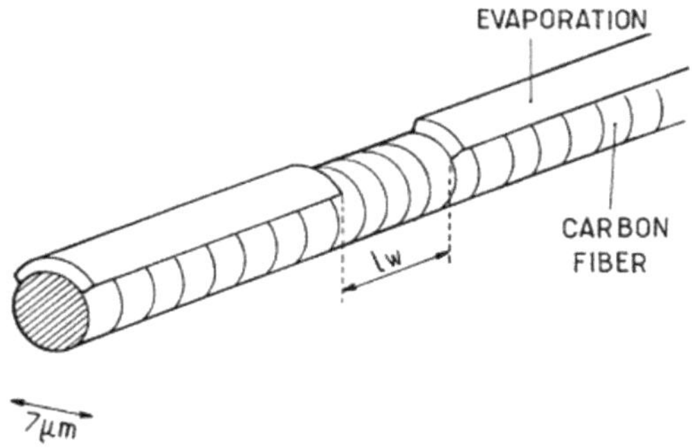

Figure 2.1.2. Cryogenic hot wire probe, showing the active region, $\ell_w =$ O(7 μm). After Tabeling *et al.* (Tabeling *et al* (1996)).

Considering a time record of velocity obtained from a probe fixed in space we may say that it is due partly to the temporal changes of the velocity at that point and partly to changes in the flow sweeping past the probe. Generally in turbulence, the latter contribution far exceeds the former, so the convention is to consider that all fluctuations measured as a function of time are due to spatial fluctuations in the presence of a mean background flow sweeping them past the probe. This is the so-called Taylor's frozen-flow hypothesis: the underlying assumption is that the small scale turbulence is "frozen", or unchanging, for short times so that a variation of the velocity in time can be converted into a spatial variation by concurrent knowledge of a mean sweeping velocity. This hypothesis, while having some limitations is adequate for many purposes. We can then obtain from time-resolved measurements the wavenumber spectrum, $E(k)$, of velocity fluctuations.

Even though the technique is straightforward, difficulties can arise when trying to convert what is common technology in room temperature measurements to the cryogenic regime. The first consideration is of the sensitivity of the device, and so in this case one has not only to find the ideal fluid for turbulence, but also the ideal materials with which to fabricate probes to measure it. Another difficulty that we had alluded to earlier is the smallness of the scales needing to be resolved: recall that as Re gets larger the smallest scale in the inertial range gets smaller; for these high-Re experiments, the corresponding Kolmogorov scale can be of order 1 μm.

The hot wires used by Tabeling *et al.* were constructed of carbon fibers 7 μm in diameter. These fibers were masked by another similar fiber stretched perpendicularly across, before evaporating on them a conducting layer (Au) to provide a good electrical connection. The sensitivity of the carbon spot left underneath the masking fiber in the case of Tabeling's wires was within a factor of ten of that corresponding to a typical Allen Bradley resistor and so was quite good. The active area was then a cylinder of diameter and length equal to 7 μm. Even this small size is larger than the Kolmogorov scale at the highest Re achieved, although the wire can be made thinner at the expense of its robustness.

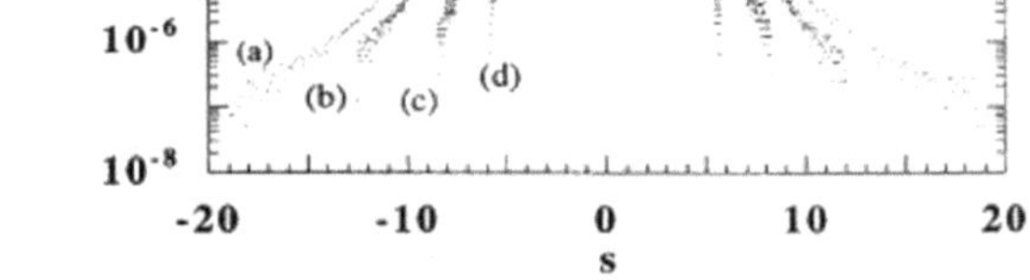

Figure 2.1.3. Log-linear plot of the PDF of longitudinal velocity increments, calculated on different scales (a) $r \approx 10\mu$m. (b) $r \approx 100\mu$m. (c) $r \approx 1$ mm. (d) $r \approx 1$ cm. Here, $\eta \approx 2.5\mu$m and $R_\lambda \approx 3000$. The large-scale Reynolds number is of the order R_λ^2. After Tabeling *et al.* (Tabeling *et al* (1996))

Fig. 2.1.3 shows the probability density function (PDF) of longitudinal *increments* of velocity calculated on different scales. These increments have the following definition:

$$\Delta V_r = V(x + r) - V(x), \tag{2.1.1}$$

where V is the local velocity, and r is the separation length, determined from the measured separation time using the Taylor frozen flow hypothesis described above. From Fig. 2.1.3 we can see that as the scales decrease toward the Kolmogorov dissipation length, the increments display more intermittency, indicated here by the presence of long "tails" in the distribution. Such intermittency refers to the existence of "patchiness" in the spatial distribution of small scales and leads a departure of the random velocity field from the self-similarity assumed by Kolmogorov's phenomenological theory. It is a general feature of turbulent flows related to the occasional and large excursions of the data (e.g., velocity increments) from the mean; hence it appears in the form of "tails" in the PDF. While there has been much effort in recent years to quantify and elucidate the nature of intermittency (Arneodo *et al* (1996); Sreenivasan and Antonia (1997)), there is, as yet, no rigorous theory to explain all the observations. Broadly speaking, intermittency means that Gaussianity is inapplicable to turbulence, though large-scale quantities do approximate Gaussian behavior under certain conditions (Tennekes and Lumley (1971)).

We can partially quantify the intermittency, regarding it as a measure of the distribution of tails, by the flatness factor given by

$$F(r) = \frac{< \Delta V_r^4 >}{< \Delta V_r^2 >^2}. \tag{2.1.2}$$

Taking the limit of this quantity as the separation distance approaches zero yields the flatness factor $\tilde{F}$ of the derivative. Clearly, the relevance and accuracy of the derivative obtained from experimental data depends on whether the velocity has been resolved and sampled adequately. By "adequate", we mean that at least the Kolmogorov scale has been resolved with good temporal fidelity. It is known from measurement that $\tilde{F}$ increases with increasing Reynolds number monotonically (Sreenivasan and Antonia (1997)). The data at higher Reynolds numbers have been obtained in atmospheric flows where resolving the Kolmogorov scale (of the order of 0.5 mm) from conventional hot wires presents no major problem. It is here that we find an unexpected result(see Fig. 2.1.4); the flatness of the measured "derivative" saturates, or even decreases, with increasing Reynolds number beyond a certain critical value, given by $Re_\lambda \simeq 700$. Here we have introduced the Taylor scale Reynolds number Re_λ, which is based on an internal scale of the flow (i.e., the Taylor microscale). The microscale Reynolds number is roughly equivalent to the square root of the Reynolds number based on the large scales which we introduced above.

In fact, this break in behavior has been interpreted by the original authors (Tabeling and Willaime (2002)) in terms of a new bifurcation at around $Re_\lambda \simeq 700$. If this is true, the measurements have revealed either some new physics that was missed in earlier experiments and analysis, or the derivative flatness is dependent on the particular method of generating turbulence, which would run counter to the the idea of scale similarity. The possibility that the tendency for the flatness to reach constancy at some high Reynolds number could merely be an experimental artifact related to the inadequate resolution of the probe (since the turbulent scales in the flow are quite small) was discussed (Emsellem *et al* (1997)) and regarded as unlikely. However, in another recent paper (Gylfason, Aqqalasomayajula and Warhaft (2004)), it has been pointed out that this anomalous tendency disappears when the resolution of measurement is improved. At the least, this debate suggests that there is a great need to resolve the velocity data better in order to know with any certainty the true behavior of the flatness.

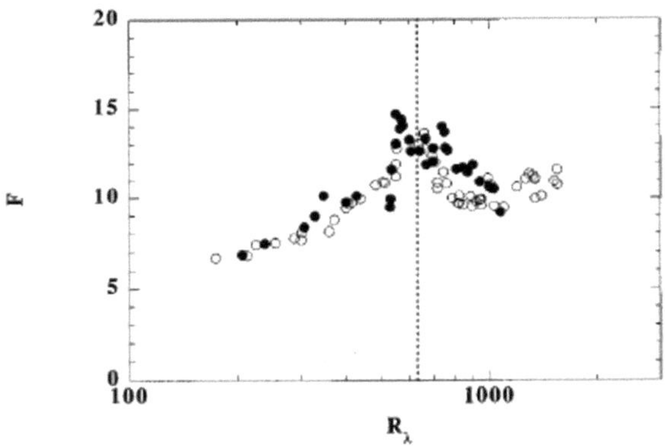

Figure 2.1.4. Flatness of velocity derivatives versus Re_λ (after Emsellem *et al* (1997)). The solid and open symbols refer to two different experimental cells, the larger one (open symbols) having a radius roughly 3 times that of the other (solid symbols).

Conventional wisdom states that the gradients can be computed from the velocity data reliably if the latter is measured with a resolution of the order of the Kolmogorov or dissipation scale. That has been the ideal to aim for in most experimental efforts, and, indeed, some at high Reynolds numbers fall short of this goal. For some time, however, there has been a growing realization (Sreenivasan and Meneveau (1988); Paladin and Vulpiani (1987); Sreenivasan (2004); Yakhot and Sreenivasan (2005)) that the Kolmogorov scale is a *fluctuating* quantity (due to a so-called multi-fractal nature of the energy dissipation), and that scales significantly smaller than the mean value must exist in turbulent flows. This feature has been verified in direct numerical simulations (Schumacher, Sreenivasan and Yakhot (2006)) in which the grid resolution was made much finer than the Kolmogorov scale. A recent paper (Yakhot and Sreenivasan (2004)) discusses how the grid-resolution depends on the moment order and the Reynolds number. The higher the order of the moment and the higher the Reynolds number, the more stringent is the required resolution in measurement and simulations. Simulations of passive scalar (Schumacher Sreenivasan and Yeung (2005)), made with resolutions much smaller than the Batchelor scale (the nominally smallest scale for passive scalars), confirm that much finer scales than the Batchelor scale indeed exist. So we see that the requirements for high-Re experiments may become even more stringent and thus the advantages of novel fluids like low temperature helium bring with them also new challenges.

We note that similar experiments were conducted (Maurer and Tabeling (1988)) in the same apparatus, where helium II replaced the gas phase fluid, as will be discussed below. Due to its high conductivity (and the resulting counterflow–see the chapter by L. Skrbek) hot wires would be completely ineffective in this case and hence Tabeling and co-workers used small pressure transducers to probe the velocity. More specifically, the output of these pressure transducers could be related to the squared velocity and hence to the turbulent energy directly. What was interesting and important in these experiments is that the authors also warmed the liquid helium above the lambda point and performed the same measurements: at all temperatures below the lambda point down to 1.4 K, where the normal component is of the order of a percent, the structure function exponents were indistinguishable from those observed in Helium I turbulence. This was the best available direct evidence that turbulence in helium II could indeed behave quasi-classically (see Figure 2.3.1), as has been discussed extensively in a recent review (Vinen and Niemela (2002)).

2.2 Thermal turbulence

Having explored an example of isothermal turbulence in helium, let us now turn our attention to thermally-generated turbulence; in particular to turbulent thermal convection, a subject in which the use of helium has played a major role in recent years. Indeed, turbulent thermal convection has long been a problem of major interest for a number of reasons: it plays a prominent role in the energy transport within stars, atmospheric and oceanic circulations, the generation of the earth's magnetic field, and countless engineering processes in which heat transport is important. Threlfall (Threlfall (1975)) cab be credited with recognizing the advantages of using low temperature helium gas to investigate high-Ra turbulent convection in the laboratory, even if his original motivation was more toward developing a better method of mixing and equilibration of critical fluids than to fluid turbulence, per se. A decade later, low temperature work by Libchaber and co-workers (Castaing *et al* (1989); Wu (1991)) brought a broader and lasting awareness of the potential of helium.

Rayleigh-Bénard convection

It is not only difficult to conceive of, but perhaps not even clear that we would benefit from trying to recreate every real problem of convection in the laboratory, including every aspect related to their complex boundary conditions, rotation, magnetic fields, ionization, chemical reactions, etc. Indeed, it is better to consider instead a simpler problem, one that contains the essential physics of real problems, but allows a straightforward treatment, both theoretically and experimentally. In the case of buoyancy-driven turbulence, this model problem is Rayleigh-Bénard convection (RBC). In RBC, a thin fluid layer of (typically) infinite lateral extent is contained between two isothermal surfaces with the bottom surface maintained slightly hotter than the other. When the expansion coefficient is positive (which, as we have already discussed, is the case usually), an instability develops because the hot fluid from below rises to the top and the colder fluid from above sinks to the bottom. The applied driving force is measured in terms of a Rayleigh number, Ra, which is a non-dimensional measure of the imposed temperature difference across the fluid layer (see Equation (1.2.6)). The Prandtl number, Pr, defined in Equation (1.2.7), is an additional dynamical parameter that has some importance in thermal turbulence because it determines the state and thickness of the viscous and thermal *boundary layers* set up on the solid surfaces. With increasing Ra the dynamical state of RBC goes from a uniform and parallel "roll" pattern at the onset ($Ra \sim 10^3$) to a turbulent state at $Ra \sim 10^7$–10^8.

For the RBC system, the conservation equations can also be written in dimensionless form, scaling lengths by L, times by L^2/κ, velocities by κ/L, pressures by $\rho_0\kappa^2/L^2$ (ρ_0 being the mean density in the layer), and temperatures, T, by ΔT. In this way, we obtain for the dimensionless momentum and energy equations respectively:

$$\frac{\partial \mathbf{u}}{\partial t} + \mathbf{u} \cdot \nabla \mathbf{u} = -\nabla p + Pr \nabla^2 \mathbf{u} + Pr Ra T \mathbf{e}_z \tag{2.2.1}$$

$$\frac{\partial T}{\partial t} + \mathbf{u} \cdot \nabla T = \nabla^2 T \tag{2.2.2}$$

where $\mathbf{e}_z$ is the unit vector in the upward vertical direction. Mass conservation is given by (1.2.2).

The apparatus used for low temperature experiments all have more or less the same general features (although some important details may differ). We refer the interested reader to the original works that we will discuss for the actual set-ups used (Niemela *et al* (2000); Chavanne *et al* (2001); Niemela and Sreenivasan (2003)). For our purposes here it is sufficient to note the following salient features. To approximate the constant temperature top and bottom boundaries, annealed oxygen-free copper is used, with a conductivity near 1 $kW\ m^{-1}\ K^{-1}$ at helium temperatures, nearly 5 orders of magnitude larger than the molecular conductivity of the contained helium gas. On the bottom plate a constant heating is applied, using a distributed heater (either a serpentine metal film or spiraling wires) while the heater at the top plate regulates the temperature as part of a feedback loop, where the plate is in contact with a cold reservoir (liquid helium bath). Temperature fluctuations within the turbulent gas are measured using small cubes of doped germanium or silicon, nominally of the order of a few hundred microns on a side. The fluid is contained laterally by thin wall stainless steel, typically only as thick as necessary to withstand the pressure differential between the fluid and the surrounding vacuum which eliminates parasitic heat inputs through conduction or convection. Radiative heat input through the vacuum is reduced through the use of shields at various graded temperatures.

Fluid experiments impose a particular restriction that is not necessarily of trivial importance, whether the flows are turbulent or laminar: the fluid must be confined laterally by walls. In our case, the walls must also be strong enough to withstand the pressure difference between the high vacuum outside and the normal pressure inside, and not too conductive compared to the fluid. The length scale associated with the lateral containment of the fluid then introduces another dynamical parameter in the form of an aspect ratio $\Gamma = D/H$, where D is the diameter for the case of cylindrical containers and H, the vertical height between the top and bottom plates, is the relevant proxy for the length scale L in Equation (1.2.6). For recent helium experiments, Γ is small of the order of unity, so the correspondence with theoretical models which banish lateral walls to infinity is fraught with difficulties. Considering that we wish to approximate $\Gamma \gg 1$, it is of some concern that most experiments have the value $\Gamma = 1/2$. It is, of course, easy to understand from the definition of the Rayleigh number in Equation (1.2.6) why such a small aspect ratio has been chosen. Maximizing the height is paramount to achieving high Ra, and to make the width of the same order or much larger than the largest feasible height is often technically difficult *and* expensive. But the effects of small Γ are nontrivial as we shall see.

Heat transport and scaling

The heat transport in convection is usually given in terms of the Nusselt number Nu

$$Nu = \frac{q}{q_{cond}} = \frac{qH}{k_f \Delta T}, \tag{2.2.3}$$

where q is the applied heat flux, q_{cond} is the heat flux in the absence of convection, given by Fourier's law, and k_f is the thermal conductivity of the fluid. Nu represents the ratio of the effective turbulent thermal conductivity of the fluid to its molecular value and can reach values of over 10^4 in helium experiments (Niemela *et al* (2000)) thus demonstrating the enormous enhancement of transport possible in cryogenic turbulent convection. It is this global quantity that is often of interest in natural and engineering turbulent processes, and the ability to predict its value in the limit of high Ra corresponding to those processes—and often out of reach of theory, simulation or ordinary laboratory experiments—is clearly useful. Since we normally expect $Nu = f(Ra, Pr, \Gamma, ...)$ to attain some limiting power-law form in fully developed turbulence as $Ra \to \infty$, we may hope to extrapolate results from experiments at the highest Ra if we have, indeed, exceeded the Ra beyond which the functional relation ceases to change (i.e., our definition of "high enough" introduced earlier).

Taking a rather simple view, it is possible to imagine two limiting cases for the scaling of Nu. In the first case, we consider that the global flux of heat is determined by processes occurring in the two thermal boundary layers at the top and bottom of the heated fluid layer. These layers, defined as the extent over which the molecular conduction remains important, are always present at the solid boundary where a heat flux is imposed, although they become thin as Ra is increased. Their thickness λ can be estimated as the distance over which the imposed temperature gradient would produce a state of marginal stability for convection; that is, when the Rayleigh number *based on their height* exceeds something of order 10^3 (roughly corresponding to the critical $Ra_c = 1100$ for one rigid boundary and one "free" boundary). Assuming that the temperature outside the thermal boundary layer is uniform (because of strong turbulent mixing), one can easily show that $\lambda = H/2Nu$. Clearly, λ can become rather small as Ra and hence Nu increases; for instance, for the highest Ra achieved at present in the laboratory (Niemela *et al* (2000)) the thermal boundary layers at the boundaries of a 1 meter column of fluid would have a thickness of the order 100 μm. In fact, this sharp gradient in temperature, realized in the limit of vanishing λ, is the source of "stress" and results in the emission of bursts of hot or cold fluid into the interior– commonly referred to as "plumes" or "thermals" (Sparrow, Husar and Goldstein (1970); Castaing *et al* (1989); Theerthan and Arakeri (1998); Du and Tong (2000))–for the marginally unstable boundary layer.

Now, in the sense that the bulk fluid outside the boundary layers is fully turbulent and "randomized", it should act more or less as a thermal short circuit. Therefore its precise nature or spatial extent should be immaterial to the flux of heat and from the definition of Nu given in 2.2.3 we can then immediately write down the asymptotic power law (Malkus (1954)) $Nu \sim Ra^{1/3}$ (where we assume that the heat flux q appearing in Nu has no implicit height dependence). This simple argument has been made more precise by others (Howard (1966)) and rigorously shown (Constantin and Doering (1999)) to hold when Pr and Ra are both very large.

There is another limit one can think of in which molecular properties lose relevance in determining heat transport—for instance, when boundary layers are mixed by a vigorous turbulent flow so that molecular conduction is supplanted entirely by small scale turbulent motion. Here, an exponent of 1/2 (modulo logarithmic corrections) has been proposed; the phenomenological theory has been developed by Kraichnan (Kraichnan (1962)). The theory of Grossmann and Lohse (Grossmann and Lohse (2001)) is consistent with both "asymptotic" scalings given above, in different regimes of Ra and Pr, but does not anticipate the 1/2-power to occur in the range covered by most available experiments. In Kraichnan's work (Kraichnan (1962)) this scaling was proposed to occur for $Ra \sim 10^{24}$ although this work made use of empirical input which has since been improved upon. The one experiment (Niemela *et al* (2000)) which could have been expected to reveal the state does not show it.

Of course, it is not merely the exponent, but also the prefactor which determines our ability to make a prediction of Nu at higher Ra, so it is of some importance to determine both from an experiment at *sufficiently* high Ra. As we saw in the introduction, low temperature helium gas provides an opportunity not only to attain the highest possible Ra in an apparatus of given height, but also to obtain many orders of magnitude of it without making mechanical changes to the apparatus (or change its aspect ratio which is often done in conventional fluids).

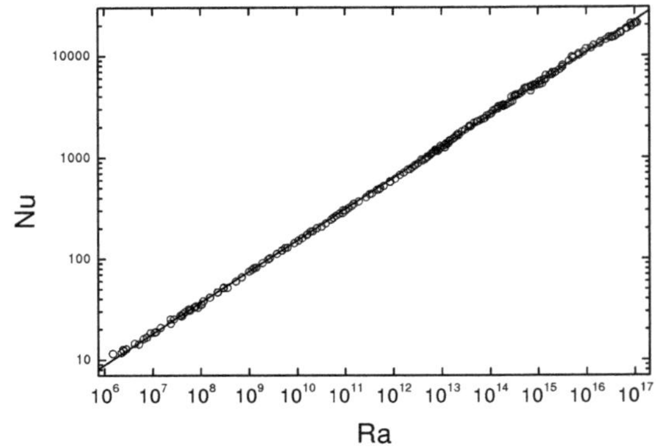

Figure 2.2.1. Log-log plot of the Nusselt number versus Rayleigh number. The line through the data is a least-square fit over the entire Ra range, and represents a $d \log Nu / d \log Ra$ slope of 0.32.

Figure 2.2.1, plotted using the data from Niemela *et al.* (Niemela *et al* (2000)), illustrates the enormous range of Ra and Nu possible in low temperature experiments of modest size. It represents the highest Ra achieved in a laboratory experiment and also the largest range of Ra in the turbulent scaling regime, both of which are a testament to the utility of cryogenic helium gas as as test fluid. The average slope over 11 decades is 0.32, quite close, in fact, to 1/3. Low temperature experiments of Chavanne *et al.* (Chavanne *et al* (2001)), and by the present authors for differing aspect ratios ranging up to $\Gamma = 4$ (Niemela and Sreenivasan (2003, 2006)), have found a scaling exponents as high as 0.4 over a limited range at the very highest Ra, but those high-Ra data have been obtained by operating very close to the critical point, introducing a number of concerns, particularly whether they are obtained under Boussinesq conditions. That the Prandtl number is also rapidly varying in this region of high Ra further complicates the interpretation: specifically, Pr strongly *increases*, which should act to strengthen and laminarize the boundary layers in precise opposition to the conditions prescribed for observing the enhanced scaling. Perhaps the best we can do is to refer to the large aspect ratio experiments (Niemela and Sreenivasan (2006)) in the region of Boussinesq conditions but also high Ra

(up to 10^{13}): here the flow more closely approximates the ideal RBC, and in particular, the lateral confinement poses a less severe restriction consistent with the observation of a significantly weakened mean wind. The resulting scaling for Nu is 0.33-0.34 over the highest *two decades* of Boussinesq turbulent convection, giving the best evidence that the 1/3 scaling is probably the correct limit for RBC flow, at least for the range of Ra presently attainable in the laboratory.

Let us briefly consider the boundary conditions in somewhat more detail. While all experiments attempt to recreate the appropriate ideal boundary conditions for RBC, this is always an approximation and it is useful to examine what kind of effect departures from "ideality" may have on the observed flows, even if such departures are small. The data shown in Fig. 2.2.1 have been corrected for two main effects, although we state at the outset that these common corrections are particulary small for low temperature apparatus. We have already pointed out the thermal advantages of cryogenic helium experiments over conventional fluid experiments in having heating plates of thermal conductivity many orders of magnitude greater than that of the fluid. However, for very high Ra, the effective thermal conductivity of the fluid, $Nu \times k$, becomes large, and the plates can no longer be regarded as strictly isothermal. The resulting "mixed" boundary conditions could then affect dynamical features such as plume production (Verzicco (2004)). This effect is is immeasurably small over almost the entire range of Ra used by Niemela *et al.* (Niemela *et al* (2000)) For the data shown in Fig. 2.2.1, we estimate the corrections to be order of a few percent only for $Ra > 5 \times 10^{16}$. The same order of magnitude correction in similarly constructed water experiments would develop at a Rayleigh number roughly 7 orders of magnitude lower!

The other correction required is that due to sidewall conduction. Some (small) part of the heat is inevitably conducted from the bottom plate to the top through the side walls instead of through the fluid. The two parallel conductances normally add together simply to form the total conductance. It has therefore been standard procedure to subtract this parallel conductance, assuming for this purpose that the temperature difference prescribed between the plates varies linearly between them. The onset of turbulence and thermal short-circuiting in the bulk of the flow makes such corrections inadequate in a range of moderately large Nu, where the temperature profile is no longer approximately linear but not so high as to make any parallel conduction absolutely negligible. Various models (Roche *et al* (2001); Ahlers (2001); Niemela and Sreenivasan (2003); Verzicco (2002); Roche *et al* (2004)) have been developed for the purpose, all of them reasonable but none perfect. The data in Fig. 2.2.1 have been corrected for plate effects using the scheme proposed by Verzicco (Verzicco (2004)) and modified in Brown *et al.* (Ahlers *et al* (2005)) and for wall effects using the scheme proposed by Roche *et al.* (Roche *et al* (2001)). The effect is small, changing the original $Nu - Ra$ scaling exponent of 0.31 to 0.32.

An interesting and largely unexpected feature of turbulence in convecting flows is the presence of a large scale *coherent* motion that arises from self-organization of small scale turbulent features (plumes). This is the so-called mean wind (Niemela *et al* (2001); Sreenivasan, Bershadskii and Niemela (2002); Cioni, Ciliberto and Sommeria (1997); Qui and Tong (2002); Villermaux (1995); Sun, Xia and Tong (2005); Brown,Nikolaenko and Ahlers (2005)), and its study has added significantly to our overall understanding of turbulent convection. The wind phenomenon has had a rather broad reach; for example, quantitative observations of occasional reversals of the mean wind flow direction have been shown using data obtained from helium experiments (Sreenivasan, Bershadskii and Niemela (2004); Hwa *et al* (2005)) to be related to standard critical systems. Furthermore, the lifetimes of the metastable states of the bi-directional mean flow have intriguing analogies with reversals of the earth's magnetic field polarity, a phenomena arising from turbulent convection within the outer core (Glatzmaier, *et al* (1999)); there is also a more quantitative statistical analogies with the lifetime of solar flare activity driven by turbulent convection in the Sun's outer layer (Bershadskii, Niemela and Sreenivasan (2004)). This latter conclusion may indicate the existence of an underlying universality class, or a more direct physical similarity in the convective processes that lead to re-shuffling of the magnetic footprints and ultimately to flare extinction. In general, the understanding of the transition from one metastable state to another is a virgin territory in the presence of large fluctuations.

2.3 The superfluid washing-machine

Before moving toward the discussion, we briefly return to the point made earlier, that the quasi-classical nature of superfluid turbulence can be observed in the Kármán flow experiments when they are extended below the lambda point (Tabeling *et al* (1996)). As we mentioned previously, in these experiments the Kolmogorov description was found to work equally well to describe the decay of turbulence energy; in fact, it was impossible to distinguish which measurements were taken above or below the superfluid transition, both giving the $-5/3$ law decay in the energy spectrum (Equation (1.2.9)). As discussed previously, measurements were made using a local pressure probe inserted into the fluid in a region close to the rotating plates and thus having a large mean flow. The pressure fluctuations measured by the probe were then interpreted to give the velocity fluctuations which are directly connected to the turbulent energy (Maurer and Tabeling (1988)). The resulting energy spectra in frequency space are shown in Fig. 2.3.1. Note that the three curves are displaced for clarity, and only the curve labelled "A" is taken for temperature above the superfluid transition. Of the other two, the curve labelled "C" was measured at 1.4K, where the normal fluid fraction of the density is only a few percent, yet all three show essentially identical energy spectra (with the canonical −5/3 roll-off). At all temperatures below the superfluid transition, the structure function exponents, for example, were identical to those observed in normal fluid turbulence.

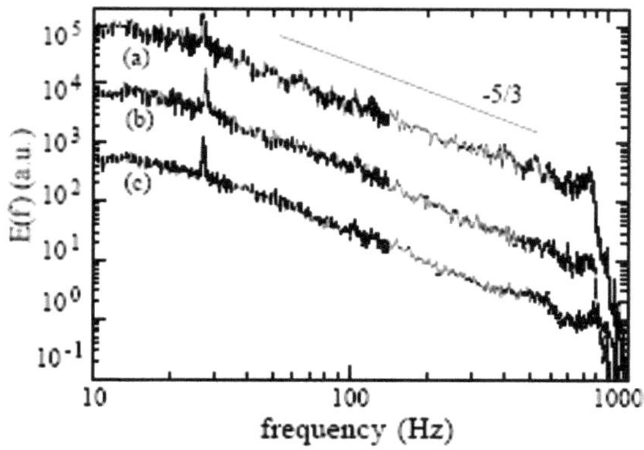

Figure 2.3.1. Results by Maurer & Tabeling (Maurer and Tabeling (1988)) showing the energy spectrum obtained under the same conditions but at different temperatures (a) 2.3 K, (b) 2.08 K, (c) 1.4 K.

3 Summary and Discussion

As we saw, very large values as well as large ranges of *Re* were possible in the Kármán flow using gaseous helium. A transition to reduced intermittency was noted above a certain threshold *Re*. This has not yet been seen in turbulence work using water and air. It is not firmly established that the observed result is due to poor probe resolution, but there are enough concerns on this score that the result will need to be substantiated by measurements in other flow configurations by using different and/or smaller probes. It is, in fact, somewhat unfortunate that this has not occurred, given the importance of the results!

There are questions on how to assess the errors due to the finite size of the probe. If the local *Re* based on the probe size is large enough, we have to worry about vortex shedding by the probe itself, leading to altered heat transfer laws for the hot wires (Corrsin (1963)). Analysis of this behavior (Emsellem *et al* (1997); Grossmann and Lohse (1993)) shows that hot wire response would likely affect flatness measurements: the viscous boundary layer set up around the probe limits the response time by the need for heat to diffuse through this layer and maintain the temperature of the probe constant (Grossmann and Lohse (1993)).

As for thermal turbulence, we saw that very near the critical point the increase in *Ra* is associated with the divergence of the specific heat. The diminishing value of thermal diffusivity following from this makes the Prandtl number grow substantially in excess of 0.7. Thus, two control parameters are simultaneously varying! The larger the fluid layer height, the better in this regard, so that *Ra* can be increased as much as possible without resorting to a close approach to the critical point, which also can lead to a large enough vertical asymmetry (due to the temperature dependence of the fluid properties) for the applicability of Boussinesq approximation to be in serious doubt.

Thus we see that it is not enough merely to raise the appropriate control parameter (*Re* or *Ra*); one has to consider also certain consequences for measurement, especially, as we pointed out initially, as the questions we ask become more refined.

In fact, the general prospect of obtaining very high *Re* and *Ra* in apparatus of small physical size has imposed severe demands on instrumentation. The combination of small sizes of the apparatus and high values of the control parameter leads to the generation of dissipation scales that are physically very small (of the order of microns or less in existing experiments) and oscillate with frequencies on the order of MHz. Resolving these scales requires nontrivial development of very small, fast, sensitive and durable probes and associated measuring equipment. Related to this is an interesting concern that has been raised at various times: do the increasingly small scales being made available in cryogenic helium flows render tenuous the continuum approximation of classical hydrodynamics? Even in the experiments already completed, does one have to worry about aspects of molecular motion? This question has addressed (Frisch (1995)) in a general context and it has been demonstrated that the ratio of the dissipation scale to the molecular mean free path *grows* with increasing Reynolds number. The hydrodynamic approximation thus, if anything, becomes better applicable at higher *Re*.

Bibliography

G. Ahlers, *Phys. Rev. E.* **63**, 015303 (2001).

R.A. Antonia, A.J. Chambers, S. Rajagopalan, K.R. Sreenivasan, C.A. Friehe, *J. Phys. Ocean.* **8**, 28 (1978).

R.A. Antonia, B.R. Satyaprakash, A.K.M.F. Hussain, *J. Fluid Mech.* **119**, 55 (1982).

A. Arneodo, C. Baudet, F. Belin, R. Benzi, B. Castaing, B. Chabaud, R. Chavarria, S. Ciliberto, R. Camussi, F. Chilla, B. Dubrulle, Y. Gagne, B. Hebral, J. Herweijer, M. Marchand, J. Maurer, J.F. Muzy, A. Naert, A. Noullez, J. Peinke, F. Roux, P. Tabeling, W. van de Water, H. Willaime, *Europhys. Lett.* **34**, 411 (1996).

G.K. Batchelor, *The Theory of Homogeneous Turbulence*, Cambridge Univesity Press (1953)

A. Bershadskii, J.J. Niemela, K.R. Sreenivasan, *Phys. Lett. A* **331**, 15 (2004).

E. Brown, A. Nikolaenko, D. Funfschilling, G. Ahlers, *Phys. Fluids* **17**, 075108 (2005).

E. Brown, A. Nikolaenko, G. Ahlers, *Phys. Rev. Lett.* **95**, 084503 (2005).

F.H. Busse, *Rep. Prog. Phys.* **41**, 1929 (1978).

B. Castaing, G. Gunaratne, F. Heslot, L. Kadanoff, A. Libchaber, S. Thomae, X.-Z. Wu, S. Zaleski, G. Zanetti, *J.* Fluid Mech. **204**, 1 (1989).

O. Chanal, B. Baguenard, O. Bethoux, B. Chabaud, *Rev. Sci. Instrum.* **68**, 2442 (1997).

X. Chavanne, F. Chilla, B. Chabaud, B. Castaing, B. Hebral, *Phys. Fluids* **13**, 1300 (2001).

S. Cioni, S. Ciliberto, J. Sommeria, *J. Fluid Mech.* **335**, 111 (1997).

P. Constantin, C.R. Doering, *J. Stat. Phys.* **94**, 159 (1999).

S. Corrsin, in *Handbuch der Physik*, eds. S. Flugge and C. Truesdell, Springer, Berlin, **8**, part 2, p. 524 (1963).

P.P. Craig, J.R. Pellam, *Phys. Rev.* **108**, 1109 (1957).

P.A. Davidson, *Turbulence: An Introduction for Scientists and Engineers.* Oxford University Press, Oxford (2004).

R.J. Donnelly (ed.), *High Reynolds number flows using liquid and gaseous helium.* Springer-Verlag, New York (1991).

Y.B. Du, P. Tong, *J. Fluid Mech.* **407**, 57 (2000).

V.Emsellem, L.P. Kadanoff, D. Lohse, P. Tabeling, Z. Wang, *Phys. Rev. E* **55**, 2672 (1997).

G. Falkovich, K.R. Sreenivasan, *Physics Today* **59**, 43 (2006).

U. Frisch, *Turbulence: The Legacy of A.N. Kolmogorov*, Cambridge University Press (1995).
G.A. Glatzmaier, R.C. Coe, L. Hongre, P.H. Roberts, *Nature* **401**, 885 (1999).
S. Grossmann, D. Lohse, *Phys. Lett. A* **174**, 449 (1993).
S. Grossmann, D. Lohse, *Phys. Rev. Lett.* **86**, 3316 (2001).
A. Gylfason, A. Ayyalasomayajula, Z. Warhaft, *J. Fluid Mech.* **501**, 213 (2004).
L.N. Howard, in *Proc.* 11th *Intern. Cong. Appl. Mech.* (ed. H. Gortler), Springer, Berlin, p. 1109 (1966).
R.C. Hwa, C.B. Yang, S. Bershadskii, J.J. Niemela, K.R. Sreenivasan, *Phys. Rev. E* **72**, 066308 (2005).
J.C. Kaimal, J.C. Wyngaard, D.A. Hsugen, O.R. Cote, Y. Izumi, S.J. Caughey, C.J. Readings, *J. Atmos. Sci.* **33**, 2152 (1976).
Y. Kaneda, T. Ishihara, M. Yokakawa, M. Itakura, A. Uno, *Phys. Fluids* **15**, L21 (2003).
A.N. Kolmogorov, *Dokl. Akad. Nauk SSSR* **30**, 9 (1941).
R.H. Kraichnan, *Phys. Fluids* **5**, 1374 (1962).
B.B. Mandelbrot, *Scientific American* **280**, 50 (1999).
W.V.R. Malkus, *Proc. R. Soc. Lond. A* **225**, 196 (1954).
M.P. Marder, *Condensed Matter Physics*, Wiley, New York (2000).
J. Maurer, P. Tabeling, *Europhys. Lett*, **43**, 29 (1998).
A.S. Monin, A.M. Yaglom, *Statistical Fluid Mechanics, vol. 2*, MIT Press, Cambridge, USA (1975).
J.J. Niemela, L. Skrbek, K.R. Sreenivasan, R.J. Donnelly, *Nature* **404**, 837 (2000).
J.J. Niemela, L. Skrbek, K.R. Sreenivasan, R.J. Donnelly, *J. Fluid Mech.* **449**, 169 (2001).
J.J. Niemela, K.R. Sreenivasan, *J. Fluid. Mech.* **481**, 355 (2003).
J.J. Niemela, K.R. Sreenivasan, *J. Fluid Mech.* **557** 411, 2006.
J.J. Niemela and K.R. Sreenivasan, *J. Low Temp. Phys.* **146**, 499 (2007).
National Research Council Report: *Condensed Matter and Materials Physics: Basic Research for Tomorrow's Technology*, 308 pages, National Academy Press (1999).
A. Oberbeck, *Annalen der Physik und Chemie* **7**, 271 (1879).
G. Paladin, A. Vulpiani, *Phys. Rev. A* **35**, 1971 (1987).
X.L. Qui, P. Tong, *Phys. Rev. E* **64**, 036304 (2002).
P. Roche, B. Castaing, B. Chabaud, B. Hebral, J. Sommeria, *Europhys. J.* **24**, 405 (2001).

P.E. Roche, B. Castaing, B. Chabaud, B. Hebral, *J. Low Temp. Phys.* **134**, 1011 (2004).
J. Schumacher, K.R. Sreenivasan, V. Yakhot, submitted (2006).
J. Schumacher, K.R. Sreenivasan, P.K. Yeung, *J. Fluid Mech.* **531**, 113 (2005).
E.M. Sparrow, R.B. Husar, R.J. Goldstein, *J. Fluid Mech.* **41**, 793 (1970).
E.M. Spiegel, *Ann. Rev. Astron. Astrophys.* **9**, 323 (1971).
K.R. Sreenivasan *Phys. Fluids* **27**, 1048 (1984).
K.R. Sreenivasan, *Phys. Fluids* **7**, 2778 (1995).
K.R. Sreenivasan, *Phys. Fluids* **10**, 528 (1998).
K.R. Sreenivasan, *Rev. Mod. Phys.* **71**, S383 (1999).
K.R. Sreenivasan, *Flow, Turbulence and Combustion* **72**, 115 (2004)
K.R. Sreenivasan, R.A. Antonia, *Annu. Rev. Fluid. Mech.* **29**, 435 (1997).
K.R. Sreenivasan, A. Bershadskii, *J. Fluid. Mech.* **554**, 477 (2006).
K.R. Sreenivasan, A. Bershadskii, J.J. Niemela, *Phys. Rev. E* **65**, 056306 (2002).
K.R. Sreenivasan, A. Bershadskii, J.J. Niemela, *Physica A* **340**, 574 (2004).
K.R. Sreenivasan, R.J. Donnelly, *Adv. Appl. Mech.* **37**, 239-276 (2001).
K.R. Sreenivasan, B. Dhruva, *Prog. Theor. Phys.* **130**, 103 (1998).
K.R. Sreenivasan, C. Meneveau *Phys. Rev. A* **38**, 6287 (1988).
C. Sun, K.-Q. Xia, and P. Tong, *Phys. Rev. E* **72**, 026302 (2005).
P. Tabeling, H. Willaime, *Phys. Rev. E* **65**, 066301 (2002).
P. Tabeling, G. Zocchi, F. Belin, J. Maurer, H. Willaime, *Phys. Rev. E* **53** 1613 (1996).
H. Tennekes, J.L. Lumley, *A First Course in Turbulence*, MIT Press, Cambridge, Ma. (1971).
S.A. Theerthan, J.H. Arakeri, *J. Fluid Mech.* **373**, 221 (1998).
D.C. Threlfall, *J. Fluid Mech.* **67**, 17 (1975).
D.J. Tritton, *Physical Fluid Dynamics*, Clarendon Press, Oxford (1988).
R. Verzicco, *J. Fluid Mech.* **473**, 201 (2002).
R. Verzicco, *Phys. Fluids* **16**, 1965 (2004).
E. Villermaux, *Phys. Rev. Lett.* **75**, 4618 (1995).
W.F. Vinen, J.J. Niemela, *J. Low Temp. Phys.* **128**, 167 (2002).
W.-Z. Wu, Ph.D. thesis, University of Chicago, Chicago (1991).
V. Yakhot, K.R. Sreenivasan, *Physica A* **343**, 147 (2004).
V. Yakhot, K.R. Sreenivasan, *J. Stat. Phys.* **121**, 823 (2005).
M.V. Zagarola, A.J. Smits, *Phys. Rev. Lett.* **78**, 239 (1997).

MIX
Papier aus verantwortungsvollen Quellen
Paper from responsible sources
FSC® C105338

FSC
www.fsc.org

If you have any concerns about our products,
you can contact us on
ProductSafety@springernature.com

In case Publisher is established outside the EU,
the EU authorized representative is:
Springer Nature Customer Service Center GmbH
Europaplatz 3, 69115 Heidelberg, Germany

Printed by Libri Plureos GmbH
in Hamburg, Germany